A HANDBOOK OF INTEGER SEQUENCES

A HANDBOOK OF INTEGER SEQUENCES

N. J. A. Sloane
Mathematics Research Center
Bell Telephone Laboratories, Inc.
Murray Hill, New Jersey

ACADEMIC PRESS, INC.
Harcourt Brace Jovanovich, Publishers
San Diego New York Berkeley Boston
London Sydney Tokyo Toronto

ACADEMIC PRESS, INC.
San Diego, California 92101

United Kingdom Edition published by
ACADEMIC PRESS LIMITED
24-28 Oval Road, London NW1 7DX

Library of Congress Catalog Card Number: 72-82647

AMS (MOS) 1970 Subject Classifications: 05A10, 05A15, 05A17, 05C05, 05C30, 10A35, 10A40, 10L99

ISBN 0-12-648550-X

PRINTED IN THE UNITED STATES OF AMERICA
89 90 91 92 93 9 8 7 6 5

To John Riordan

CONTENTS

PREFACE

In spite of the large number of existing mathematical tables, until now there has been no table of sequences of integers. Thus someone coming across the sequence 1, 2, 5, 15, 52, 203, 877, 4140, . . . would have had difficulty in finding out that these are the Bell numbers, and that they have been extensively studied. This handbook remedies this situation. The main table contains a list of some 2300 sequences of integers, collected from all branches of mathematics and science. The sequences are arranged in numerical order, and for each one a brief description and a reference is given.

The first part of the book describes how to use the table, gives methods for analyzing unknown sequences, and contains an illustrated description of the most important sequences.

Who will use this handbook? Anyone who has ever been confronted with a strange sequence, whether in an intelligence test in high school, e.g.,

1, 8, 11, 69, 88, 96, 101, 111, 181, 609, . . .

(guess![1]), or in solving a mathematical problem, e.g.,

1, 2, 5, 14, 42, 132, 429, 1430, . . .

(the Catalan numbers), or from a counting problem, e.g.,

1, 1, 2, 4, 9, 20, 48, 115, 286, 719, . . .

(the number of rooted trees with *n* points), or in physics, e.g.,

1, 0, 3, 22, 192, 2046, 24853, . . .

[1]For many more terms and the explanation, see the main table.

(coefficients of the partition function for a cubic lattice), or in chemistry, e.g.,

1, 1, 1, 2, 3, 5, 9, 18, 35, 75, 159, . . .

(the number of distinct hydrocarbons of the methane series), or in electrical engineering, e.g.,

3, 7, 46, 4336, 134281216, . . .

(the number of Boolean functions of n variables), will find this handbook useful.

Besides identifying sequences, the handbook will serve as an index to the literature for locating references on a particular problem, and for quickly finding numbers like 7^{12}, the number of partitions of 30, the 18th Catalan number, or the expansion of π to 60 decimal places. It might also be useful to have around when the first signals arrive from Betelgeuse (sequence ***2311*** for example would be a friendly beginning).

ACKNOWLEDGMENTS

This book was begun at Cornell University in the years 1965–1969, and finished at Bell Telephone Laboratories from 1969 to 1972. During that time I have been sustained by the support and encouragement of Richard Guy of the University of Alberta, Ron Graham and Henry Pollak of Bell Telephone Laboratories, John Riordan of Rockefeller University, and Ann Snitow of Rutgers University. Most of the sequences were found by searching through the stacks of the libraries of Cornell University, Brown University, and Bell Telephone Laboratories, and I thank the staffs of these excellent libraries for their patience and cooperation. Other sequences were suggested by friends and correspondents, to all of whom I am most grateful. E. R. Berlekamp, J. J. Cannon, D. G. Cantor, B. Ganter, F. Harary, D. E. Knuth, Shen Lin, W. F. Lunnon, R. C. Read, P. R. Stein, and J. W. Wrench, Jr. have been especially helpful. Finally I thank Eleanor Potter and Herman P. Robinson for a thorough reading of the manuscript.

The table was produced by first recording the sequences on punched cards, and (except when the sequence was generated by the author) comparing a listing of the cards with the original tables. These cards were then stored on magnetic tape, and the table has been typeset automatically from this tape. My thanks are due to the staff of the Bell Laboratories computation center at Murray Hill, especially the keypunch operators, for their untiring assistance.

ABBREVIATIONS

Abbreviations of the references are listed in the bibliography

$[a]$	the largest integer $\leq a$
$a ** b$	a^b
$C(i, j)$	the binomial coefficient $\binom{i}{j}$
$\exp(a)$	e^a
gf	generating function
LCM	least common multiple
REF	reference(s)
seq.	sequence

CHAPTER I

DESCRIPTION OF THE BOOK

It is the fate of those who toil at the lower employments of life, to be driven rather by the fear of evil, than attracted by the prospect of good; to be exposed to censure, without hope of praise; to be disgraced by miscarriage, or punished for neglect, where success would have been without applause, and diligence without reward.

Among these unhappy mortals is the writer of dictionaries; whom mankind have considered, not as the pupil, but the slave of science, the pionier of literature, doomed only to remove rubbish and clear obstructions from the paths of Learning and Genius, who press forward to conquest and glory, without bestowing a smile on the humble drudge that facilitates their progress. Every other authour may aspire to praise; the lexicographer can only hope to escape reproach, and even this negative recompense has yet been granted to very few.

Samuel Johnson, Preface to the "Dictionary," 1755

1.1 DESCRIPTION OF A TYPICAL ENTRY

The main table is a list of about 2300 sequences of integers. A typical entry is:

256 1, 2, 3, 5, 8, 13, 21, 34, 55, 89, 144, 233, 377, 610, 987, 1597, 2584, 4181, 6765, 10946, 17711, 28657, 46368, 75025, 121393, 196418, 317811, 514229, 832040, 1346269
FIBONACCI NUMBERS A(N) = A(N − 1) + A(N − 2). REF HW1 148. REC 11 20 62. HO1.

and consists of the following items:

256	the sequence identification number
1, 2, 3, 5, 8, 13, 21, . . .	the sequence itself
FIBONACCI NUMBERS A(N) = A(N − 1) + A(N − 2)	a name or descriptive phrase (in this case a recurrence) for the sequence
REF	references

HW1 148 G. H. Hardy and E. M. Wright, "An Introduction to the Theory of Numbers," Oxford Univ. Press, 3rd ed., page 148, 1954

REC 11 20 62 *Recreational Mathematics Magazine,* Volume 11, page 20, 1962.

HO1 V. E. Hoggatt, Jr., "Fibonacci and Lucas Numbers," Houghton Mifflin, Boston, 1969

1.2 ARRANGEMENT

The entries are arranged in lexicographic order, so that sequences beginning 1, 2, 1 come before those beginning 1, 2, 2, and so on.

1.3 NUMBER OF TERMS GIVEN

Whenever possible enough terms are given to fill two lines. If fewer terms are given, it is because they have never been calculated so far as the author knows. (He would be very pleased to be corrected.) Finding the next term in the following sequences is known to be difficult (others of a similar type can be located via the index):

11, 48, 66–68, 124, 125, 129, 142, 143, 149, 150, 181, 189, 195, 226, 246, 248, 271, 304, 309, 317, 321–325, 329, 330, 358, 373, 380, 393, 435, 450, 465, 477, 516, 559–561, 580, 581, 595, 596, 614, 615, 621, 648, 650, 730, 731, 745, 757, 782, 788, 809, 812, 911, 954, 972, 994, 998, 1052, 1099, 1115, 1133, 1167, 1210, 1244, 1245, 1339, 1340, 1403, 1404, 1467, 1518, 1537, 1803, 2248, 2342.

These sequences all represent unsolved problems.

1.4 REFERENCES

To conserve space, journal references are extremely abbreviated. They usually give the exact page on which the sequence may be found, but neither the author nor the title of the article. To find out more the reader must go to a library; this book is meant to used in conjunction with a library. Quite a small one will do. A considerable fraction of the sequences will be found in the following nine great works:

Indexes	Dickson [D12] Lehmer [LE1] Fletcher, Miller, Rosenhead, and Comrie [FMR]
Tables	Davis [DA2] Abramowitz and Stegun [AS1] David, Kendall, and Barton [DKB]
Combinatorics	Riordan [R1] David and Barton [DB1] Comtet [CO1]

and in four journals:

American Mathematical Monthly [AMM]
Fibonacci Quarterly [FQ]
Journal of Combinatorial Theory [JCT]
Mathematics of Computation [MTAC]

Unusual sequences may send the reader to more exotic sources, but in any case he should first check Chapter III where additional information about some of the commoner sequences is given, and the index to see if other sequences (and hence references) of a similar type are listed.

Journal references usually give volume, page, and year, in that order. (See the example at beginning of this chapter.) Years after 1899 are abbreviated, by dropping the 19. Earlier years are not abbreviated. Sometimes to avoid ambiguity we use the more expanded form of: journal name (series number), volume number (issue number), page number, year.

References to books give volume (if any) and page. (See the example at the beginning of this chapter.)

The references do not attempt to give the discoverer of a sequence, but rather the most extensive table of the sequence that has been published.

1.5 WHAT SEQUENCES ARE INCLUDED?

Rule 1 The sequence must consist of nonnegative integers. (Sequences alternating in sign have been replaced by their absolute values. Interesting sequences of fractions have been entered by numerators and denominators separately. Some sequences of real numbers have been replaced by their integer parts, others by the nearest integers.)

Rule 2 The sequence must be infinite.

A few, like the Mersenne primes, have been given the benefit of the doubt.

Rule 3 The first two terms must be 1, n, where n is between 2 and 999.

An initial 1 has been silently inserted before the first term if this is greater than 1, and extra 1's and 0's at the beginning have been silently deleted. (See the beginning of Chapter II for examples.)

Rule 3 Enough terms must be known to separate the sequence from its neighbors in the table.

Rule 4 The sequence should have appeared in the scientific literature, and must be well-defined and interesting.

The selection has inevitably been subjective, but the goal has been to include a broad variety of sequences and as many as possible.

1.6 HOW ARE ARRAYS OF NUMBERS TREATED?

Arrays of numbers (binomial coefficients, Stirling numbers of the first kind, etc.) have been entered by rows, columns, or diagonals, whichever seemed appropriate.

1.7 SUPPLEMENTS

It is planned to issue supplements to the Handbook from time to time, containing new sequences and corrections and extensions to the original sequences. Readers wishing to receive these supplements should notify the author.

CHAPTER

II

HOW TO HANDLE A STRANGE SEQUENCE

2.1 HOW TO SEE IF A SEQUENCE IS IN THE TABLE

Obtain as many terms of the sequence as possible. The initial terms are handled as follows: Recall that the sequence must begin 1, n, where n is between 2 and 999. Find the first term in the sequence that is greater than 1, and replace all the terms that come before it by a single 1. Then look it up in the table. The initial 1 is just a marker, and need not be in the original sequence. For example, if the sequence begins

1, 2, 3, 5, 8, 13, . . .	see under	1, 2, 3, 5, 8, 13, . . .
2, 3, 5, 8, 13, . . .	see under	1, 2, 3, 5, 8, 13, . . .
−1, 1, 0, 1, 1, 2, 3, 5, 8, . . .	see under	1, 2, 3, 5, 8, . . .
1, 0, 0, 2, 24, 552, 21280, . . .	see under	1, 2, 24, 522, 21280, . . .

2.2 IF THE SEQUENCE IS NOT IN THE TABLE

(i) Try changing or redefining the sequence. Some typical changes are inserting or deleting an initial term (e.g., seq. ***46*** occurs as both 1, 2, 1, 2, 3, 6, 9, 18, . . . and 1, 2, 3, 6, 9, 18, . . .); adding or subtracting 1 or 2 from all the terms (e.g., seq. ***309*** occurs as both 1, 2, 3, 6, 20, 168, . . . and 1, 4, 18, 166, . . .); and multiplying all the terms by 2 or dividing by any common factor.

(ii) If all these methods fail, and it seems certain that the sequence is not in this handbook, please send the sequence and anything that is known about it, including appropriate references, to the author for possible inclusion in later editions.[1]

[1]Address: Mathematics Research Center, Bell Telephone Laboratories, Inc., Murray Hill, New Jersey 07974.

2.3 FINDING THE NEXT TERM

Suppose the beginning of a sequence is given as

0	1	2	3	4	5	6	7
a_0	a_1	a_2	a_3	a_4	a_5	a_6	a_7

and a rule or explanation for it is desired. If nothing is known about the history of the sequence or if it is an arbitrary sequence, nothing can be said and any continuation is possible. (Any $n + 1$ points can be fitted by an nth degree polynomial.)

But the sequences normally encountered, and those in this handbook, are distinguished in that they have been produced in some intelligent and systematic way. Occasionally such sequences have a simple explanation, and if so, the methods given below may help to find it. These methods can be divided roughly into two classes: those which look for a systematic way of generating the nth term a_n from the terms $a_0, \ldots, a_{n-1}$ before it, e.g., $a_n = a_{n-1} + a_{n-2}$, i.e., methods which seek an internal explanation; and those which look for a systematic way of going from n to a_n, e.g., a_n is the number of divisors of n, or the number of trees with n nodes, or the nth prime number, i.e., methods which seek an external explanation. The former methods are described in the rest of this chapter, the latter in Chapter III.

In practice it is usually clear for one reason or another when a correct explanation for a sequence has been found.

(For the related problems of defining the complexity of a sequence, and extrapolating a sequence of real numbers, see the interesting work of Martin-Lof [IC 9 602 66] and Fine [IC 16 331 70 and FI1].)

2.4 LOOK FOR A RECURRENCE

Let the sequence be $a_0, a_1, a_2, a_3, \ldots$. Is there a systematic way of getting the nth term a_n from the preceding terms $a_{n-1}, a_{n-2}, \ldots$? A rule for doing this, such as $a_n = a_{n-1}^2 - a_{n-2}$, is called a *recurrence*, and of course provides a method for getting as many terms of the sequence as desired.

In studying sequences and recurrences it is useful to define a *generating function* (gf) associated with the sequence, usually an ordinary gf:

$$A(x) = a_0 + a_1x + a_2x^2 + a_3x^3 + \cdots,$$

but sometimes an exponential gf:

$$E(x) = a_0 + a_1\frac{x}{1!} + a_2\frac{x^2}{2!} + a_3\frac{x^3}{3!} + \cdots.$$

(These are formal power series having the sequence as coefficients; questions of convergence do not arise.)

Once a recurrence has been found for the sequence, techniques for solving it will be found in the works by Riordan [R1 19], Batchelder [BAT], and Levy and Lessman [LE2].

For example, consider seq. ***256***, the Fibonacci numbers: 1, 1, 2, 3, 5, 8, 13, 21, 34, ... These are generated by the recurrence $a_n = a_{n-1} + a_{n-2}$, and from this it is not difficult to obtain the generating function

$$1 + x + 2x^2 + 3x^3 + 5x^4 + \cdots = \frac{1}{1 - x - x^2},$$

and the explicit formula for the nth term

$$a_n = \frac{1}{\sqrt{5}}\left[\left(\frac{1+\sqrt{5}}{2}\right)^{n+1} - \left(\frac{1-\sqrt{5}}{2}\right)^{n+1}\right].$$

2.4.1 Method of Differences

This is the standard method for finding recurrences. In simple cases, it will even find an explicit formula for the nth term of a sequence, e.g., if this is a polynomial (such as $a_n = n^2 + 1$) or a simple exponential (such as $a_n = 2^n + n + 1$).

If the sequence is

$$a_0,\ a_1,\ a_2,\ a_3,\ a_4,\ \ldots,$$

its first differences are the numbers

$$\Delta a_0 = a_1 - a_0, \qquad \Delta a_1 = a_2 - a_1, \qquad \Delta a_2 = a_3 - a_2, \qquad \ldots,$$

its second differences are

$$\Delta^2 a_0 = \Delta a_1 - \Delta a_0, \qquad \Delta^2 a_1 = \Delta a_2 - \Delta a_1, \qquad \Delta^2 a_2 = \Delta a_3 - \Delta a_2, \qquad \ldots,$$

and so on. The 0th differences are the original sequence: $\Delta^0 a_0 = a_0$, $\Delta^0 a_1 = a_1$, $\Delta^0 a_2 = a_2$, . . . ; and the mth differences are

$$\Delta^m a_n = \Delta^{m-1} a_{n+1} - \Delta^{m-1} a_n$$

or, in terms of the original sequence,

$$\Delta^m a_n = \sum_{i=0}^{m} (-1)^i \binom{m}{i} a_{m+n-i}. \qquad (1)$$

Therefore if the differences of some order can be identified, Eq. (1) gives a recurrence for the sequence.

Furthermore, if the differences a_k, Δa_k, $\Delta^2 a_k$, $\Delta^3 a_k$, . . . are known for some fixed value of k, then a formula for the nth term is given by

$$a_{n+k} = \sum_{m=0}^{n} \binom{n}{m} \Delta^m a_k . \tag{2}$$

Example (i) Seq. ***1562***

n	1		2		3		4		5		6		7		8
a_n	1		5		12		22		35		51		70		92
Δa_n		4		7		10		13		16		19		22	
$\Delta^2 a_n$			3		3		3		3		3		3		
$\Delta^3 a_n$				0		0		0		0		0			

Since $\Delta^2 a_n = 3$, $\Delta a_{n+1} - \Delta a_n = 3$, or $a_{n+2} - 2a_{n+1} + a_n = 3$, a recurrence for the sequence. An explicit formula is obtained from Eq. (2) with $k = 1$:

$$a_{n+1} = 1 + 4\binom{n}{1} + 3\binom{n}{2} = 1 + 4n + \frac{3}{2}n(n-1) = \frac{1}{2}(n+1)(3n+2).$$

In general, if the mth differences are zero, a_n is a polynomial in n of degree $m - 1$.

Example (ii) Seq. ***1382***

n	1		2		3		4		5		6		7
a_n	1		4		11		26		57		120		247
Δa_n		3		7		15		31		63		127	
$\Delta^2 a_n$			4		8		16		32		64		

Here $\Delta^2 a_n = 2^{n+1}$, $\Delta a_n = 2^{n+1} - 1$, and $a_n = 2^{n+1} - n - 2$. Equation (2) gives the same answer.

Example (iii) Seq. ***552*** (the Pell numbers)

n	1		2		3		4		5		6		7
a_n	1		2		5		12		29		70		169
Δa_n		1		3		7		17		41		99	
$\Delta^2 a_n$			2		4		10		24		58		
$\frac{1}{2}\Delta^2 a_n$			1		2		5		12		29		

Since $\frac{1}{2}\Delta^2 a_n = a_n$, Eq. (1) gives the recurrence $a_{n+2} - 2a_{n+1} - a_n = 0$. Calculating further differences shows that $\Delta^m a_1 = 2^{[m/2]}$ and so Eq. (2) gives the formula

$$a_{n+1} = \sum_{m=0}^{n} \binom{n}{m} 2^{[m/2]}.$$

Example (iv) Seq. ***469***

n	1	2	3	4	5	6	7	8
a_n	1	2	4	10	26	76	232	764
Δa_n	1	2	6	16	50	156	532	
$n^{-1}\,\Delta a_n$	1	1	2	4	10	26	76	

Notice that Δa_n is divisible by n, and in fact $n^{-1}\,\Delta a_n = a_{n-1}$, so that $a_{n+1} = a_n + na_{n-1}$. Again Eq. (2) gives a formula for a_n.

2.4.2 Other Methods of Attack

Is the sequence close to a known sequence, such as the powers of 2? If so, try subtracting off the known sequence. For example, seq. ***1382*** (again): 1, 4, 11, 26, 57, 120, 247, 502, 1013, 2036, 4083, ... The last four numbers are close to powers of 2: 512, 1024, 2048, 4096; and then it is easy to find $a_n = 2^n - n - 1$.

Is a simple recurrence such as $a_n = \alpha a_{n-1} + \beta a_{n-2}$ likely? For this to happen, the ratio $\rho_n = a_{n+1}/a_n$ of successive terms must approach a constant as n increases. Use the values a_2 to a_5 to determine α and β and then see if a_6, a_7, . . . are generated correctly.

If the ratio ρ_n has first differences which are approximately constant, this suggests a recurrence of the type $a_n = \alpha n a_{n-1} + \cdots$. For example, seq. ***704***: 1, 2, 7, 30, 157, 972, 6961, 56660, 516901, . . . has successive ratios 2, 3.5, 4.29, 5.23, 6.19, 7.16, 8.14, 9.12, . . . with differences approaching 1, suggesting $a_n \approx na_{n-1} + {?}$. Subtracting na_{n-1} from a_n, we obtain the original sequence 0, 1, 2, 7, 30, 157, 972, . . . again, so $a_n = na_{n-1} + a_{n-2}$.

This example illustrates the principle that whenever $\rho_n = a_{n+1}/a_n$ seems to be close to a recognizable sequence r_n, one should try to analyze the sequence $b_n = a_{n+1} - r_n a_n$.

A recurrence of the form $a_n = na_{n-1} +$ (small term) can be identified by the fact that the 10th term is approximately 10 times the 9th. For example, seq. ***766***: 0, 1, 2, 9, 44, 265, 1854, 14833, 133496, 1334961, . . . , $a_n = na_{n-1} + (-1)^n$.

The recurrence $a_n = a_{n-1}^2 + \cdots$ is characterized by the fact that each term is about twice as long as the one before. For example, seq. ***331***: 1, 2, 3, 7, 43, 1807, 3263443, 10650056950807, . . . , and $a_n = a_{n-1}^2 - a_{n-1} + 1$.

2.4.3 Factorizing

Does the sequence, or one obtained from it by some simple operation, have many factors?

Example (i) Seq. ***1614***: 1, 5, 23, 119, 719, 5039, 40319, ... As it stands, the sequence cannot be factored, since 719 is prime, but the addition of 1 to all the terms gives the highly composite sequence $2, 6 = 2 \cdot 3$, $24 = 2 \cdot 3 \cdot 4$, $120 = 2 \cdot 3 \cdot 4 \cdot 5, \ldots$, which are the factorial numbers (see Section 3.13).

The presence of only small primes may also suggest binomial coefficients:

Example (ii) Seq. ***577*** (the Catalan numbers): $1, 2, 5, 14 = 2 \cdot 7$, $42 = 2 \cdot 3 \cdot 7$, $132 = 4 \cdot 3 \cdot 11$, $429 = 3 \cdot 11 \cdot 13$, $1430 = 2 \cdot 5 \cdot 11 \cdot 13$, $4862 = 2 \cdot 11 \cdot 13 \cdot 17, \ldots$ and

$$a_n = \frac{1}{n+1}\binom{2n}{n}$$

(see Section 3.5).

Sequences arising in number theory are sometimes *multiplicative,* i.e., have the property that $a_{mn} = a_m a_n$ whenever m and n have no common factor. For example, seq. ***86***: 1, 2, 2, 3, 2, 4, 2, 4. . . . , the number of divisors of n.

2.4.4 Self-Generating Sequences

This section describes some recurrences of a simple yet unusual type. They have been called (rather arbitrarily) *self-generating.*

In the first two examples let $A = \{a_0 = 1, a_1, a_2, \ldots\}$ be a sequence of 1's and 2's.

(i) If every 1 in A is replaced by 1, 2 and every 2 by 2, 1 a new sequence A' is obtained. Imposing the condition that $A = A'$ forces A to be seq. ***71***: 1, 2, 2, 1, 2, 1, 1, ... Sequences ***21*** and ***36*** are of the same type.

(ii) Let $A'' = \{b_0, b_1, b_2, \ldots\}$, where b_n is the length of the nth run in A. (A *run* is a maximal string of identical symbols.) The condition $A = A''$ forces A to be seq. ***70***: 1, 2, 2, 1, 1, 2, 1, 2, 2, 1, ...

In the remaining examples, $A = \{a_0 = 1, a_1, a_2, \ldots\}$ is a nondecreasing sequence of integers.

(iii) Let c_n be the number of times n occurs in A, for $n = 1, 2, \ldots$. If $c_n = n$, A is seq. ***89***: 1, 2, 2, 3, 3, 3, 4, 4, 4, 4, ... If $c_n = a_{n-1}$, A is seq. ***91***: 1, 2, 2, 3, 3, 4, 4, 4, 5, 5, 5, 6, 6, 6, 6, ... (Seq. ***965*** is related to the latter sequence.)

(iv) The condition that $a_{n+1} - a_n$ be the smallest positive integer not equal to $a_i - a_j$ for any $i,j \leq n$ forces a to be seq. ***416***: 1, 2, 4, 8, 13, 21, ... The conditions $a_0 = 1$, $a_2 = 2$, and that a_n be the smallest integer which can be written uniquely as the sum of two distinct preceding terms force A to be seq. ***201***: 1, 2, 3, 4, 6, 8, 11, 13, ... Sequences ***231***, ***254***, ***425***, and ***909*** have similar explanations.

CHAPTER

III

ILLUSTRATED DESCRIPTION OF SOME IMPORTANT SEQUENCES

While Chapter II studied ways of getting the *n*th term of a sequence from the preceding terms, this chapter considers externally generated sequences, such as the sequences in which the *n*th term is the number of graphs with *n* nodes or the *n*th triangular number. An informal and illustrated description is given of some of the most important such sequences.

3.1 GRAPHS AND TREES

Stated informally, a *graph* consists of a finite set of points (or nodes) some of which are joined by lines (or edges). Figure 1 illustrates seq. ***479***, the number of graphs with *n* nodes.

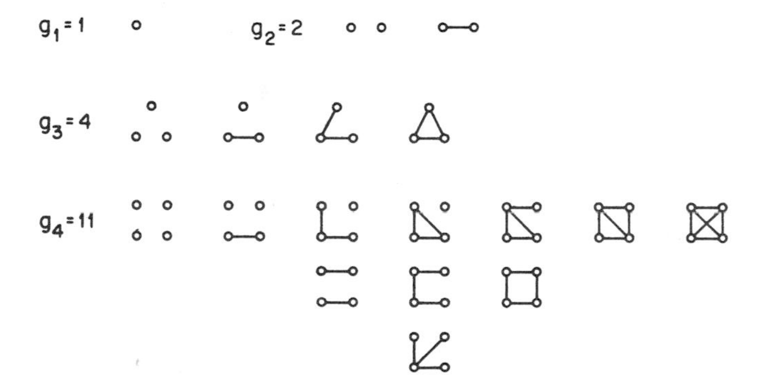

Fig. 1. Seq. ***479***, graphs or reflexive symmetric relations.

A *digraph*, or directed graph, is a graph with arrows on the edges (Fig. 2, seq. ***1229***). Figure 3 shows seq. ***1069***, digraphs of functions, i.e., digraphs with exactly one arrow directed out of each node.

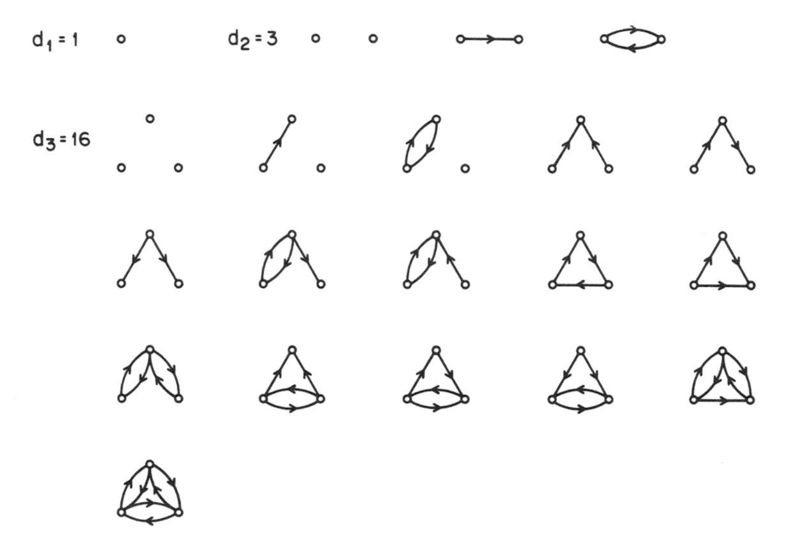

Fig. 2. Seq. ***1229***, digraphs or reflexive relations.

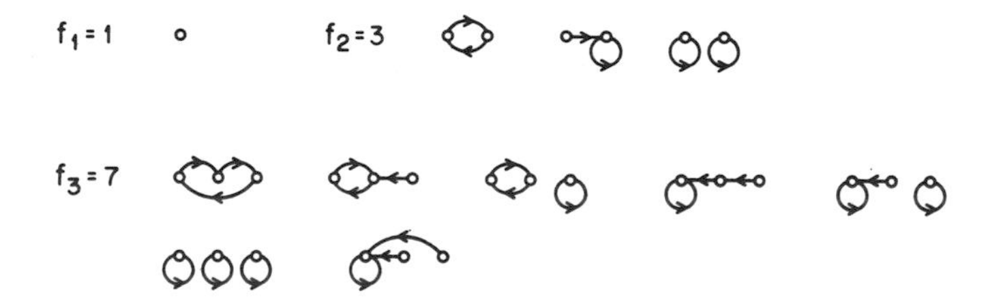

Fig. 3. Seq. ***1069***, functional digraphs.

A *tree* is a connected graph containing not closed paths (Fig. 4, seq. ***299***). A *rooted* tree is a tree with a distinguished node called Eve, or the root. Figure 5 illustrates seq. ***454***, the number of rooted trees with n nodes. The generating function (gf) of this sequence is

$$r(x) = x + x^2 + 2x^3 + 4x^4 + 9x^5 + \cdots$$

and satisfies

$$r(x) = x \exp[r(x) + \tfrac{1}{2}r(x^2) + \tfrac{1}{3}r(x^3) + \cdots].$$

The generating function for trees,

$$t(x) = x + x^2 + x^3 + 2x^4 + 3x^5 + 6x^6 + \cdots,$$

is then given by

$$t(x) = r(x) - \tfrac{1}{2}r^2(x) + \tfrac{1}{2}r(x^2)\,.$$

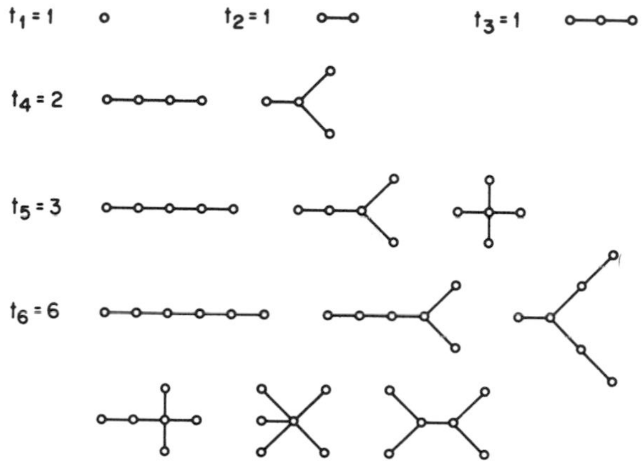

$t_1 = 1$ $t_2 = 1$ $t_3 = 1$

$t_4 = 2$

$t_5 = 3$

$t_6 = 6$

Fig. 4. Seq. ***299***, trees.

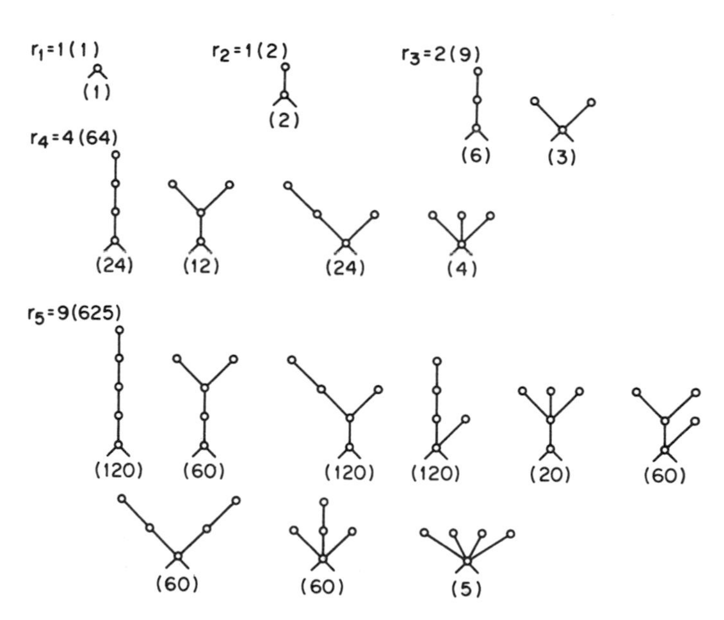

Fig. 5. Seq. ***454***, rooted trees. (The numbers in parentheses give seq. ***771***, labeled rooted trees.)

Any of these graphs may be *labeled* by (if there are n nodes) attaching the numbers from 1 to n to the nodes. For example in Fig. 5, the numbers in parentheses give the number of ways of labeling each tree, and then the total number of labeled rooted trees with n nodes is n^{n-1}, seq. ***771***. Usually when graphs are mentioned in the main table they are unlabeled unless stated otherwise.

The degree of a node is the number of edges meeting it. Figure 6 shows seq. ***118***, series-reduced trees, or trees without nodes of degree 2.

For further information about the preceding sequences and for the enumeration of other kinds of graphs, see Riordan [R1] and Harary [HA5].

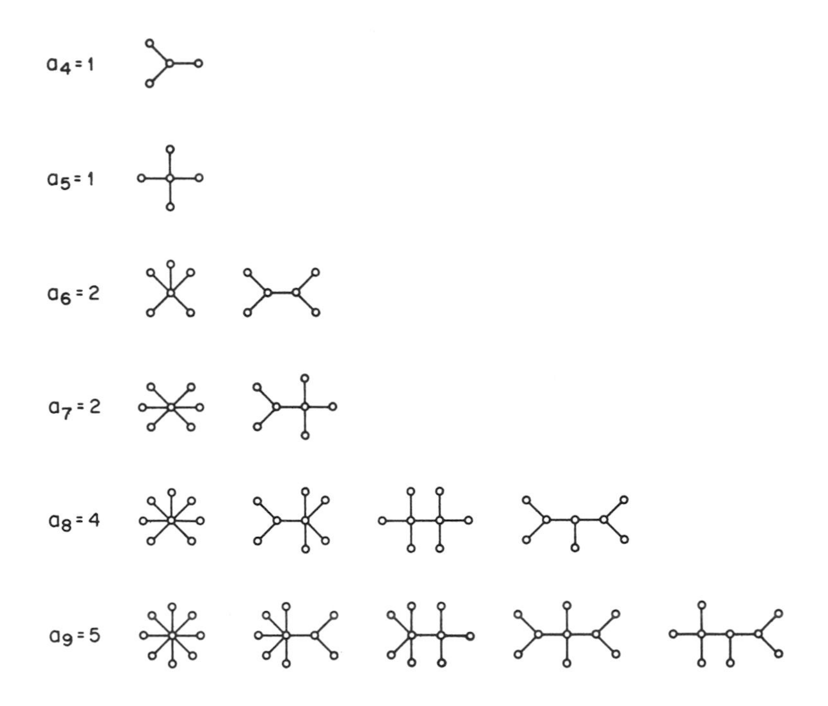

Fig. 6. Seq. ***118***, series-reduced trees.

3.2 RELATIONS

A *relation* R on a set S is any subset of $S \times S$, and xRy means $(x, y) \in R$ or "x is related to y." A relation is *reflexive* if xRx for all x in S, *symmetric* if $xRy \Rightarrow yRx$, *antisymmetric* if xRy and $yRx \Rightarrow x = y$, and *transitive* if xRy and $yRz \Rightarrow xRz$.

The most important types of relations are:

(1) unrestricted, or digraphs with loops of length 1 allowed (seq. ***784***: 2, 10, 104, 3044, 291968, . . .);

(2) symmetric, or graphs with loops of length 1 allowed (seq. ***646***: 2, 6, 20, 90, 544, 5096, 79264, . . .);

(3) reflexive, or digraphs (Fig. 2, seq. ***1229*** again);

(4) reflexive symmetric, or graphs (Fig. 1, seq. ***479*** again);

(5) reflexive transitive, or topologies (Fig. 7, seq. ***1133***: 1, 3, 9, 33, 139, 718, 4535, ?. For the connection between digraphs and topologies, see Birkhoff [B11 117]);

(6) reflexive symmetric transitive, or partitions (Fig. 20, p. 24, seq. ***244***);

(7) reflexive antisymmetric transitive, or partially ordered sets (Fig. 8, seq. ***588***: 1, 2, 5, 16, 63, 318, 2045, ?).

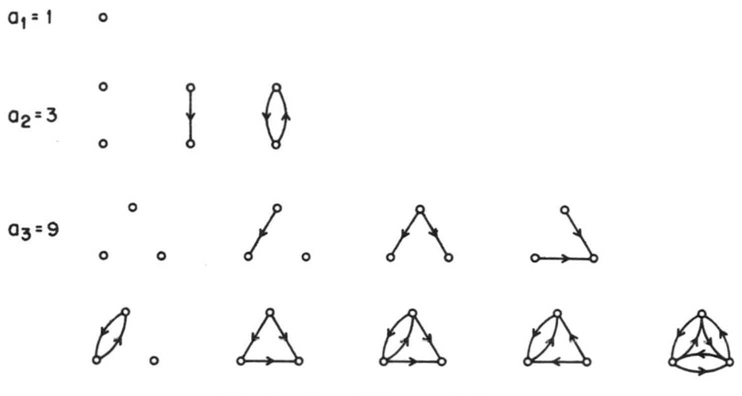

Fig. 7. Seq. ***1133***, topologies.

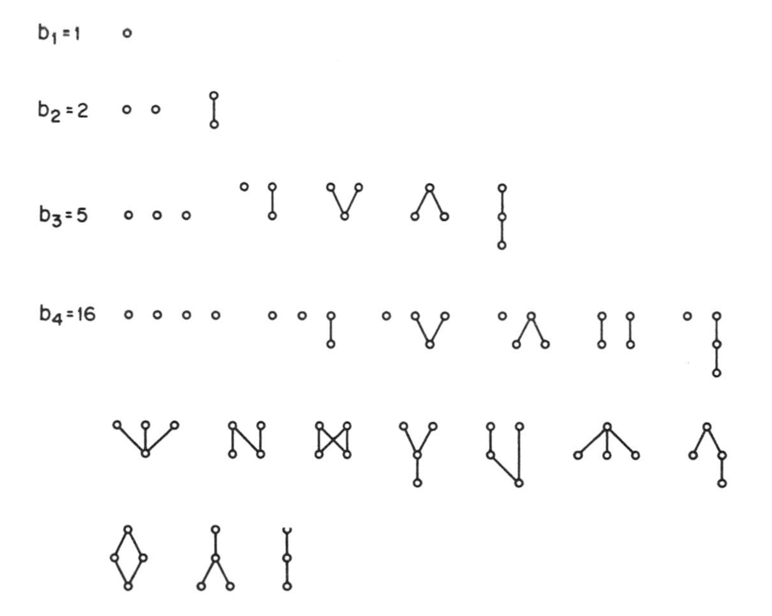

Fig. 8. Seq. ***588***, partially ordered sets.

This assumes that the graphs are unlabeled, i.e., that the elements of the set S are indistinguishable. If the elements of S are labeled 1 through n, the corresponding numbers are:

(1) 2^{n^2};
(2) $2^{n(n+1)/2}$;
(3) $2^{n(n-1)}$;
(4) $2^{\binom{n}{2}}$

(these four [(1)–(4)] are not in the table, but the sequences of their exponents are);

(5) seq. ***1476***: 1, 4, 29, 355, 6942, 209527, 9535241, ?;

(6) seq. ***585***: the Bell numbers or the number of equivalence relations on a set of n objects (see Fig. 22, p. 25);

(7) seq. ***1244***: 1, 3, 19, 219, 4231, 130023, 6129859, ?.

3.3 GEOMETRIES

The numbers of topologies were shown in Fig. 7; the following are also basic geometrical sequences:

A *linear space* is a system of (abstract) points and lines such that every two points lie on a unique line, and every line contains at least two points. A *geometry* is a system of points, lines, planes, . . . with an analogous definition. Figure 9 shows seq. ***462***: 1, 1, 2, 4, 9, 26, 101, 950, ?, the number of geometries with n points. (See Crapo and Rota [JM2 49 127 70]. The * denotes 5 points in general position in 4-dimensional space.) The planar figures in Fig. 9 form seq. ***271***: 1, 1, 2, 3, 5, 10, 24, 69, 384, ?, the number of linear spaces (Doyen [BSM 19 424 67]).

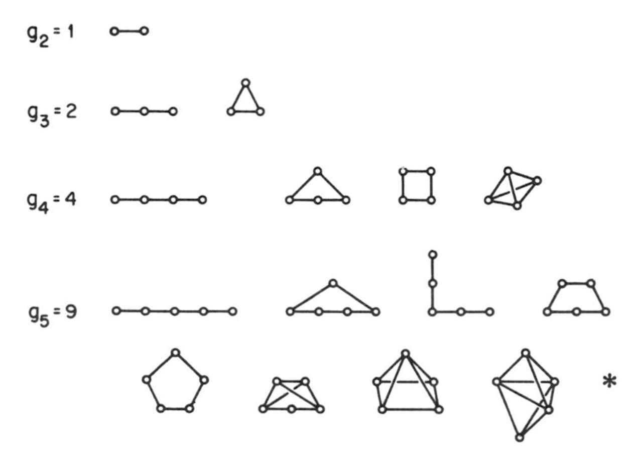

Fig. 9. Seq. ***462***, geometries (for * see text).

3.4 COMBINATIONS AND FIGURATE NUMBERS

The most basic combinatorial number is the *binomial coefficient*

$$\binom{n}{r} = \frac{n(n-1)(n-2)\cdots(n-r+1)}{1\cdot 2\cdot 3\cdots r} = \frac{n!}{r!(n-r)!},$$

which is the number of selections, or combinations, of n unlike things taken r at a time, has gf

$$(1+x)^n = \sum_{r=0}^{n} \binom{n}{r} x^r,$$

and is the $(r+1)$th term in the $(n+1)$th row of Pascal's triangle

$$\begin{array}{ccccccccccccc}
 & & & & & & 1 & & & & & & \\
 & & & & & 1 & & 1 & & & & & \\
 & & & & 1 & & 2 & & 1 & & & & \\
 & & & 1 & & 3 & & 3 & & 1 & & & \\
 & & 1 & & 4 & & 6 & & 4 & & 1 & & \\
 & 1 & & 5 & & 10 & & 10 & & 5 & & 1 & \\
1 & & 6 & & 15 & & 20 & & 15 & & 6 & & 1 \\
 & & & & & & \vdots & & & & & &
\end{array}$$

These are also called *figurate* numbers since they are the numbers of points in certain figures. For example, $\binom{n}{2}$ and $\binom{n}{3}$ are the *triangular* and *tetrahedral* numbers (Fig. 10, seqs. ***1002***, ***1363***).

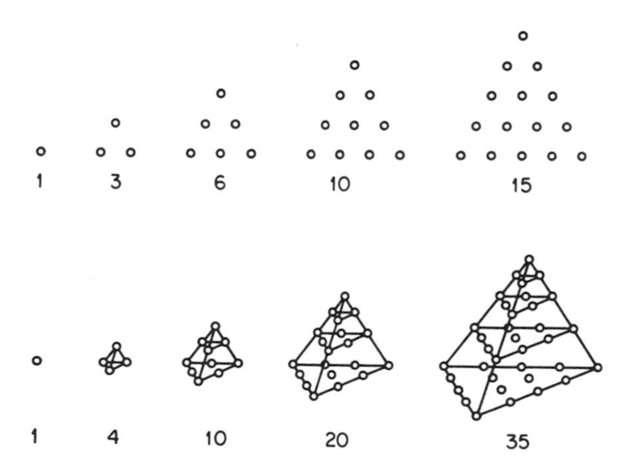

Fig. 10. Seqs. ***1002*** and ***1363***, the triangular and tetrahedral numbers.

Other examples of figurate numbers are the *polygonal* numbers $P(r, s) = \frac{1}{2}r(rs - s + 2)$. Figure 11 shows seq. ***1350***, the *square* numbers $P(r, 2) = r^2$; and seq. ***1562***, the *pentagonal* numbers $P(r, 3) = \frac{1}{2}r(3r - 1)$.

Many other figurate numbers, including cubes, fourth powers, etc., will be found in the table. For further pictures see Hogben [HO3].

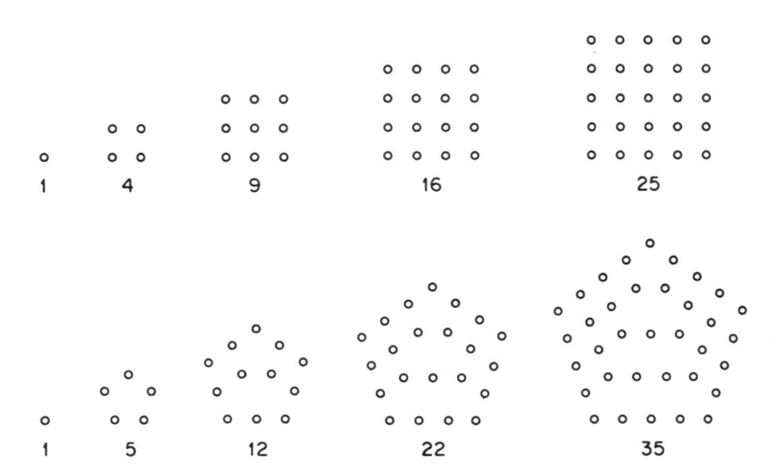

Fig. 11. Seqs. ***1350*** and ***1562***, the square and pentagonal numbers.

3.5 CATALAN NUMBERS AND DISSECTIONS

Next to the figurate numbers, the Catalan numbers are the most frequently occurring combinatorial numbers. (Gould [GO4] lists over 240 references.) They are defined by

$$c_n = \frac{1}{n+1}\binom{2n}{n},$$

and form seq. ***557***: 1, 2, 5, 14, 42, 132, 429, 1430, 4862, 16796, ... A gf is

$$1 + x + 2x^2 + 5x^3 + \cdots = (2x)^{-1}[1 - (1 - 4x)^{1/2}].$$

Some of the interpretations of c_n are:

(1) The number of ways of dissecting a convex polygon of $n+2$ sides into n triangles by drawing nonintersecting diagonals (Fig. 12a).

(2) The number of ways of completely parenthesizing a product of $n + 1$ letters (so that there are two factors inside each set of parentheses). The examples for $n = 1, 2, 3$ (arranged to show the correspondence with the dissections of Fig. 12a) are:

$n = 1$ (ab); $n = 2$ $a(bc)$, $(ab)c$;
$n = 3$ $(ab)(cd)$, $a((bc)d)$, $((ab)c)d$, $a(b(cd))$, $(a(bc))d$.

(3) The number of bifurcated rooted planar trees with $n + 1$ endpoints. (A planar tree is one which has been drawn on a plane, and bifurcated means that each edge splits in two at each node. See Fig. 12b. The trees are drawn to show the correspondence with the dissections and the parentheses.)

(4) In an election with two candidates A and B, each receiving n votes, c_n is the number of ways the votes can come in so that A is never behind B (Feller [FE1 1 71] and Comtet [CO1 1 94]).

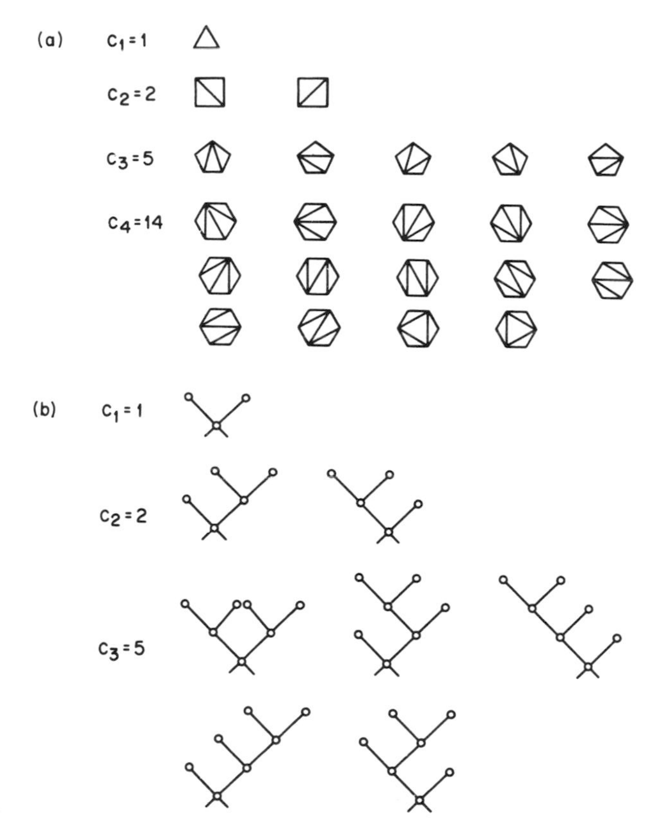

Fig. 12. Seq. *577*, the Catalan numbers.

Figure 13 illustrates seq. ***942***, the number of different dissections of a polygon when two dissections are considered to be the same if a rotation or reflection sends one into the other.

Figure 14 illustrates seq. ***391***, giving $\frac{1}{2}n(n + 1) + 1$, the maximum number of pieces obtained by slicing a pancake with n slices. The numbers of n-sided polygons in the nth diagram of Fig. 14 form seq. ***1181***: 0, 0, 1, 3, 12, 70, 465, 3507, 30016, . . . (Robinson [AMM 58 462 51]). Seq. ***491***: 2, 4, 8, 15, 26, 42, 64, 93, 130, 176, . . . gives $(n+2)(n+3)/6$, the maximum number of pieces obtained with n slices of a cake.

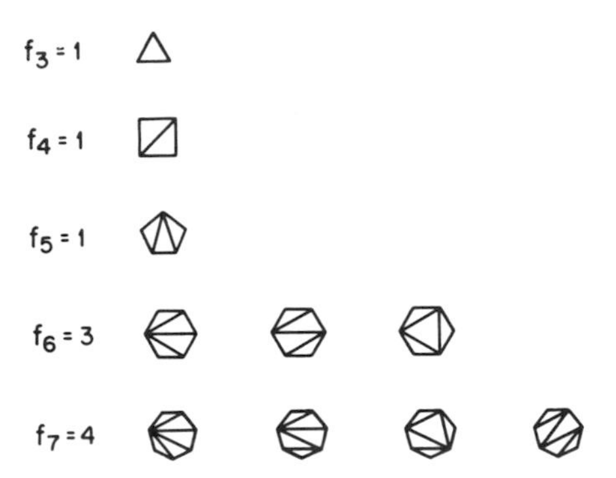

Fig. 13. Seq. ***942***, dissections of a polygon.

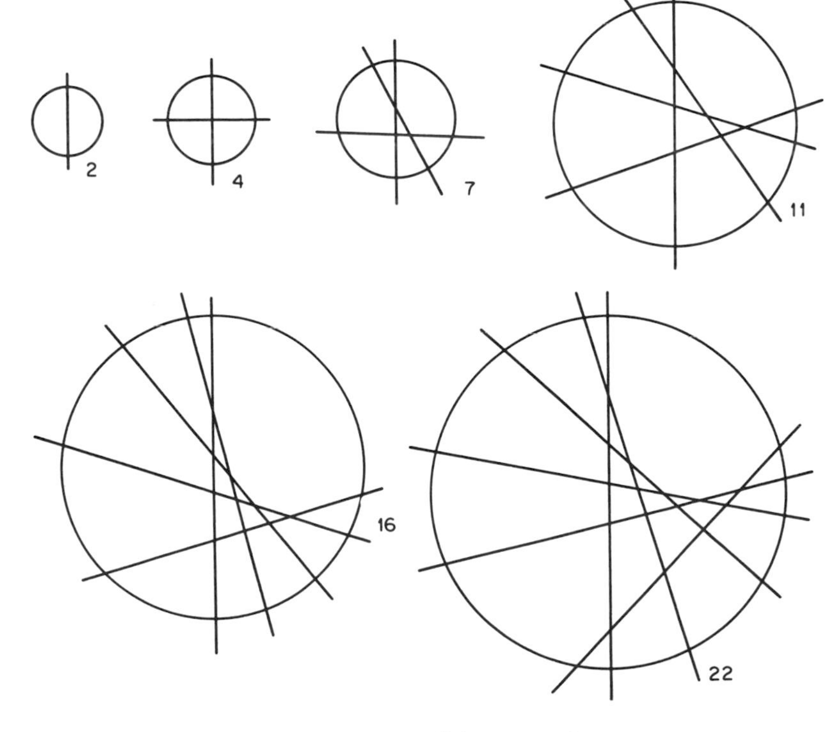

Fig. 14. Seq. ***391***, slicing a pancake.

3.6 NECKLACES AND IRREDUCIBLE POLYNOMIALS

Figure 15 illustrates seq. ***203***, T_n, the number of different necklaces that can be made from beads of two colors, when the necklaces can be rotated but not turned over. This is also the number of irreducible binary polynomials whose degree divides n, an important sequence in digital circuitry; and has the formula $T_n = \Sigma\ \phi(d) 2^{n/d}$, where $\phi(d)$ is the Euler totient function (seq. ***111***, Section 3.14) and the sum is over all divisors d of n. (See Berlekamp [BE2 70] and Golomb [CMA 1 358 69].) If turning over is allowed, the number of different necklaces is given by seq. ***202***: 2, 3, 4, 6, 8, 13, 18, 30, 46, 78, ... (See Gilbert and Riordan [IJM 5 657 61].)

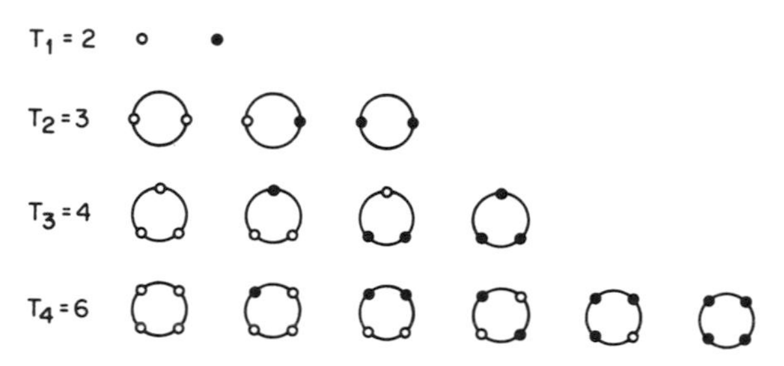

Fig. 15. Seq. ***203***, necklaces.

3.7 KNOTS

Figure 16 shows seq. ***322***: 0, 0, 1, 1, 2, 3, 7, 18, 41, 123, 367, ?, the number of knots with n crossings, in which the crossings alternate. (See Tait [TA1 1 334] and Conway [JL2 343].)

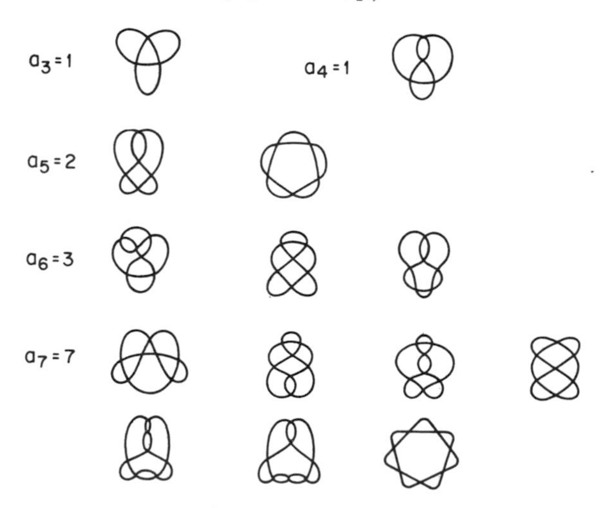

Fig. 16. Seq. ***322***, knots.

3.8 STAMPS

Figure 17 shows seq. ***576***: 1, 1, 2, 5, 14, 39, 120, 358, 1176, 3527, . . . (six more terms are known), the number of ways of folding a strip of stamps.

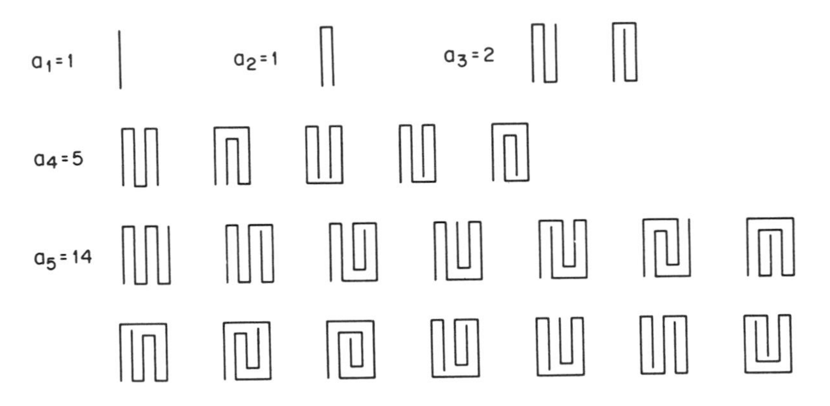

Fig. 17. Seq. ***576***, folding a strip of stamps.

3.9 POLYOMINOES

A *polyomino* with *p* squares is a connected set of *p* squares from a chessboard pattern. Polyominoes are *free* if they can be rotated and turned over (Fig. 18), and *fixed* otherwise. Unless otherwise stated, all polyominoes are free. Polyominoes may also be formed from triangles, rectangles, cubes (Fig. 19), etc. In no case is a formula known for the general term. (See Golomb [GO2].)

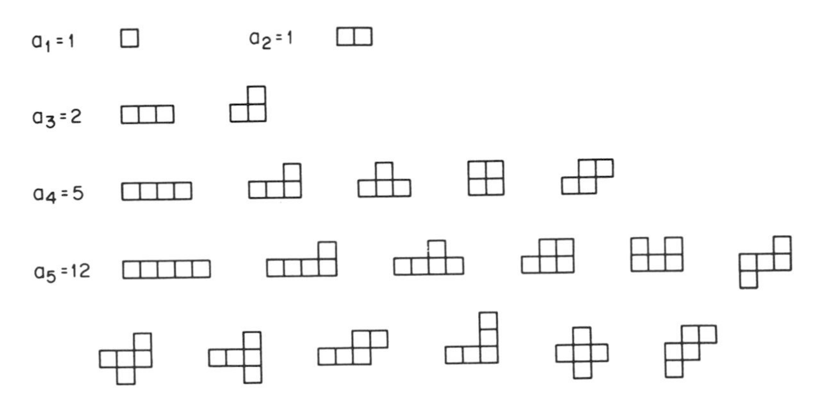

Fig. 18. Seq. ***561***, square polyominoes.

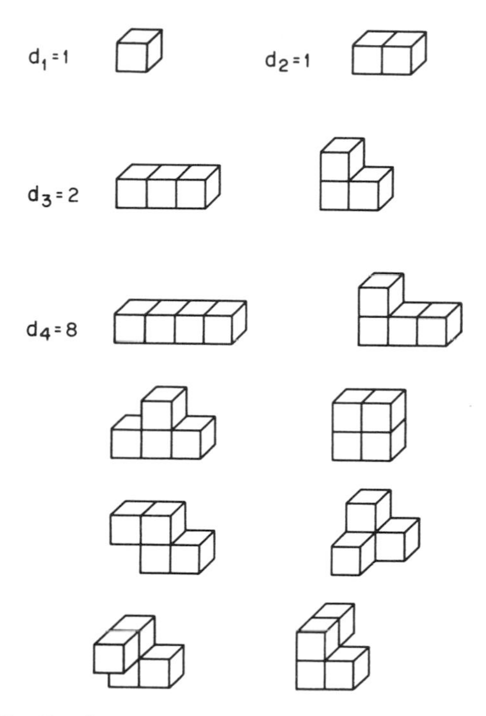

Fig. 19. Seq. *731*, polyominoes made from cubes.

3.10 BOOLEAN FUNCTIONS

A Boolean (or switching) function is a function $f(x_1, \ldots, x_n)$, where each variable x_i is 0 or 1, and f takes on the values 0 or 1.

These arise in the design of logical circuits, when the names of the variables do not matter. So it is natural to say that two such functions are equivalent if they differ only in the names of the variables (so that $x_1 + x_2x_3$ is equivalent to $x_2 + x_1x_3$), and to ask for the number of inequivalent functions. The answers to this (which is seq. ***1405***: 4, 12, 80, 3984, . . .) and to many similar questions (allowing complementation of the variables, etc.) are given by the Pólya counting theory (Section 3.11).

Two generalizations that will be found in the table are (i) Post functions, which are functions $f(x_1, \ldots, x_n)$, where each x_i and f can take any value from 0 to $m - 1$; and (ii) switching networks, which are n-input, k-output networks such that each of the outputs is a Boolean function of the n inputs. For details see Harrison [HA2, MU3 85].

3.11 PÓLYA COUNTING THEORY

A large number of counting problems involving graphs, necklaces, Boolean functions, and patterns of various kinds have been solved by the theorems of Redfield, Pólya, and De Bruijn. (See Riordan [R1 131], De Bruijn [BE6 144], Harrison [HA2 127, MU3 85], and Harary [HA5 178].)

3.12 PARTITIONS

The following are the most important sequences of partitions.

The main such sequence is number ***244***: 1, 2, 3, 5, 7, 11, . . . , giving the number of partitions of n into integer parts (Fig. 20). A gf is

$$1 + x + 2x^2 + 3x^3 + 5x^4 + \cdots = \prod_{i=1}^{\infty} (1 - x^i)^{-1}.$$

(See Gupta [RS2] and David *et al.* [DKB 273].)

Those partitions of n in which all parts are distinct form seq. ***100***: 1, 1, 2, 2, 3, 4, 5, . . . with gf

$$1 + x + x^2 + 2x^3 + 2x^4 + \cdots = \prod_{i=1}^{\infty} (1 + x^i).$$

The partitions of the even numbers into parts which are powers of two form the binary partition function $b(n)$, seq. ***378***: 1, 2, 4, 6, 10, 14, 20, 26, 36, 46, . . . , with recurrence $b(n) = b(n-1) + b([\frac{1}{2}n])$.

$p(1) = 1$	1
$p(2) = 2$	$2,\ 1^2$
$p(3) = 3$	$3,\ 21,\ 1^3$
$p(4) = 5$	$4,\ 31,\ 2^2,\ 21^2,\ 1^4$
$p(5) = 7$	$5,\ 41,\ 32,\ 31^2,\ 2^21,\ 21^3,\ 1^5$
$p(6) = 11$	$6,\ 51,\ 42,\ 41^2,\ 3^2,\ 321,\ 31^3,\ 2^3,\ 2^21^2,\ 21^4,\ 1^6$
$p(7) = 15$	$7,\ 61,\ 52,\ 51^2,\ 43,\ 421,\ 41^3,\ 3^21,\ 32^2,\ 321^2,\ 31^4,\ 2^31,\ 2^21^3,\ 21^5,\ 1^7$

Fig. 20. Seq. ***244***, the number of partitions of n.

Figure 21 illustrates the number of planar partitions of n, seq. ***1016***, with gf

$$1 + x + 3x^2 + 6x^3 + \cdots = \prod_{i=1}^{\infty} (1 - x^i)^{-i}.$$

Figure 22 shows $S(n, k)$, the *Stirling numbers of the second kind,* or the number of partitions of a set of n labeled objects into k parts.

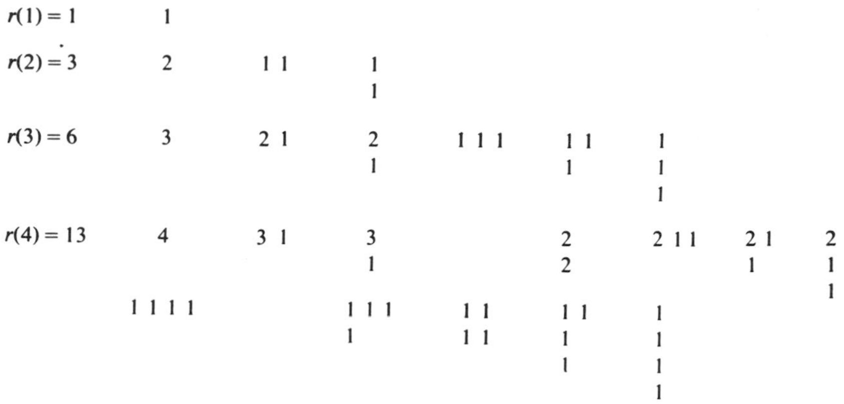

$r(1) = 1$ 1

$r(2) = 3$ 2   1 1   1
1

$r(3) = 6$ 3   2 1   2
1   1 1 1   1 1
1   1
1
1

$r(4) = 13$ 4   3 1   3
1   2
2   2 1 1   2 1
1   2
1
1

1 1 1 1   1 1 1
1   1 1
1 1   1 1
1
1   1
1
1
1

Fig. 21. Seq. ***1016***, planar partitions.

n \ k	1	2	3	4	Total
1	1				1
2	12	1, 2			2
3	123	1, 23 2, 13 3, 12	1, 2, 3		5
4	1234	1, 234 2, 134 3, 124 4, 123 12, 34 13, 24 14, 23	1, 2, 34 1, 3, 24 1, 4, 23 2, 3, 14 2, 4, 13 3, 4, 12	1, 2, 3, 4	15

Fig. 22. $S(n, k)$, the Stirling numbers of the second kind, and seq. ***585***, the Bell numbers.

The numbers continue:

							row sums $B(n)$
1							1
1	1						2
1	3	1					5
1	7	6	1				15
1	15	25	10	1			52
1	31	90	65	15	1		203
1	63	301	350	140	21	1	877
			⋮				⋮

A gf for $S(n, k)$ is

$$x^n = \sum_{k=0}^{n} S(n, k)\, x(x-1) \cdots (x-k+1).$$

Both the columns and diagonals of this array will be found in the main table.

The row sums are the *Bell numbers* $B(n)$, seq. ***585***. $B(n)$ is also the number of equivalence relations on a set of n objects (Section 3.2) and has gf

$$1 + x + 2\frac{x^2}{2!} + 5\frac{x^3}{3!} + \cdots = e^{e^x - 1}.$$

(See Abramowitz and Stegun [AS1 835], David *et al.* [DKB 223], and Comtet [CO1 2 38].)

3.13 PERMUTATIONS

A *permutation* of n objects is any rearrangement of them, and is specified either by a table:

1	2	3	4	5
3	5	4	1	2

or by a product of cycles: (134)(25), both of which mean replace 1 by 3, 3 by 4, 4 by 1, 2 by 5, and 5 by 2.

Figure 23 shows $s(n, k)$, the *Stirling numbers of the first kind,* or the numbers of permutations of n objects containing k cycles. The numbers continue:

							row sums $n!$
1							1
1	1						2
2	3	1					6
6	11	6	1				24
24	50	35	10	1			120
120	274	225	85	15	1		720
720	1764	1624	735	175	21	1	5040
			⋮				⋮

A gf for $s(n, k)$ is

$$x(x-1) \cdots (x-n+1) = \sum_{k=0}^{n} (-1)^{n-k} s(n, k) x^k.$$

Both the columns and diagonals of this array will be found in the main table. The row sums are the *factorial* numbers $n!$, seq. ***659***, the total number of permutations of n objects. References are as given above for the Stirling numbers of the second kind.

Factorial n is the product $1 \cdot 2 \cdot 3 \cdots n$ of the first n numbers. The products of the first n even numbers, $(2n)!! = 2 \cdot 4 \cdot 6 \cdots (2n) = 2^n \cdot n!$, seq. ***742***: 2, 8, 48, 384, 3840, 46080, . . . , and of the first n odd numbers, $(2n-1)!! = 1 \cdot 3 \cdot 5 \cdots (2n-1) = (2n)!/(2^n \cdot n!)$, seq. ***1217***: 1, 3, 15, 105, 945, 10395, . . . , are called *double factorials.*

n \ k	1	2	3	4	Total
1	(1)				1
2	(12)	(1)(2)			2
3	(123) (132)	(1)(23) (2)(13) (3)(12)	(1)(2)(3)		6
4	(1234) (1243) (1324) (1342) (1423) (1432)	(1)(234) (1)(243) (2)(134) (2)(143) (3)(124) (3)(142) (4)(123) (4)(132) (12)(34) (13)(24) (14)(23)	(1)(2)(34) (1)(3)(24) (1)(4)(23) (2)(3)(14) (2)(4)(13) (3)(4)(12)	(1)(2)(3)(4)	24

Fig. 23. $s(n, k)$, the Stirling numbers of the first kind; and seq. ***659***, the factorial numbers.

$D_2 = 1$

1 2
2 1

$D_3 = 2$

1 2 3   1 2 3
2 3 1   3 1 2

$D_4 = 9$

1 2 3 4   1 2 3 4   1 2 3 4
2 1 4 3   2 3 4 1   2 4 1 3

1 2 3 4   1 2 3 4   1 2 3 4
3 1 4 2   3 4 1 2   3 4 2 1

1 2 3 4   1 2 3 4   1 2 3 4
4 1 2 3   4 3 1 2   4 3 2 1

Fig. 24. Seq. ***766***, derangements.

Figure 24 shows D_n, the number of *derangements* of n objects, or the permutations in which every object is moved from its original position (seq. ***766***). These are also called subfactorial or *rencontres* numbers, and have the recurrence $D_n = nD_{n-1} + (-1)^n$. (See Riordan [R1 57].)

Figure 25 illustrates seq. ***587***, the *Euler numbers* E_n, or the number of permutations of n objects which first rise and then alternately fall and rise. (Only the second rows of the permutations are shown.)

The even numbered Euler numbers form seq. ***1667***: 1, 5, 61, 1385, 50521, . . . , and have gf

$$1 + 1\frac{x^2}{2!} + 5\frac{x^4}{4!} + 61\frac{x^6}{6!} + \cdots = \sec x.$$

(Often these are called the Euler numbers instead of seq. ***587***.)

The odd numbered Euler numbers form seq. ***829***: 1, 2, 16, 272, 7936, 353792, . . . , and are called the *tangent numbers* $T_n = E_{2n-1}$. They have gf

$$x + 2\frac{x^3}{3!} + 16\frac{x^5}{5!} + \cdots = \tan x.$$

$E_1 = 1$	1				
$E_2 = 1$	1 2				
$E_3 = 2$	1 3 2	2 3 1			
$E_4 = 5$	1 3 2 4	1 4 2 3	2 3 1 4	2 4 1 3	3 4 1 2
$E_5 = 16$	1 3 2 5 4	1 4 2 5 3	1 4 3 5 2	1 5 2 4 3	
	1 5 3 4 2	2 3 1 5 4	2 4 1 5 3	2 4 3 5 1	
	2 5 1 4 3	2 5 3 4 1	3 4 1 5 2	3 4 2 5 1	
	3 5 1 4 2	3 5 2 4 1	4 5 1 3 2	4 5 2 3 1	

Fig. 25. Seq. ***587***, the Euler numbers.

The *Bernoulli numbers* B_n are defined by

$$B_n = \frac{2nE_{2n-1}}{2^{2n}(2^{2n}-1)},$$

and form the sequence

$$\frac{1}{6}, \frac{1}{30}, \frac{1}{42}, \frac{1}{30}, \frac{5}{66}, \frac{691}{2730}, \frac{7}{6}, \frac{3617}{510}, \ldots$$

with gf

$$1 - \frac{x}{2} + \frac{1}{6}\frac{x^2}{2!} - \frac{1}{30}\frac{x^4}{4!} + \frac{1}{42}\frac{x^6}{6!} - \cdots = \frac{x}{e^x - 1}.$$

The numerators and denominators form seqs. ***1677*** and ***1746***.

Finally the *Genocchi numbers* are defined by $G_n = 2^{2-2n}\, n\, E_{2n-1}$, and form seq. ***1233***: 1, 1, 3, 17, 155, 2073, 38227, . . ., with gf

$$1\frac{x}{2!} + 1\frac{x^3}{4!} + 3\frac{x^5}{6!} + 17\frac{x^7}{8!} + \cdots = \tan\frac{1}{2}x .$$

The Euler, tangent, Bernoulli, and Genocchi numbers arise in all branches of mathematics. For applications and properties see Jordan [JO2], David and Barton [DB1], Comtet [CO1] and Gould [AMM 79 44 72]; for tables see Fletcher *et al.* [FMR 1 65] and Knuth and Buckholtz [MTAC 21 663 67].

3.14 SEQUENCES FROM NUMBER THEORY

The table contains many number-theoretic sequences, of which the following are typical:

(1) The prime numbers, lucky numbers, and other sequences generated by sieves (seqs. ***241***, ***377***, ***1035***, ***1048***);

(2) the Euler totient function $\phi(n)$: the number of integers not exceeding and relatively prime to n (seq. ***111***);

(3) from the Goldbach conjecture: the number of ways of writing $2n$ as a sum of two primes (various sequences—see index);

(4) quadratic partitions of primes: a prime of the form $4n + 1$ has a unique representation as $a^2 + b^2$ with $a \geqslant b$. Sequences ***169*** and ***33*** give a and b;

(5) the number of integers less than or equal to 2^n expressible in the form $u^2 + v^2$, where u and v are integers (seq. ***265***);

(6) Mersenne primes: the numbers n such that $2^n - 1$ is prime (seq. ***248***);

(7) from Euler's proof that there are an infinity of primes: let $p_1 = 2$, $p_2, \ldots, p_n$ be primes, and define p_{n+1} to be the smallest (largest) prime factor of $p_1 p_2 \cdots p_n + 1$ (seqs. ***329***, ***330***);

(8) Beatty sequences: if α, β are positive irrational numbers such that $(1/\alpha) + (1/\beta) = 1$, then the *Beatty sequences*

$$[\alpha], [2\alpha], [3\alpha], \ldots \qquad \text{and} \qquad [\beta], [2\beta], [3\beta], \ldots$$

together contain all the positive integers without repetition, where $[x]$ denotes the greatest integer less than or equal to x. (See Honsberger [HO2].) For example, $\alpha = \frac{1}{2}(1 + \sqrt{5}) = 1.61803\ldots$ gives seqs. ***917***: 1, 3, 4, 6, 8, 9, . . . and ***509***: 2, 5, 7, 10, 13, 15, ...

The following test for Beatty sequences is due to R. L. Graham. If $a_1, a_2, \ldots$ is a Beatty sequence, then the values of $a_1, \ldots, a_{n-1}$ determine

a_n to within 1. Look at the sums $a_1 + a_{n-1}, a_2 + a_{n-2}, \ldots, a_{n-1} + a_1$. If all these sums have the same value, V say, then a_n must equal V or $V + 1$; but if they take on the two values V and $V + 1$, and no others, then a_n must equal $V + 1$. If anything else happens, it is not a Beatty sequence. For example, in seq. ***917***, $a_1 + a_1 = 2$ so a_2 must be 2 or 3 (it is 3); $a_1 + a_2 = 4$ so a_3 must be 4 or 5 (it is 4); $a_1 + a_3 = 5$ and $a_2 + a_2 = 6$, so a_4 must be 6 (it is); and so on.

For further information about number-theoretic sequences see the comprehensive works of Dickson [DI2] and Lehmer [LE1].

3.15 PUZZLE SEQUENCES

This section describes some sequences with simple yet unexpected generating principles. They have all been given as puzzles at one time or another. Of course all of the sequences given in Chapters II and III make good puzzles.

(1) Sequences related to well-known constants (e.g., seq. ***1291***: 1, 4, 1, 4, 2, 1, 3, 5, 6, 2, 3, . . . , the decimal expansion of $\sqrt{2}$) or to other common sequences (seq. ***2127***: 1, 15, 29, 12, 26, 12, 26, 9, . . . is related to the calendar—guess!). See also seqs. ***684***, ***880***, ***1679***, ***1812***, etc.

(2) Sequences depending on the binary expansions of numbers (e.g., seq. ***41***: 1, 2, 1, 2, 2, 3, 1, 2, 2, . . . gives the number of 1's in the binary expansion of $n + 1$; see also seqs. ***360***, ***388***).

(3) Sequences depending on the English words or Arabic numerals used to describe them (e.g., seq. ***2218***: 1, 21, 21000, 101, 121, 1101, . . . , the smallest number requiring n words in English; see also seqs. ***1818***, ***1897***).

(4) The terms not in some well-known sequence (e.g., seq. ***1319***: 4, 6, 7, 9, 10, 11, 12, 14, 15, . . . , the non-Fibonacci numbers).

(5) Sequences obtained by *bisecting* (i.e., taking every other term of) well-known sequences (e.g., seq. ***1067***: 1, 3, 7, 18, 47, 123, 322, . . . , a bisection of seq. ***924***, the Lucas numbers; see also seqs. ***569***, ***1101***).

(6) Sequences obtained by alternating the terms of two sequences (seq. ***889***: 3, 2, 1, 7, 4, 1, 1, 8, 5, 2, 9, . . . , mixing π and e, is the only example given).

The following pleasing puzzles are not in the table because they are finite or are not integers.

(7) $\frac{1}{4}, \frac{1}{2}$, 1, 3, 6, 12, 24, 30, 120, 240, 1200, 2400, English money in 1950.

(8) 3, 8, 8, 4, 89, 75, 30, 28, ?, planetary diameters in thousands of statute miles.

(9) 8, 5, 4, 9, 1, 7, 6, 3, 2, 0; or 8, 8000000000, . . . , 18, 18000000000, . . . , 18000000, . . . , 18000, . . . , 80, . . . , 88, . . . , 85, . . . , 84, . . . , 11, . . . , 15, . . . , 5, . . . , 4, . . . , the numbers arranged in alphabetical order (in English).

(10) 12, 13, 14, 15, 20, 22, 30, 110, 1100, the number 12 written to the bases 10, 9, 8, . . . , 2.

(11) 14, 18, 23, 28, 34, 42, 50, 59, 66, 72, 79, 86, 96, 103, 110, 116, 125, 137, 145, 157, 168, 181, 191, 207, 215, 225, 231, 238, 242, the local stops on the New York IRT subway.

(12) 1714, 1727, 1760, 1820, 1910, 1936, dates of the accessions of the Georges to the English throne.

(13) 1732, 1735, 1743, 1751, 1758, 1767, 1767, 1782, 1773, 1790, 1795, 1784, 1800, 1804, 1791, 1809, 1808, 1822, 1822, 1831, 1830, 1837, 1833, 1837, 1843, 1858, 1857, 1856, 1865, 1872, 1874, 1882, 1884, 1890, 1917, 1908, 1913, dates of birth of presidents of the U.S.A.

(14) The integers 1, 2, 3, . . . drawn next to a mirror. (See Fig. 26.)

(15) O, T, T, F, F, S, S, E, N, T, E, T, T, F, F, S, S, E, N, T, T, T, T, . . . , the initial letters of the English names for the numbers.

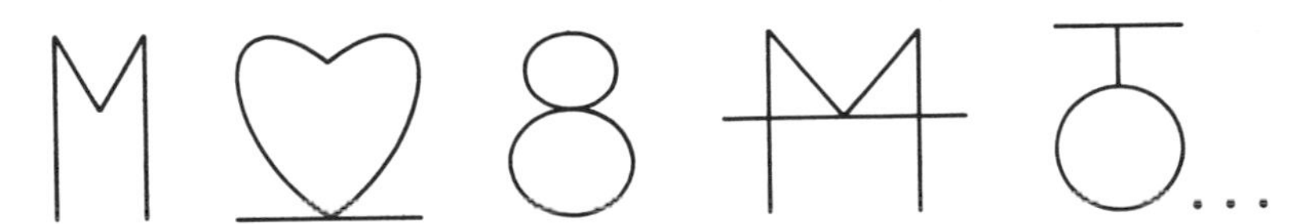

Fig. 26. A puzzle.

3.16 SEQUENCES FROM LATTICE STUDIES IN PHYSICS

In the last twenty years physicists have studied a number of basic combinatorial problems related to crystal lattices. Typical problems are to find the number of self-avoiding paths of length *n* on a given lattice, or the number of ways a particular graph can be drawn on the lattice. A number of such sequences will be found in the main table. For further information see Montroll [BE6 96], Sykes *et al.* [JMP 7 1557 66], Kasteleyn [HA1 43], Percus [PE3], and Domb [ACP 15 229 69].

THE MAIN TABLE OF SEQUENCES

SEQUENCES BEGINNING 1, 2, 0 AND 1, 2, 1

1 1, 2, 0, 0, 1, 1, 2, 0, 0, 2, 0, 0, 1, 2, 1, 0, 2, 0, 0, 0, 0, 3, 2, 0, 0, 2, 0, 0, 1, 0, 2, 0, 1, 2, 0, 0, 2, 2, 0, 0, 0, 2, 0, 0, 0, 1, 3, 0, 2, 2, 0, 0, 0, 0, 2, 0, 0, 2, 0, 0, 1, 4, 0, 0, 2, 0, 0, 0, 1
RELATED TO THE DIVISORS OF N. REF QJM 20 164 1884.

2 1, 2, 0, 1, 0, 0, 1, 2, 0, 0, 1, 0, 0, 2, 0, 2, 0, 0, 0, 1, 0, 1, 2, 0, 0, 2, 0, 0, 0, 0, 1, 2, 0, 2, 0, 0, 0, 2, 0, 0, 0, 0, 1, 3, 0, 0, 2, 0, 0, 0, 0, 2, 0, 0, 0, 2, 0, 2, 1, 0, 0, 2, 0, 0, 0, 0, 0, 2, 0
RELATED TO THE DIVISORS OF N. REF MES 31 67 01. LE1 8.

3 1, 2, 0, 1, 1, 1, 0, 0, 3, 2, 1, 1, 2, 1, 2, 1, 2, 1, 0, 2, 3, 3, 1, 0, 2, 4, 1, 2, 4, 1, 1, 3, 5, 4, 1, 2, 3, 4, 4, 3, 5, 3, 1, 4, 8, 6, 1, 2, 7, 6, 4, 8, 6, 3, 4, 6, 10, 8, 4, 5, 6, 10, 10, 7, 10, 7, 4
RELATED TO PARTITIONS INTO PRIMES. REF PNISI 21 187 55.

4 1, 2, 0, 2, 1, 1, 2, 1, 2, 2, 2, 2, 3, 2, 4, 3, 4, 4, 4, 5, 5, 5, 6, 5, 6, 7, 6, 9, 7, 9, 9, 9, 11, 11, 11, 13, 12, 14, 15, 15, 17, 16, 18, 19, 20, 21, 23, 22, 25, 26, 27, 30, 29, 32, 32, 35
PARTITIONS INTO DISTINCT PRIMES. REF PNISI 21 186 55. PURB 107 285 57.

5 1, 2, 0, 2, 2, 4, 3, 2, 8, 1, 8, 8, 12, 11, 4, 25, 4, 24, 21, 40, 31, 16, 82, 14
FROM SYMMETRIC FUNCTIONS. REF PLMS 23 309 23.

6 1, 2, 0, 2, 3, 2, 6, 4, 9, 14, 11, 26, 29, 34, 62, 68, 99, 140, 169, 252, 322, 430, 607, 764, 1059, 1424, 1845, 2546
FROM SYMMETRIC FUNCTIONS. REF PLMS 23 315 23.

7 1, 2, 0, 3, 1, 1, 0, 3, 3, 2, 2, 4, 0, 5, 2, 2, 3, 3, 0, 1, 1, 3, 0, 2, 1, 1, 0, 4, 5, 3, 7, 4, 0, 1, 1, 2, 0, 3, 1, 1, 0, 3, 3, 2, 2, 4, 4, 5, 5, 2, 3, 3, 0, 1, 1, 3, 0, 2, 1, 1, 0, 4, 5, 3, 7, 4, 8, 1, 1
THE GAME OF DAWSONS KAYLES. REF PCPS 52 517 56.

8 1, 2, 0, 4, 9, 18, 17, 0, 24, 35, 36, 12, 40, 11, 0, 13, 56, 30, 79, 45, 39, 67, 100, 0, 133, 83, 48, 53, 104, 138, 7, 163, 100, 26, 0, 28, 116, 217, 9, 248, 104, 17, 80, 79, 8, 139
THE MINIMAL SEQUENCE. REF AT1 177.

9 1, 2, 0, 8, 24, 72, 240, 896, 3640, 15688, 70512
ENERGY FUNCTION FOR SQUARE LATTICE. REF PHA 28 926 62.

10 1, 2, 0, 9, 35, 230, 1624, 13209, 120287
LOGARITHMIC NUMBERS. REF MST 31 77 63.

11 1, 2, 0, 12, 14, 90, 192, 792, 2148, 7716, 23262, 79512, 252054
MAGNETISATION FOR CUBIC LATTICE. REF PHA 22 936 56.

12 1, 2, 1, 0, 1, 2, 3, 2, 1, 0, 1, 2, 3, 4, 3, 2, 3, 2, 1, 0, 1, 2, 3, 4, 5, 6, 5, 4, 3, 2, 3, 2, 1, 0, 1, 2, 3, 4, 5, 4, 5, 6, 5, 6, 5, 6, 7, 6, 5, 4, 3, 2, 3, 2, 3, 2, 3, 2, 1, 2, 3, 4, 3, 4, 5, 6, 7, 6, 7
LIOUVILLES FUNCTION. REF PURB 3 48 50.

13 1, 2, 1, 0, 2, 0, 1, 3, 0, 2, 2, 0, 0, 0, 1, 2, 3, 2, 0, 0, 2, 0, 2, 1, 0, 4, 0, 0, 0, 0, 1, 4, 2, 0, 3, 0, 2, 0, 0, 2, 0, 2, 2, 0, 0, 0, 2, 1, 1, 4, 0, 0, 4, 0, 0, 4, 0, 2, 0, 0, 0, 0, 1, 0, 4, 2, 2, 0, 0
RELATED TO THE DIVISORS OF N. REF MES 31 86 01. LE1 9.

14 1, 2, 1, 0, 2, 1, 0, 2, 1, 3, 2, 1, 3, 2, 4, 3, 0, 4, 3, 0, 4, 3, 0, 4, 1, 2, 3, 1, 2, 4, 1, 2, 4, 1, 2, 4, 1, 5, 4, 1, 5, 4, 1, 5, 4, 1, 0, 2, 1, 0, 2, 1, 5, 2, 1, 3, 2, 1, 3, 2, 4, 3, 2, 4, 3, 2, 4, 3, 2
GRUNDYS GAME. REF PCPS 52 525 56.

15 1, 2, 1, 0, 4, 2, 3, 1, 0, 6, 3, 2, 180, 4, 1, 0, 8, 4, 39, 2, 12, 42, 5, 1, 0, 10, 5, 24, 1820, 2, 273, 3, 4, 6, 1, 0, 12, 6, 4, 3, 320, 2, 531, 30, 24, 3588, 7, 1, 0, 14, 7, 90, 9100, 66
SOLUTIONS OF PELLIANS. REF DE1. CAY 13 430. LE1 55.

16 1, 2, 1, 1, 1, 1, 1, 1, 2, 2, 1, 2, 2, 2, 1, 1, 2, 1, 2, 2, 3, 3, 2, 3, 2, 2, 2, 1, 2, 1, 3, 3, 3, 4, 3, 3, 2, 3, 3, 1, 2, 3, 4, 4, 3, 4, 3, 4, 3, 3, 3, 3, 3, 4, 5, 5, 3, 3, 4, 4, 3, 2, 4, 3, 4, 4, 4, 6, 5
PARTITIONS INTO NOT MORE THAN 5 PENTAGONAL NUMBERS. REF LINR 1 4 46. MTAC 2 301 47.

17 1, 2, 1, 1, 1, 1, 2, 2, 1, 1, 1, 2, 1, 1, 1, 1, 2, 1, 2, 1, 1, 3, 1, 1, 1, 1, 2, 2, 1, 2, 1, 2, 1, 1, 1, 1, 2, 2, 1, 2, 2, 2, 1, 1, 1, 1, 3, 1, 1, 1, 1, 2, 1, 1, 1, 1, 2, 2, 2, 1, 1, 2, 2, 1, 1, 1, 2, 2, 1
LEONARDO LOGARITHMS. REF HM1. MTAC 23 460 69. ACA 16 109 69.

18 1, 2, 1, 1, 1, 1, 2, 2, 1, 2, 2, 1, 1, 2, 2, 3, 2, 2, 2, 2, 1, 1, 3, 3, 3, 3, 2, 2, 2, 1, 3, 4, 2, 4, 3, 3, 2, 2, 3, 4, 3, 2, 4, 2, 2, 2, 4, 5, 3, 5, 3, 5, 3, 1, 4, 5, 3, 3, 4, 3, 4, 2, 4, 6, 4, 4, 4, 5, 2
PARTITIONS INTO NOT MORE THAN 4 SQUARES. REF LINR 1 10 46. MTAC 2 301 47.

19 1, 2, 1, 1, 1, 2, 1, 2, 1, 2, 2, 1, 1, 2, 1, 2, 2, 2, 1, 2, 1, 2, 1, 2, 1, 3, 1, 1, 2, 2, 2, 2, 1, 2, 2, 2, 1, 3, 1, 2, 2, 2, 1, 2, 1, 2, 2, 2, 1, 2, 2, 2, 2, 2, 1, 3, 1, 2, 2, 1, 2, 3, 1, 2, 2, 3, 1, 2, 1
NUMBER OF DISTINCT PRIMES DIVIDING N + 4. REF AS1 844.

20 1, 2, 1, 1, 1, 3, 2, 1, 1, 2, 1, 1, 1, 5, 1, 2, 1, 2, 1, 1, 1, 3, 2, 1, 3, 2, 1, 1, 1, 7, 1, 1, 1, 4, 1, 1, 1, 3, 1, 1, 1, 2, 2, 1, 1, 5, 2, 2, 1, 2, 1, 3, 1, 3, 1, 1, 1, 2, 1, 1, 2, 11, 1, 1, 1, 2, 1, 1, 1
NUMBER OF ABELIAN GROUPS OF ORDER N + 2. REF MZT 56 21 52.

21 1, 2, 1, 1, 2, 1, 2, 1, 2, 1, 1, 2, 1, 2, 1, 1, 2, 1, 2, 1, 1, 2, 1, 2, 1, 2, 1, 1, 2, 1, 2, 1, 1, 2, 1, 2, 1, 2, 1, 1, 2, 1, 2, 1, 1, 2, 1, 2, 1, 2, 1, 1, 2, 1, 2, 1, 1, 2, 1, 2, 1, 1, 2, 1, 2, 1, 2, 1, 1
A SELF-GENERATING SEQUENCE. REF MMAG 36 180 63.

22 1, 2, 1, 1, 2, 1, 2, 1, 6, 1, 2, 3, 2, 1, 1, 1, 1, 2, 8, 2, 1, 8, 2, 1, 2, 1, 3, 4, 18, 1, 2, 1, 1, 10, 3, 1, 2, 1, 1, 1, 2, 2, 1, 2, 1, 6, 1, 3, 8, 2, 10, 5, 16, 2, 1, 2, 3, 4, 3, 1, 3, 2, 2, 1, 11, 16, 1
FERMAT QUOTIENTS. REF BE4 35 666 13. KR1 1 131. LE1 10.

23 1, 2, 1, 1, 2, 1, 2, 3, 2, 1, 3, 2, 2, 3, 2, 2, 4, 2, 3, 4, 2, 3, 5, 1, 4, 5, 2, 3, 5, 1, 3, 5, 3, 3, 5, 3, 5, 7, 3, 5, 7, 4, 4, 7, 3, 3, 7, 4, 3, 9, 5, 3, 7, 5, 3, 8, 5, 4, 8, 5, 3, 7, 5, 3, 9, 4, 3, 12, 6
DECOMPOSITIONS OF 2N INTO SUM OF 2 LUCKY NUMBERS. REF MMAG 29 119 59. ST4. MTAC 19 332 65.

24 1, 2, 1, 1, 2, 1, 4, 2, 0, 1, 0, 7, 3, 8, 30, 8, 23, 2, 14, 10, 28, 2, 18, 4, 24, 8, 12, 4, 0, 28, 33, 28, 44, 40, 2, 22, 16, 18, 25, 7, 22, 12, 12, 10, 30, 8, 42, 27, 34, 32, 47, 27, 39, 25
EXPANSION OF AN INFINITE PRODUCT. REF JRAM 221 218 66.

25 1, 2, 1, 1, 2, 2, 1, 1, 0, 3, 2, 1, 2, 1, 1, 2, 2, 2, 2, 2, 1, 3, 1, 0, 1, 3, 2, 2, 2, 1, 3, 2, 0, 2, 1, 1, 4, 2, 1, 3, 2, 2, 2, 2, 1, 4, 2, 1, 1, 3, 3, 3, 3, 4, 1, 5, 3, 2, 4, 3, 2, 4, 2, 1, 2, 4, 2, 3, 1
PARTITIONS INTO A PRIME AND A SQUARE. REF AM3 47 30 43.

26 1, 2, 1, 1, 2, 2, 1, 2, 1, 2, 3, 2, 2, 1, 2, 1, 2, 2, 2, 2, 1, 3, 2, 2, 2, 2, 1, 4, 2, 1, 2, 2, 2, 1, 2, 4, 2, 2, 2, 1, 3, 2, 2, 2, 2, 2, 2, 2, 1, 2, 4, 1, 4, 2, 2, 1, 4, 2, 2, 2, 2, 2, 2, 1, 2, 3, 4, 2, 2
RELATED TO THE DIVISORS OF N. REF PLMS 15 106 1884.

27 1, 2, 1, 1, 2, 2, 1, 2, 2, 1, 3, 2, 1, 2, 3, 2, 2, 2, 1, 4, 3, 2, 2, 2, 2, 3, 3, 1, 4, 4, 2, 2, 3, 2, 3, 4, 2, 3, 3, 2, 4, 3, 2, 4, 4, 2, 4, 4, 1, 4, 5, 1, 2, 3, 4, 6, 4, 3, 2, 5, 2, 3, 3, 3, 6, 5, 2, 2, 5
PARTITIONS INTO NOT MORE THAN 3 TRIANGULAR NUMBERS. REF LINR 1 12 46. MTAC 2 301 47.

28 1, 2, 1, 2, 1, 2, 2, 0, 2, 2, 1, 0, 0, 2, 3, 2, 2, 0, 0, 2, 2, 0, 0, 2, 1, 0, 2, 2, 2, 2, 1, 2, 0, 2, 2, 2, 2, 0, 2, 0, 4, 0, 0, 0, 1, 2, 0, 0, 2, 0, 2, 2, 1, 2, 0, 2, 2, 0, 0, 2, 0, 2, 0, 2, 2, 0, 4, 0, 0
THE SQUARE OF EULERS PRODUCT. REF PLMS 21 190 1889.

29 1, 2, 1, 2, 1, 2, 2, 3, 4, 1, 5, 3, 6, 8, 3, 4, 8, 3, 0, 2, 8, 4, 4, 13, 9, 5, 18, 2, 2, 8, 3, 10, 0, 4, 2, 19, 14, 7, 8, 0, 20, 4, 1, 8, 2, 15, 7, 8, 26, 10, 26, 18, 10, 2, 10, 28, 29
RELATED TO EULERS PRODUCT. REF JRAM 179 128 38.

30 1, 2, 1, 2, 1, 2, 3, 1, 3, 2, 1, 2, 3, 3, 1, 3, 2, 1, 3, 2, 3, 4, 2, 1, 2, 1, 2, 7, 2, 3, 1, 5, 1, 3, 3, 2, 3, 3, 1, 5, 1, 2, 1, 6, 6, 2, 1, 2, 3, 1, 5, 3, 3, 3, 1, 3, 2, 1, 5, 7, 2, 1, 2, 7, 3, 5, 1, 2, 3
DIFFERENCES BETWEEN CONSECUTIVE PRIMES. REF AS1 870 (DIVIDED BY 2).

31 1, 2, 1, 2, 1, 3, 2, 2, 1, 3, 1, 2, 2, 4, 1, 3, 1, 3, 2, 2, 1, 4, 2, 2, 3, 3, 1, 3, 1, 5, 2, 2, 2, 4, 1, 2, 2, 4, 1, 3, 1, 3, 3, 2, 1, 5, 2, 3, 2, 3, 1, 4, 2, 4, 2, 2, 1, 4, 1, 2, 3, 6, 2, 3, 1, 3, 2, 3, 1
NUMBER OF PRIMES DIVIDING N + 2. REF AS1 844.

32 1, 2, 1, 2, 1, 3, 2, 2, 1, 4, 1, 2, 2, 5, 1, 4, 1, 4, 2, 2, 1, 7, 2, 2, 3, 4, 1, 5, 1, 7, 2, 2, 2, 9, 1, 2, 2, 7, 1, 5, 1, 4, 4, 2, 1, 12, 2, 4, 2, 4, 1, 7, 2, 7, 2, 2, 1, 11, 1, 2, 4, 11, 2, 5, 1, 4, 2, 5
NUMBER OF WAYS OF FACTORIZING N + 2. REF EUR 18 17 55.

33 1, 2, 1, 2, 1, 4, 2, 5, 3, 5, 4, 1, 3, 7, 4, 7, 6, 2, 9, 7, 1, 2, 8, 4, 1, 10, 9, 5, 2, 12, 11, 9, 5, 8, 7, 10, 6, 1, 3, 14, 12, 7, 4, 10, 5, 11, 10, 14, 13, 1, 8, 5, 17, 16, 4, 13, 6, 12, 1, 5, 15
QUADRATIC PARTITIONS OF PRIMES. REF CU2 1. AMM 56 526 49.

34 1, 2, 1, 2, 1, 4, 3, 13, 5, 1, 1, 8, 1, 2, 4, 1, 1, 40, 1, 11, 3, 7, 1, 7, 1, 1, 5, 1, 49, 4, 1, 65, 1, 4, 7, 11, 1, 399, 2, 1, 3, 2, 1, 2, 1, 5, 3, 2, 1, 10, 1, 1, 1, 1, 2, 1, 1, 3, 1, 4, 1, 1, 2, 5, 1
CONTINUED FRACTION EXPANSION OF EULERS CONSTANT. REF LE4. MTAC 25 403 71.

35 1, 2, 1, 2, 1, 5, 2, 2, 1, 5, 1, 2, 1, 14, 1, 5, 1, 5, 2, 2, 1, 15, 2, 2, 5, 4, 1, 4, 1, 51, 1, 2, 1, 14, 1, 2, 2, 14, 1, 6, 1, 4, 2, 2, 1, 52, 2, 5, 1, 5, 1, 15, 2, 13, 2, 2, 1, 13, 1, 2, 4, 267, 1, 4
NUMBER OF GROUPS OF ORDER N + 2. REF AJM 52 617 30. HS1. CM1 134.

36 1, 2, 1, 2, 2, 1, 2, 1, 2, 2, 1, 2, 2, 1, 2, 1, 2, 2, 1, 2, 1, 2, 2, 1, 2, 2, 1, 2, 1, 2, 2, 1, 2, 2, 1, 2, 1, 2, 2, 1, 2, 1, 2, 2, 1, 2, 2, 1, 2, 1, 2, 2, 1, 2, 1, 2, 2, 1, 2, 2, 1, 2, 1, 2, 2, 1, 2, 2, 1
A SELF-GENERATING SEQUENCE. REF AMM 64 198 57.

37 1, 2, 1, 2, 2, 1, 3, 2, 2, 3, 1, 3, 3, 2, 4, 2, 3, 3, 1, 4, 3, 3, 5, 2, 4, 4, 2, 5, 3, 3, 4, 1, 4, 4, 3, 6, 3, 5, 5, 2, 6, 4, 4, 6, 2, 5, 5, 3, 6, 3, 4, 4, 1, 5
REPRESENTATIONS AS A SUM OF FIBONACCI NUMBERS. REF FQ 4 305 66.

38 1, 2, 1, 2, 2, 2, 1, 2, 2, 3, 2, 1, 1, 2, 2, 3, 3, 2, 1, 2, 2, 2, 1, 1, 1, 2, 3, 4, 4, 3, 2, 1, 1, 2, 1, 0, 0, 1, 2, 3, 3, 3, 2, 3, 3, 3, 3, 2, 2, 3, 3, 2, 2, 1, 0, 1, 1, 2, 1, 1, 1, 0, 1, 2, 2, 1, 2, 3, 3
MERTENS FUNCTION. REF WIEN 106(2A) 843 1897. LE1 7.

39 1, 2, 1, 2, 2, 2, 2, 3, 2, 4, 3, 4, 4, 4, 5, 5, 5, 6, 5, 6, 7, 6, 9, 7, 9, 9, 9, 11, 11, 11, 13, 12, 14, 15, 15, 17, 16, 18, 19, 20, 21, 23, 22, 25, 26, 27, 30, 29, 32, 32, 35, 37, 39, 40, 42
PARTITIONS INTO DISTINCT PRIMES. REF PNISI 21 186 55. PURB 107 285 57.

40 1, 2, 1, 2, 2, 2, 2, 3, 3, 3, 2, 3, 2, 4, 4, 2, 3, 4, 3, 4, 5, 4, 3, 5, 3, 4, 6, 3, 5, 6, 2, 5, 6, 5, 5, 7, 4, 5, 8, 5, 4, 9, 4, 5, 7, 3, 6, 8, 5, 6, 8, 6, 7, 10, 6, 6, 12, 4, 5, 10, 3, 7, 9, 6, 5, 8, 7, 8
DECOMPOSITIONS OF 2N INTO SUM OF 2 ODD PRIMES. REF GR1 19. LE1 80.

41 1, 2, 1, 2, 2, 3, 1, 2, 2, 3, 2, 3, 3, 4, 1, 2, 2, 3, 2, 3, 3, 4, 2, 3, 3, 4, 3, 4, 4, 5, 1, 2, 2, 3, 2, 3, 3, 4, 2, 3, 3, 4, 3, 4, 4, 5, 2, 3, 3, 4, 3, 4, 4, 5, 3, 4, 4, 5, 4, 5, 5, 6, 1, 2, 2, 3, 2, 3, 3
NUMBER OF ONES IN BINARY EXPANSION OF N+1. REF FQ 4 374 66. ANY 175 177 70.

42 1, 2, 1, 2, 3, 1, 3, 1, 4, 1, 1, 2, 4, 5, 5, 1, 2, 3, 6, 3, 1, 5, 2, 4, 1, 7, 5, 3, 5, 7, 1, 5, 7, 3, 1, 4, 5, 6, 8, 1, 2, 7, 9, 4, 5, 3, 5, 2, 1, 9, 5, 6, 7, 10, 11, 3, 1, 4, 11, 6, 7, 8, 9, 7, 1, 4, 9, 5
QUADRATIC PARTITIONS OF PRIMES. REF CU2 1. LE1 55.

43 1, 2, 1, 2, 3, 3, 3, 1, 1, 3, 4, 2, 1, 3, 4, 1, 5, 3, 5, 5, 2, 4, 5, 3, 4, 2, 6, 1, 7, 7, 1, 3, 7, 5, 4, 5, 7, 8, 6, 8, 7, 7, 6, 3, 7, 9, 7, 9, 8, 1, 3, 9, 5, 6, 3, 7, 10, 1, 6, 4, 10, 7, 9, 5, 9, 2, 11
QUADRATIC PARTITIONS OF PRIMES. REF CU2 1. LE1 55.

44 1, 2, 1, 2, 3, 4, 8, 8, 18, 18, 38, 28, 142, 72, 234, 360, 669, 520, 2606, 1608, 7293
NECKLACES. REF IJM 8 269 64.

45 1, 2, 1, 2, 3, 6, 8, 16, 24, 42, 69, 124, 208, 378, 668, 1214, 2220, 4110, 7630, 14308, 26931
NECKLACES. REF IJM 5 663 61.

46 1, 2, 1, 2, 3, 6, 9, 18, 30, 56, 99, 186, 335, 630, 1161, 2182, 4080, 7710, 14532, 27594, 52377, 99858, 190557, 364722, 698870, 1342176, 2580795, 4971008, 9586395
IRREDUCIBLE POLYNOMIALS, OR NECKLACES. REF IJM 5 663 61. JSIAM 12 288 64.

47 1, 2, 1, 2, 5, 1, 1, 2, 1, 1, 3, 10, 2, 1, 3, 2, 24, 1, 3, 2, 3, 1, 1, 1, 90, 2, 1, 12, 1, 1, 1, 1, 5, 2, 6, 1, 6, 3, 1, 1, 2, 5, 2, 1, 2, 1, 1, 4, 1, 2, 2, 3, 2, 1, 1, 4, 1, 1, 2, 5, 2, 1, 1, 3, 29, 8, 3
CONTINUED FRACTION EXPANSION OF KHINTCHINES CONSTANT. REF MTAC 20 446 66.

48 1, 2, 1, 2, 5, 17, 92, 994, 28262, 2700791
THRESHOLD FUNCTIONS. REF PGEC 19 821 70.

49 1, 2, 1, 2, 9, 96, 2690, 226360, 64646855, 68339572672
THRESHOLD FUNCTIONS. REF PGEC 19 821 70.

50 1, 2, 1, 2, 1382, 4, 3617, 87734, 349222, 310732, 472728182, 2631724, 13571120588, 13785346041608, 7709321041217, 303257395102
MULTIPLES OF BERNOULLI NUMBERS. REF RO2 331. FMR 1 74.

51 1, 2, 1, 3, 1, 2, 1, 4, 1, 2, 1, 3, 1, 2, 1, 5, 1, 2, 1, 3, 1, 2, 1, 4, 1, 2, 1, 3, 1, 2, 1, 6, 1, 2, 1, 3, 1, 2, 1, 4, 1, 2, 1, 3, 1, 2, 1, 5, 1, 2, 1, 3, 1, 2, 1, 4, 1, 2, 1, 3, 1, 2, 1, 7, 1, 2, 1, 3, 1
NUMBER OF TWOS DIVIDING 2N. REF MMAG 40 164 67.

52 1, 2, 1, 3, 1, 3, 2, 5, 1, 6, 3, 2, 2, 8, 3, 9, 2, 3, 5, 11, 1, 10, 6, 9, 3, 14, 2, 15, 4, 5, 8, 6, 3, 18, 9, 6, 2, 20, 3, 21, 5, 6, 11, 23, 2, 21, 10, 8, 6, 26, 9, 10, 3, 9, 14, 29, 2, 30, 15, 3
REDUCED TOTIENT FUNCTION. REF CAU (2) 12 43. LE1 7. (DIVIDED BY 2.)

53 1, 2, 1, 3, 1, 4, 2, 3, 1, 8, 1, 3, 3, 8, 1, 8, 1, 8, 3, 3, 1, 20, 2, 3, 4, 8, 1, 13, 1, 16, 3, 3, 3, 26, 1, 3, 3, 20, 1, 13, 1, 8, 8, 3, 1, 48, 2, 8, 3, 8, 1, 20, 3, 20, 3, 3, 1, 44, 1, 3, 8, 32, 3, 13
PERFECT PARTITIONS. REF R1 124.

54 1, 2, 1, 3, 2, 1, 4, 1, 2, 5, 7, 4, 2, 6, 5, 1, 2, 3, 11, 6, 1, 5, 3, 10, 7, 12, 1, 2, 9, 4, 13, 7, 6, 5, 9, 14, 16, 8, 11, 2, 7, 3, 4, 10, 1, 17, 19, 2, 8, 14, 16, 1, 13, 20, 9, 3, 8, 5, 6, 11, 14
QUADRATIC PARTITIONS OF PRIMES. REF CU2 1. LE1 55.

55 1, 2, 1, 3, 2, 1, 5, 2, 1, 4, 6, 3, 2, 7, 4, 3, 1, 7, 4, 9, 1, 8, 5, 10, 4, 7, 3, 2, 5, 8, 12, 2, 1, 9, 11, 8, 4, 7, 2, 1, 14, 6, 9, 5, 11, 13, 2, 14, 16, 4, 11, 8, 3, 2, 7, 10, 17, 12, 11, 1, 7, 13
QUADRATIC PARTITIONS OF PRIMES. REF CU2 1. LE1 55.

56 1, 2, 1, 3, 2, 3, 1, 4, 3, 5, 2, 5, 3, 4, 1, 5, 4, 7, 3, 8, 5, 7, 2, 7, 5, 8, 3, 7, 4, 5, 1, 6, 5, 9, 4, 11, 7, 10, 3, 11, 8, 13, 5, 12, 7, 9, 2, 9, 7, 12, 5, 13, 8, 11, 3, 10, 7, 11, 4, 9, 5, 6, 1, 7
A(2N) = A(N), A(2N + 1) = A(N + 1) + A(N). REF ELM 2 95 47.

57 1, 2, 1, 3, 2, 3, 2, 1, 4, 5, 4, 1, 6, 3, 5, 7, 6, 7, 2, 8, 1, 7, 3, 6, 8, 5, 6, 3, 9, 8, 5, 4, 10, 11, 2, 11, 6, 4, 10, 12, 9, 12, 11, 1, 9, 13, 2, 7, 13, 4, 12, 13, 14, 11, 7, 9, 10, 4, 15, 14, 9, 6
QUADRATIC PARTITIONS OF PRIMES. REF CU2 1. KNAW 54 14 51.

58 1, 2, 1, 3, 2, 3, 2, 5, 2, 6, 3, 4, 4, 8, 3, 9, 4, 6, 5, 11, 4, 10, 6, 9, 6, 14, 4, 15, 8, 10, 8, 12, 6, 18, 9, 12, 8, 20, 6, 21, 10, 12, 11, 23, 8, 21, 10, 16, 12, 26, 9, 20, 12, 18, 14, 29, 8
EULER TOTIENT FUNCTION. REF AS1 840. MTAC 23 682 69. (DIVIDED BY 2.)

59 1, 2, 1, 3, 2, 4, 4, 5, 7, 7, 11, 11, 16, 18, 23, 29, 34, 45, 52, 68, 81, 102, 126, 154, 194, 235, 296, 361, 450, 555, 685, 851, 1046, 1301, 1601, 1986, 2452, 3032, 3753, 4633
A(N) = A(N - 2) + A(N - 5). REF MMAG 41 17 68.

60 1, 2, 1, 3, 3, 2, 1, 2, 1, 4, 4, 4, 1, 4, 1, 4, 3, 2, 1, 4, 3, 5, 4, 2, 1, 3, 1, 3, 5, 2, 3, 3, 1, 4, 5, 2, 1, 3, 1, 5, 2, 4, 1, 2, 5, 3, 5, 2, 1, 2, 5, 2, 3, 2, 1, 3, 1, 6, 2, 3, 5, 2, 1, 4, 6, 5, 1, 3, 1
ITERATES OF A NUMBER-THEORETIC FUNCTION. REF MTAC 23 181 69

61 1, 2, 1, 3, 3, 2, 4, 1, 3, 5, 5, 3, 6, 1, 7, 3, 6, 5, 3, 7, 6, 9, 9, 5, 8, 4, 10, 9, 7, 3, 11, 3, 9, 1, 11, 12, 8, 10, 12, 9, 11, 5, 9, 13, 3, 6, 1, 13, 3, 2, 10, 8, 15, 15, 7, 9, 13, 1, 15, 14, 15
QUADRATIC PARTITIONS OF PRIMES. REF CU2 1. LE1 55.

62 1, 2, 1, 3, 4, 3, 7, 7, 8, 14, 15, 21, 28, 33, 47, 58, 76, 103, 125, 169, 220, 277, 373
FROM SYMMETRIC FUNCTIONS. REF PLMS 23 314 23.

63 1, 2, 1, 3, 16, 380, 1227756, 400507805615570
NONDEGENERATE BOOLEAN FUNCTIONS. REF PGEC 12 464 63.

64 1, 2, 1, 4, 2, 2, 2, 2, 4, 1, 2, 4, 2, 2, 2, 4, 2, 1, 2, 2, 1, 2, 2, 4, 4, 2, 2, 1, 2, 1, 2, 2, 4, 2, 2, 4, 1, 2, 2, 2, 2, 2, 1, 2, 2, 2, 2, 2, 4, 2, 2, 4, 2, 2, 2, 2, 1, 1, 2, 4, 1, 2, 2, 4, 2, 2, 2, 2, 2
RELATED TO FIBONACCI NUMBERS. REF HM1. MTAC 23 459 69. ACA 16 109 69.

65 1, 2, 1, 4, 2, 5, 4, 7, 8, 5, 2, 7, 10, 1, 10, 8, 2, 7, 4, 13, 1, 14, 8, 14, 11, 7, 14, 13, 16, 8, 11, 16, 17, 7, 2, 19, 4, 17, 19, 11, 1, 14, 5, 10, 22, 16, 4, 23, 20, 8, 23, 13, 10, 5, 16, 22
QUADRATIC PARTITIONS OF PRIMES. REF CU2 1. KNAW 54 14 51.

66 1, 2, 1, 4, 7, 24, 62, 216, 710, 2570, 9215, 34146, 126853, 477182
COLORED POLYOMINOES. REF LU2.

67 1, 2, 1, 4, 10, 36, 108, 392, 1363, 5000, 18223, 67792, 252938, 952540
ONE-SIDED COLORED POLYOMINOES. REF LU2.

68 1, 2, 1, 6, 12, 46, 92, 341, 1787, 9233
QUEENS PROBLEM. REF SL1 47. WE2 238.

SEQUENCES BEGINNING 1, 2, 2

69 1, 2, 2, 1, 0, 2, 1, 2, 2, 3, 2, 1, 3, 3, 3, 1, 2, 4, 3, 3, 3, 5, 3, 4, 3, 5, 4, 3, 1, 2, 5, 5, 4, 4, 5, 5, 5, 6, 5, 7, 4, 7, 5, 3, 5, 9, 5, 7, 7, 9, 6, 5, 5, 8, 5, 8, 7, 11, 6, 7, 4, 9, 7, 11, 7, 10, 9, 9
GENUS OF MODULAR GROUP. REF NBS 67B 62 63.

70 1, 2, 2, 1, 1, 2, 1, 2, 2, 1, 2, 2, 1, 1, 2, 1, 1, 2, 2, 1, 2, 1, 1, 2, 1, 2, 2, 1, 1, 2, 1, 1, 2, 1, 2, 2, 1, 2, 2, 1, 1, 2, 1, 2, 2, 1, 2, 1, 1, 2, 1, 1, 2, 2, 1, 2, 2, 1, 1, 2, 1, 2, 2, 1, 2, 2, 1, 1, 2
A SELF-GENERATING SEQUENCE. REF AMM 73 681 66.

71 1, 2, 2, 1, 2, 1, 1, 2, 2, 1, 1, 2, 1, 2, 2, 1, 2, 1, 1, 2, 1, 2, 2, 1, 1, 2, 2, 1, 2, 1, 1, 2, 2, 1, 1, 2, 1, 2, 2, 1, 1, 2, 2, 1, 2, 1, 1, 2, 1, 2, 2, 1, 2, 1, 1, 2, 2, 1, 1, 2, 1, 2, 2, 1, 2, 1, 1, 2, 1
A SELF-GENERATING SEQUENCE.

72 1, 2, 2, 1, 2, 1, 2, 4, 2, 4, 1, 4, 2, 3, 6, 6, 4, 3, 4, 4, 2, 2, 6, 4, 8, 4, 1, 4, 5, 2, 6, 4, 4, 2, 3, 6, 8, 8, 8, 1, 8, 4, 7, 4, 10, 8, 4, 5, 4, 3, 4, 10, 6, 12, 2, 4, 8, 8, 4, 14, 4, 5, 8, 6, 3, 6, 12
DIVISOR CLASSES OF QUADRATIC FIELDS. REF BO1 425.

73 1, 2, 2, 1, 2, 2, 2, 3, 2, 2, 4, 2, 2, 4, 2, 3, 4, 4, 2, 3, 4, 2, 6, 3, 2, 6, 4, 3, 4, 4, 4, 6, 4, 2, 6, 4, 4, 8, 4, 3, 6, 4, 4, 5, 4, 4, 6, 6, 4, 6, 6, 4, 8, 4, 2, 9, 4, 6, 8, 4, 4, 8, 8, 3, 8, 8, 4, 7, 4
CLASS NUMBERS. REF NBS 67B 62 63.

74 1, 2, 2, 1, 2, 3, 1, 3, 1, 2, 4, 3, 4, 2, 1, 3, 5, 5, 4, 2, 5, 4, 5, 1, 6, 4, 2, 1, 4, 7, 7, 6, 7, 2, 4, 6, 5, 7, 8, 8, 3, 7, 1, 5, 6, 3, 2, 8, 9, 7, 6, 8, 9, 4, 2, 5, 8, 1, 10, 10, 3, 7, 5, 2, 8, 10, 9, 7
QUADRATIC PARTITIONS OF PRIMES. REF CU2 1. LE1 55.

75 1, 2, 2, 1, 4, 1, 4, 2, 2, 3, 2, 2, 4, 3, 2, 4, 2, 4, 4, 4, 2, 5, 2, 6, 2, 5, 0, 4, 2, 6, 4, 4, 2, 7, 0, 8, 2, 3, 2, 6, 2, 8, 4, 6, 2, 7, 2, 10, 2, 8, 0, 6, 2, 8, 2, 6, 0, 7, 2, 12, 4, 5, 2, 10, 0, 12, 2
DECOMPOSITIONS OF N INTO SUM OF 2 PRIMES. REF PLMS 23 315 23.

76 1, 2, 2, 2, 0, 2, 1, 3, 2, 3, 1, 4, 2, 4, 3, 5, 4, 7, 3, 6, 5, 8, 6, 10, 6, 10, 9, 12, 9, 15, 11, 16, 14, 18, 14, 22, 19, 25, 22, 27, 23, 33, 29, 36, 33, 40, 38, 49, 43, 53, 51, 61, 57, 71
RESTRICTED PARTITIONS. REF JNSM 9 220 69.

77 1, 2, 2, 2, 2, 2, 3, 2, 3, 3, 3, 4, 3, 2, 4, 3, 4, 4, 3, 3, 5, 4, 4, 6, 4, 3, 6, 3, 4, 7, 4, 5, 6, 3, 5, 7, 6, 5, 7, 5, 5, 9, 5, 4, 10, 4, 5, 7, 4, 6, 9, 6, 6, 9, 7, 7, 11, 6, 6, 12, 4, 5, 10, 4, 7, 10, 6
DECOMPOSITIONS OF 2N INTO SUM OF 2 PRIMES. REF AFAS 23(2) 117 1894. LE1 79. ST4. MTAC 19 332 65.

78 1, 2, 2, 2, 2, 3, 3, 3, 4, 5, 5, 5, 6, 7, 8, 8, 9, 11, 12, 12, 14, 16, 17, 18, 20, 23, 25, 26, 29, 33, 35, 37, 41, 46, 49, 52, 57, 63, 68, 72, 78, 87, 93, 98, 107, 117, 125, 133, 144
PARTITIONS INTO DISTINCT ODD PARTS. REF PLMS 42 553 36. CJM 4 383 52. GU4.

79 1, 2, 2, 2, 2, 3, 4, 4, 4, 5, 6, 6, 6, 8, 9, 10, 10, 12, 13, 14, 14, 16, 19, 20, 21, 23, 26, 27, 28, 31, 34, 37, 38, 43, 46, 49, 50, 55, 60, 63, 66, 71, 78, 81, 84, 90, 98, 104, 107, 116
PARTITIONS INTO SQUARES.

80 1, 2, 2, 2, 4, 4, 4, 7, 7, 8, 12, 12, 16, 21, 21, 31, 37, 38, 58, 65, 71, 106, 114, 135, 191, 201, 257, 341, 359, 485, 605, 652, 904, 1070, 1202, 1664, 1894, 2237, 3029, 3370
A GENERALIZED FIBONACCI SEQUENCE. REF FQ 4 244 66.

81 1, 2, 2, 2, 4, 6, 6, 8, 11, 13, 17, 24, 28, 36
ROTATABLE PARTITIONS. REF JLMS 43 504 68.

82 1, 2, 2, 2, 4, 6, 10, 18, 32, 56, 102, 186, 341, 630, 1170, 2184, 4096, 7710, 14563, 27594, 52428, 99864, 190650, 364722, 699050, 1342177, 2581110, 4971026, 9586980
(2N)/N.**

83 1, 2, 2, 3, 2, 2, 3, 2, 5, 2, 3, 2, 6, 3, 5, 2, 2, 2, 2, 7, 5, 3, 2, 3, 5, 2, 5, 2, 6, 3, 3, 2, 3, 2, 2, 6, 5, 2, 5, 2, 2, 2, 19, 5, 2, 3, 2, 3, 2, 6, 3, 7, 7, 6, 3, 5, 2, 6, 5, 3, 3, 2, 5, 17, 10, 2, 3
LEAST POSITIVE PRIMITIVE ROOTS. REF AS1 864.

84 1, 2, 2, 3, 2, 2, 3, 2, 5, 2, 3, 2, 7, 3, 5, 2, 2, 2, 2, 7, 5, 3, 2, 3, 5, 2, 5, 2, 11, 3, 3, 2, 3, 2, 2, 7, 5, 2, 5, 2, 2, 2, 19, 5, 2, 3, 2, 3, 2, 7, 3, 7, 7, 11, 3, 5, 2, 43, 5, 3, 3, 2, 5, 17, 17, 2
LEAST POSITIVE PRIME PRIMITIVE ROOTS. REF RS5 2.

85 1, 2, 2, 3, 2, 3, 4, 2, 2, 7, 2, 6, 9, 2, 2, 3, 2, 4, 2, 5, 2, 3, 3, 5, 2, 2, 3, 6, 3, 9, 3, 3, 4, 2, 5, 5, 4, 2, 2, 3, 2, 2, 5, 2, 2, 4, 9, 3, 6, 3, 2, 7, 3, 3, 2, 2, 2, 5, 3, 6, 2, 7, 2, 10, 2, 5, 10, 3
LEAST NEGATIVE PRIMITIVE ROOTS. REF AS1 864.

86 1, 2, 2, 3, 2, 4, 2, 4, 3, 4, 2, 6, 2, 4, 4, 5, 2, 6, 2, 6, 4, 4, 2, 8, 3, 4, 4, 6, 2, 8, 2, 6, 4, 4, 4, 9, 2, 4, 4, 8, 2, 8, 2, 6, 6, 4, 2, 10, 3, 6, 4, 6, 2, 8, 4, 8, 4, 4, 2, 12, 2, 4, 6, 7, 4, 8, 2, 6
NUMBER OF DIVISORS OF N. REF AS1 840.

87 1, 2, 2, 3, 2, 4, 2, 4, 4, 4, 2, 6, 2, 4, 4, 6, 2, 8, 2, 6, 4, 4, 2, 8, 6, 4, 6, 6, 2, 8, 2, 8, 4, 4, 4, 12, 2, 4, 4, 8, 2, 8, 2, 6, 8, 4, 2, 12, 8, 12, 4, 6, 2, 12, 4, 8, 4, 4, 2, 12, 2, 4, 8, 12, 4, 8
RELATED TO A MODULAR GROUP. REF NBS 67B 62 63.

88 1, 2, 2, 3, 3, 3, 4, 3, 4, 5, 4, 5, 4, 4, 6, 5, 6, 6, 5, 6, 4, 5, 7, 6, 8, 7, 6, 8, 6, 7, 8, 6, 7, 5, 5, 8, 7, 9, 9, 8, 10, 7, 8, 10, 8, 10, 8, 7, 10, 8, 9, 9, 7, 8, 5, 6, 9, 8, 11, 10, 9, 12, 9, 11, 13
REPRESENTATIONS AS A SUM OF FIBONACCI NUMBERS. REF FQ 4 304 66.

89 1, 2, 2, 3, 3, 3, 4, 4, 4, 4, 5, 5, 5, 5, 5, 6, 6, 6, 6, 6, 6, 7, 7, 7, 7, 7, 7, 7, 8, 8, 8, 8, 8, 8, 8, 8, 9, 9, 9, 9, 9, 9, 9, 9, 9, 10, 10, 10, 10, 10, 10, 10, 10, 10, 10, 11, 11, 11, 11, 11, 11
A SELF-GENERATING SEQUENCE. REF MMAG 38 186 65. KN1 1 43.

90 1, 2, 2, 3, 3, 4, 4, 4, 4, 5, 5, 6, 6, 6, 6, 7, 7, 8, 8, 8, 8, 9, 9, 9, 9, 9, 9, 10, 10, 11, 11, 11, 11, 11, 11, 12, 12, 12, 12, 13, 13, 14, 14, 14, 14, 15, 15, 15, 15, 15, 15, 16, 16, 16, 16
NUMBER OF PRIMES NOT GREATER THAN N+1. REF AS1 870.

91 1, 2, 2, 3, 3, 4, 4, 4, 5, 5, 5, 6, 6, 6, 6, 7, 7, 7, 7, 8, 8, 8, 8, 9, 9, 9, 9, 9, 10, 10, 10, 10, 10, 11, 11, 11, 11, 11, 12, 12, 12, 12, 12, 12, 13, 13, 13, 13, 13, 13, 14, 14, 14, 14, 14
A SELF-GENERATING SEQUENCE. REF AMM 74 740 67.

92 1, 2, 2, 3, 3, 4, 4, 4, 5, 5, 6, 6, 6, 6, 7, 7, 7, 8, 8, 8, 8, 9, 9, 9, 10, 10, 10, 10, 10, 11, 11, 11, 11, 12, 12, 12, 12, 12, 13, 13, 13, 13, 13, 14, 14, 14, 14, 15, 15, 15, 15, 15, 15, 16
INTEGER PART OF SQUARE ROOT OF PRIMES. REF AS1 2.

93 1, 2, 2, 3, 3, 4, 5, 6, 7, 9, 10, 12, 14, 17, 19, 23, 26, 30, 35, 40, 46, 52, 60, 67, 77, 87, 98, 111, 124, 140, 157, 175, 197, 219, 244, 272, 302, 336, 372, 413, 456, 504, 557
PARTITIONS INTO PRIME PARTS. REF PNISI 21 183 55.

94 1, 2, 2, 3, 3, 5, 6, 8, 8, 12, 13, 17, 19, 26, 28, 37, 40, 52, 58, 73, 79, 102, 113, 139, 154, 191, 210, 258, 284, 345, 384, 462, 509, 614, 679, 805, 893, 1060, 1171, 1382
PARTITIONS INTO NON-PRIME PARTS. REF JNSM 9 91 69.

95 1, 2, 2, 3, 3, 5, 6, 8, 9, 11, 14, 19, 22
ROTATABLE PARTITIONS. REF JLMS 43 504 68.

96 1, 2, 2, 3, 4, 1, 8, 1, 10, 9, 16, 18, 12, 42, 4, 58, 38, 82, 88, 54, 188, 18, 248, 151, 334, 338, 260, 760, 120
FROM SYMMETRIC FUNCTIONS. REF PLMS 23 309 23.

97 1, 2, 2, 3, 4, 3, 4, 4, 3, 4, 4, 5, 5, 4, 6, 5, 6, 6, 6, 4, 6, 7, 6, 6, 5, 7, 6, 10, 4, 7, 8, 5, 5, 6, 7, 6, 6, 6, 8, 6, 6, 6, 5, 5, 6, 7, 7, 7, 6, 7, 6, 5, 7, 6, 7, 9, 7, 7, 7, 9, 5, 7, 10, 7, 7
CONSECUTIVE QUADRATIC NONRESIDUES. REF BAMS 32 284 26.

98 1, 2, 2, 3, 4, 5, 6, 7, 8, 10, 11, 13, 14, 16, 18, 20, 22, 26, 29
DENUMERANTS. REF R1 152.

99 1, 2, 2, 3, 4, 5, 6, 7, 8, 11, 12, 15, 16, 19, 22, 25, 28, 34, 40
DENUMERANTS. REF R1 152.

100 1, 2, 2, 3, 4, 5, 6, 8, 10, 12, 15, 18, 22, 27, 32, 38, 46, 54, 64, 76, 89, 104, 122, 142, 165, 192, 222, 256, 296, 340, 390, 448, 512, 585, 668, 760, 864, 982, 1113, 1260, 1426
PARTITIONS INTO DISTINCT PARTS. REF AS1 836.

101 1, 2, 2, 3, 4, 5, 7, 9, 11, 15, 18, 23, 30, 37, 47, 58, 71, 90, 110, 136, 164, 201, 248, 300, 364, 436, 525, 638, 764, 919, 1090, 1297, 1549, 1845, 2194, 2592, 3060, 3590
MAXIMUM OF A PARTITION FUNCTION. REF JIMS 6 112 42. PSPM 19 172 71.

102 1, 2, 2, 3, 4, 5, 7, 9, 12, 16, 21, 28, 37, 49, 65, 86, 114, 151, 200, 265, 351, 465, 616, 816, 1081, 1432, 1897, 2513, 3329, 4410, 5842, 7739, 10252, 13581, 17991, 23833
A(N) = A(N − 2) + A(N − 3). REF JA2 90. MMAG 41 17 68.

103 1, 2, 2, 3, 4, 6, 9, 14, 22, 35, 56, 90, 145, 234, 378, 611, 988, 1598, 2585, 4182, 6766, 10947, 17712, 28658, 46369, 75026, 121394, 196419, 317812, 514230, 832041
FIBONACCI NUMBERS + 1. REF JA2 97.

104 1, 2, 2, 3, 5, 6, 9, 13, 14, 15, 20
RELATED TO ZARANKIEWICZS PROBLEM. REF TI1 126.

105 1, 2, 2, 3, 6, 0, 6, 7, 9, 7, 7, 4, 9, 9, 7, 8, 9, 6, 9, 6, 4, 0, 9, 1, 7, 3, 6, 6, 8, 7, 3, 1, 2, 7, 6, 2, 3, 5, 4, 4, 0, 6, 1, 8, 3, 5, 9, 6, 1, 1, 5, 2, 5, 7, 2, 4, 2, 7, 0, 8, 9, 7, 2, 4, 5, 4, 1, 0, 5
SQUARE ROOT OF 5. REF RS4 XVIII. MTAC 22 234 68.

106 1, 2, 2, 3, 7, 15, 12, 30, 8, 32, 162, 21
FROM SEDLACEKS PROBLEM ON SOLUTIONS OF X + Y = Z. REF GU8.

107 1, 2, 2, 3, 7, 25, 121, 721, 5041, 40321, 362881, 3628801, 39916801, 479001601, 6227020801, 87178291201, 1307674368001, 20922789888001, 355687428096001
FACTORIAL N + 1. REF AS1 833.

108 1, 2, 2, 4, 2, 4, 2, 4, 6, 2, 6, 4, 2, 4, 6, 6, 2, 6, 4, 2, 6, 4, 6, 8, 4, 2, 4, 2, 4, 14, 4, 6, 2, 10, 2, 6, 6, 4, 6, 6, 2, 10, 2, 4, 2, 12, 12, 4, 2, 4, 6, 2, 10, 6, 6, 6, 2, 6, 4, 2, 10, 14, 4, 2, 4
DIFFERENCES BETWEEN CONSECUTIVE PRIMES. REF AS1 870.

109 1, 2, 2, 4, 2, 4, 4, 8, 2, 4, 4, 8, 4, 8, 8, 16, 2, 4, 4, 8, 4, 8, 8, 16, 4, 8, 8, 16, 8, 16, 16, 32, 2, 4, 4, 8, 4, 8, 8, 16, 4, 8, 8, 16, 8, 16, 16, 32, 4, 8, 8, 16, 8, 16, 16, 32, 8, 16, 16, 32
RELATED TO BINARY EXPANSION OF N. REF GO3.

110 1, 2, 2, 4, 2, 6, 2, 6, 4, 10, 2, 12, 6, 4, 4, 16, 6, 18, 4, 6, 10, 22, 2, 20, 12, 18, 6, 28, 4, 30, 8, 10, 16, 12, 6, 36, 18, 12, 4, 40, 6, 42, 10, 12, 22, 46, 4, 42, 20, 16, 12, 52, 18, 20
REDUCED TOTIENT FUNCTION. REF CAU (2) 12 43. LE1 7.

111 1, 2, 2, 4, 2, 6, 4, 6, 4, 10, 4, 12, 6, 8, 8, 16, 6, 18, 8, 12, 10, 22, 8, 20, 12, 18, 12, 28, 8, 30, 16, 20, 16, 24, 12, 36, 18, 24, 16, 40, 12, 42, 20, 24, 22, 46, 16, 42, 20, 32, 24
EULER TOTIENT FUNCTION. REF AS1 840. MTAC 23 682 69.

112 1, 2, 2, 4, 4, 6, 7, 10, 11, 16, 17, 23, 26, 33, 37, 47, 52, 64, 72, 86, 96, 115, 127, 149, 166, 192, 212, 245, 269, 307, 338, 382, 419, 472, 515, 576, 629, 699, 760, 843, 913
EXPANSION OF A GENERATING FUNCTION. REF CAY 10 415.

113 1, 2, 2, 4, 4, 7, 8, 12, 14, 21, 24, 34, 41, 55, 66, 88, 105, 137, 165, 210, 253, 320, 383, 478, 574, 708, 847, 1039, 1238, 1507, 1794, 2167, 2573, 3094, 3660, 4378, 5170
PARTITIONS WITH NO PART OF SIZE 1. REF TA1 1 334. AS1 836.

114 1, 2, 2, 4, 4, 8, 9, 18, 23, 44, 63, 122, 190, 362, 612, 1162, 2056, 3912, 7155, 13648
NECKLACES. REF IJM 5 662 61.

115 1, 2, 2, 4, 4, 8, 10, 20, 30, 56, 94, 180, 316, 596, 1096, 2068, 3856, 7316, 13798, 26272
NECKLACES. REF IJM 5 662 61.

116 1, 2, 2, 4, 5, 7, 9, 13, 16, 22, 27, 36, 44, 57, 70, 89, 108, 135, 163, 202, 243, 297, 355, 431, 513, 617, 731, 874, 1031, 1225, 1439, 1701, 1991, 2341, 2731, 3197, 3717
PARTITIONS INTO PARTS PRIME TO 3. REF PSPM 8 145 65.

117 1, 2, 2, 4, 5, 9, 12, 21, 30, 51, 76, 127, 195, 322, 504, 826, 1309, 2135, 3410, 5545, 8900, 14445, 23256, 37701, 60813, 98514, 159094, 257608, 416325, 673933, 1089648
PACKING A BOX WITH DOMINOES. REF AMM 69 61 62.

118 1, 2, 2, 4, 5, 10, 14, 26, 42, 78, 132, 249, 445, 842, 1561, 2988, 5671, 10981, 21209, 41472, 81181, 160176, 316749, 629933, 1256070, 2515169, 5049816, 10172638
SERIES-REDUCED TREES. REF AM1 101 150 59. HA5 232. CA3.

119 1, 2, 2, 4, 6, 6, 11, 16, 20, 28, 41, 51, 70, 93, 122
PLANAR PARTITIONS. REF MA2 2 332.

120 1, 2, 2, 4, 6, 8, 18, 20, 56, 48, 178, 132, 574, 348, 1870, 1008
FOLDING A STRIP OF STAMPS. REF JCT 5 151 68.

121 1, 2, 2, 4, 6, 10, 16, 30, 52, 94
SHIFT REGISTERS. REF GO1 172.

122 1, 2, 2, 4, 6, 11, 18, 37, 66, 135, 265
BORON TREES. REF CAY 9 451.

123 1, 2, 2, 4, 6, 12, 20, 39, 71, 137, 261, 511, 995, 1974, 3915, 7841, 15749, 31835, 64540, 131453, 268498, 550324, 1130899, 2330381, 4813031, 9963288, 20665781
SERIES-REDUCED PLANTED TREES. AM1 101 150 59. CA3.

124 1, 2, 2, 4, 7, 12, 16, 32
COVERING NUMBERS. REF JLMS 44 60 69.

125 1, 2, 2, 4, 8, 4, 16, 12, 48, 80, 136, 420, 1240, 2872, 7652, 18104, 50184
QUEENS PROBLEM. REF PSAM 10 93 60.

126 1, 2, 2, 4, 8, 13, 25, 44, 83, 152, 286, 538, 1020, 1942, 3725, 7145, 13781, 26627, 51572, 100099, 194633, 379037, 739250, 1443573, 2822186, 5522889
POPULATION OF U2 + 16V**2. REF MTAC 20 567 66.**

127 1, 2, 2, 4, 10, 16, 28, 48, 76, 110, 144, 182, 222, 264, 310, 356, 408, 468, 536, 610, 684, 762, 842, 924, 1010, 1096, 1188, 1288, 1396, 1510, 1624, 1742, 1862
PERIODIC DIFFERENCES. REF TCPS 2 220 1827.

128 1, 2, 2, 4, 10, 28, 84, 264, 858, 2860, 9724, 33592, 117572, 416024, 1485800, 5348880, 19389690, 70715340
FROM BINOMIAL COEFFICIENTS. REF TH1 164. FMR 1 55.

129 1, 2, 2, 4, 12, 22, 58, 158, 448, 1342, 4199, 13384
POLYTOPES BY NUMBER OF EDGES. REF JCT 7 157 69.

130 1, 2, 2, 5, 4, 7, 7, 11, 9, 8, 6, 9, 4, 6, 22, 10, 4, 8, 4
PRIMITIVE GROUPS. REF JL2 178.

131 1, 2, 2, 5, 5, 12, 12, 27, 28, 64, 67, 147, 158, 348, 373, 799, 879, 1886, 2069, 4335, 4864
SQUARE FILAMENTS. REF PL2 1 337 70.

132 1, 2, 2, 6, 6, 18, 16, 48, 60, 176, 144, 630, 756, 1800, 2048, 7710, 7776, 27594, 24000, 84672
RELATED TO EULERS TOTIENT FUNCTION. REF BE1 296.

133 1, 2, 2, 6, 8, 18, 30, 67, 127
BORON TREES. REF CAY 9 451.

134 1, 2, 2, 6, 9, 17, 30, 54, 98, 183, 341, 645, 1220, 2327, 4451, 8555, 16489, 31859, 61717, 119779, 232919, 453584, 884544, 1727213, 3376505, 6607371
POPULATION OF U2 + 12V**2. REF MTAC 20 567 66.**

135 1, 2, 2, 6, 9, 20, 37, 86, 183, 419
HYDROCARBONS. REF BS1 201.

136 1, 2, 2, 6, 14, 34, 82, 198, 478, 1154, 2786, 6726, 16238, 39202, 94642, 228486, 551614, 1331714, 3215042, 7761798, 18738638, 45239074, 109216786, 263672646
A(N) = 2A(N - 1) + A(N - 2). REF AJM 1 187 1878.

137 1, 2, 2, 6, 16, 20, 132, 28, 1216, 936, 23540, 34782, 138048, 469456, 1601264, 9112560, 18108928, 182135008, 161934624, 3804634784, 404007680, 83297957568
FROM PERMUTATIONS OF ORDER 2. REF CJM 7 168 55.

138 1, 2, 2, 6, 38, 390, 6062, 134526
COLORED GRAPHS. REF CJM 22 596 70.

139 1, 2, 2, 7, 10, 20, 36, 65, 118, 221, 409, 776, 1463, 2788, 5328, 10222, 19714, 38054, 73685, 142944, 277838, 540889, 1054535, 2058537, 4023278
POPULATION OF U2 + 10V**2. REF MTAC 20 563 66.**

140 1, 2, 2, 8, 8, 112, 656, 5504, 49024, 491264
RELATED TO LATIN RECTANGLES. REF R1 210.

141 1, 2, 2, 8, 12, 88, 176, 2752, 8784
SELF-COMPLEMENTARY ORIENTED GRAPHS. REF KNAW 73 443 70.

142 1, 2, 2, 8, 72, 1536, 86080, 14487040, 8274797440, 17494930604032
THRESHOLD FUNCTIONS. REF PGEC 19 821 70.

143 1, 2, 2, 9, 11, 37, 79, 249, 671, 2182, 6692
POLYTOPES BY NUMBER OF EDGES. REF JCT 7 157 69.

144 1, 2, 2, 10, 28, 207, 1288, 10366, 91296
HIT POLYNOMIALS. REF RI3.

145 1, 2, 2, 10, 218, 64594, 4294642034, 18446744047940725978, 340282366920938463334247399005993378250
NONDEGENERATE BOOLEAN FUNCTIONS. REF HA2 170.

146 1, 2, 2, 10, 52246, 2631645209645100680142
INVERTIBLE BOOLEAN FUNCTIONS. REF PGEC 13 350 64.

147 1, 2, 2, 17, 1, 91
QUEENS PROBLEM. REF SL1 49.

148 1, 2, 2, 18, 66, 374, 1694, 9822, 51698
BAXTER PERMUTATIONS. REF MA4 2 25 67.

149 1, 2, 2, 20, 38, 146, 368, 1070, 2824, 7680, 19996
SUSCEPTIBILITY FOR SQUARE LATTICE. REF PHA 28 924 62.

150 1, 2, 2, 22, 563, 1676257
TYPES OF LATIN SQUARES. REF R1 210. FY1 22. JCT 5 177 68.

SEQUENCES BEGINNING 1, 2, 3

151 1, 2, 3, 0, 2, 5, 8, 5, 0, 9, 2, 9, 9, 4, 0, 4, 5, 6, 8, 4, 0, 1, 7, 9, 9, 1, 4, 5, 4, 6, 8, 4, 3, 6, 4, 2, 0, 7, 6, 0, 1, 1, 0, 1, 4, 8, 8, 6, 2, 8, 7, 7, 2, 9, 7, 6, 0, 3, 3, 3, 2, 7, 9, 0, 0, 9, 6, 7, 5
NATURAL LOGARITHM OF 10. REF RS4 2.

152 1, 2, 3, 0, 11, 0, 17, 15, 14, 51
A PARTITION FUNCTION. REF JNSM 9 103 69.

153 1, 2, 3, 0, 25, 152, 1350, 12644, 131391, 1489568, 18329481, 243365514, 3468969962, 52848096274, 857073295427, 14744289690560, 268202790690465
FROM DISCORDANT PERMUTATIONS. REF KYU 10 13 56.

154 1, 2, 3, 1, 1, 4, 5, 1, 3, 1, 3, 1, 1, 8, 15, 3, 7, 4, 5, 2, 3, 3, 6, 2, 3, 2, 3, 1, 1, 16, 19, 7, 10, 5, 15, 4, 5, 7, 15, 3, 7, 4, 5, 2, 3, 5, 13, 3, 5, 4, 7, 1, 3, 3, 5, 2, 3, 1, 3, 1, 1, 32, 47, 11
A PROBLEM IN PARITY. REF IJ1 11 163 69.

155 1, 2, 3, 1, 2, 3, 4, 2, 1, 2, 3, 3, 2, 3, 4, 1, 2, 2, 3, 2, 3, 3, 4, 3, 1, 2, 3, 4, 2, 3, 4, 2, 3, 2, 3, 1, 2, 3, 4, 2, 2, 3, 3, 3, 2, 3, 4, 3, 1, 2, 3, 2, 2, 3, 4, 3, 3, 2, 3, 4, 2, 3, 4, 1, 2, 3, 3, 2, 3
LEAST NUMBER OF SQUARES TO REPRESENT N.

156 1, 2, 3, 1, 4, 3, 2, 1, 4, 2, 6, 4, 1, 2, 7, 1, 4, 3, 2, 1, 4, 6, 7, 4, 1, 2, 8, 5, 4, 7, 2, 1, 8, 6, 7, 4, 1, 2, 3, 1, 4, 7, 2, 1, 8, 2, 7, 4, 1, 2, 8, 1, 4, 7, 2, 1, 4, 2, 7, 4, 1, 2, 8, 1, 4, 7, 2, 1, 8
THE GAME OF KAYLES. REF PCPS 52 516 56.

157 1, 2, 3, 1, 5, 4, 3, 3, 9, 2, 11, 5
A NUMBER-THEORETIC FUNCTION. REF MTS 67 11 58.

158 1, 2, 3, 2, 0, 1, 7, 2, 6, 8, 22, 7, 0, 33, 3, 14, 51, 46, 19, 12, 94, 42, 23, 113, 150, 54, 48, 345, 116, 109, 403, 498, 140, 219, 1057, 326, 259, 1271, 1641, 308, 656, 3396
FROM SYMMETRIC FUNCTIONS. REF PLMS 23 297 23.

159 1, 2, 3, 2, 1, 2, 2, 4, 2, 2, 1, 0, 4, 2, 3, 2, 2, 4, 0, 2, 2, 0, 4, 2, 3, 0, 2, 6, 2, 2, 1, 2, 0, 2, 2, 2, 2, 4, 2, 0, 4, 4, 4, 0, 1, 2, 0, 4, 2, 0, 2, 2, 5, 2, 0, 2, 2, 4, 4, 2, 0, 2, 4, 2, 2, 0, 4, 0, 0
THE SQUARE OF EULERS PRODUCT. REF PLMS 21 190 1889.

160 1, 2, 3, 2, 2, 4, 4, 4, 4, 4, 3, 5, 4, 3, 5, 5, 6, 6, 4, 6, 7, 4, 4, 7, 7, 6, 5, 5, 7, 8, 6, 5, 4, 7, 6, 6, 6, 6, 6, 6, 6, 4, 7, 6, 7, 7, 7, 5, 6, 6, 6, 7, 6, 7, 8, 7, 10, 7, 9, 9, 7, 10, 5, 5
CONSECUTIVE QUADRATIC RESIDUES. REF BAMS 32 284 26.

161 1, 2, 3, 2, 3, 4, 4, 4, 5, 6, 5, 4, 6, 4, 7, 8, 3, 6, 8, 6, 7, 10, 8, 6, 10, 6, 7, 12, 5, 10, 12, 4, 10, 12, 9, 10, 14, 8, 9, 16, 9, 8, 18, 8, 9, 14, 6, 12, 16, 10, 11, 16, 12, 14, 20, 12, 11, 24
DECOMPOSITIONS OF 2N INTO SUM OF 2 ODD PRIMES. REF FVS 4(4) 7 27. LE1 80.

162 1, 2, 3, 2, 5, 2, 3, 7, 2, 11, 13, 2, 3, 5, 17, 19, 2, 23, 7, 29, 3, 31, 2, 37, 41, 43, 47, 5, 53, 59, 2, 11, 61, 3, 67, 71, 73, 79, 13, 83, 89, 2, 97, 101, 103, 107, 7, 109, 113, 17, 127
RELATED TO HIGHLY COMPOSITE NUMBERS. REF RAM 115.

163 1, 2, 3, 2, 5, 5, 7, 10, 12, 17, 22, 29, 39, 51, 68, 90, 119, 158, 209, 277, 367, 486, 644, 853, 1130, 1497, 1983, 2627, 3480, 4610, 6107, 8090, 10717, 14197, 18807, 24914
A(N) = A(N − 2) + A(N − 3). REF AMM 15 209 08. JA2 90. FQ 6(3) 68 68.

164 1, 2, 3, 3, 3, 5, 7, 6, 6, 10, 12, 11, 13, 17, 20, 21, 21, 27, 34, 33, 36, 46, 51, 53, 58, 68, 78, 82, 89, 104, 118, 123, 131, 154, 171, 179, 197, 221, 245, 262, 279, 314, 349, 369
MOCK THETA NUMBERS. REF TAMS 72 495 52.

165 1, 2, 3, 3, 5, 9, 16, 28, 50, 89, 159, 285, 510, 914, 1639, 2938, 5269, 9451, 16952, 30410, 54555, 97871, 175588, 315016, 565168, 1013976, 1819198, 3263875, 5855833
BINARY CODES. REF PGIT 17 309 71.

166 1, 2, 3, 4, 3, 5, 3, 6, 1, 2, 6, 7, 4, 5, 8, 3, 9, 7, 6, 9, 1, 2, 6, 11, 4, 10, 9, 3, 12, 9, 12, 13, 8, 3, 14, 12, 13, 6, 1, 2, 12, 11, 5, 15, 16, 9, 3, 13, 8, 15, 12, 17, 16, 6, 14, 15, 10, 3, 17
QUADRATIC PARTITIONS OF PRIMES. REF CU2 1. LE1 55. MTAC 23 459 69.

167 1, 2, 3, 4, 5, 3, 7, 4, 6, 5, 11, 4, 13, 7, 5, 6, 17, 6, 19, 5, 7, 11, 23, 4
N DIVIDES FACTORIAL A(N). REF AMM 25 210 18.

168 1, 2, 3, 4, 5, 5, 7, 6, 6, 7, 11, 7, 13, 9, 8, 8, 17, 8, 19, 9, 10, 13, 23, 9, 10, 15, 9, 11, 29, 10, 31, 10, 14, 19, 12, 10, 37, 21, 16, 11, 41, 12, 43, 15, 11, 25, 47, 11, 14, 12, 20, 17
SUM OF PRIMES DIVIDING N. REF MTAC 23 181 69.

169 1, 2, 3, 4, 5, 6, 5, 7, 6, 8, 8, 9, 10, 10, 8, 11, 10, 11, 13, 10, 12, 14, 15, 13, 15, 16, 13, 14, 16, 17, 13, 14, 16, 18, 17, 18, 17, 19, 20, 20, 15, 17, 20, 21, 19, 22, 20, 21, 19, 20
QUADRATIC PARTITIONS OF PRIMES. REF CU2 1. AMM 56 526 49.

170 1, 2, 3, 4, 5, 6, 7, 1, 2, 3, 4, 5, 6, 7, 8, 2, 3, 4, 5, 6, 7, 8, 9, 3, 4, 5, 1, 2, 3, 4, 5, 4, 5, 6, 2, 3, 4, 5, 6, 5, 6, 7, 3, 4, 5, 6, 7, 6, 7, 8, 4, 5, 6, 2, 3, 4, 5, 6, 5, 6, 7, 3, 4, 1, 2, 3, 4, 5, 6
LEAST NUMBER OF CUBES TO REPRESENT N. REF JRAM 14 279 1835. LE1 81.

171 1, 2, 3, 4, 5, 6, 7, 6, 6, 10, 11, 12, 13, 14, 15, 8, 17, 12, 19, 20, 21, 22, 23, 18, 10, 26, 9, 28, 29, 30, 31, 10, 33, 34, 35, 24, 37, 38, 39, 30, 41, 42, 43, 44, 30, 46, 47, 24, 14
MOSAIC NUMBERS. REF BAMS 69 446 63. CJM 17 1010 65.

172 1, 2, 3, 4, 5, 6, 7, 8, 9, 10, 11, 12, 13, 14, 15, 1, 2, 3, 4, 5, 6, 7, 8, 9, 10, 11, 12, 13, 14, 15, 16, 2, 3, 4, 5, 6, 7, 8, 9, 10, 11, 12, 13, 14, 15, 16, 17, 3, 4, 5, 6, 7, 8, 9, 10, 11, 12
LEAST NUMBER OF FOURTH POWERS TO REPRESENT N. REF JRAM 46 3 1853. LE1 82.

173 1, 2, 3, 4, 5, 6, 7, 8, 9, 10, 11, 12, 13, 14, 15, 16, 17, 18, 19, 20, 21, 22, 23, 24, 25, 26, 27, 28, 29, 30, 31, 32, 33, 34, 35, 36, 37, 38, 39, 40, 41, 42, 43, 44, 45, 46, 47, 48, 49
THE NATURAL NUMBERS.

174 1, 2, 3, 4, 5, 6, 7, 8, 9, 10, 11, 12, 13, 14, 15, 16, 17, 18, 19, 20, 21, 22, 23, 24, 25, 26, 27, 28, 29, 30, 31, 32, 33, 34, 35, 36, 37, 38, 39, 42, 43, 45, 46, 47, 48, 50, 51, 52, 53
N2 + N + 41 IS PRIME.**

175 1, 2, 3, 4, 5, 6, 7, 8, 9, 10, 11, 12, 15, 20, 22, 24, 30, 33, 36, 40, 44, 48, 50, 55, 60, 66, 70, 77, 80, 88, 90, 99, 100, 101, 102, 104, 105, 110, 111, 112, 115, 120, 122, 124, 126
DIVISIBLE BY EACH DIGIT. REF JRM 1 217 68.

176 1, 2, 3, 4, 5, 6, 7, 8, 9, 10, 12, 13, 15, 16, 18, 21, 22, 24, 25, 28, 30, 33, 37, 40, 42, 45, 48, 57, 58, 60, 70, 72, 78, 85, 88, 93, 102, 105, 112, 120, 130, 133, 165, 168, 177, 190
THE SUITABLE NUMBERS OF EULER. REF BO1 427.

177 1, 2, 3, 4, 5, 6, 7, 8, 9, 10, 12, 14, 15, 16, 18, 20, 21, 24, 25, 27, 28, 30, 32, 35, 36, 40, 42, 45, 48, 49, 50, 54, 56, 60, 63, 64, 70, 72, 75, 80, 81, 84, 90, 96, 98, 100, 105, 108
CONTAIN NO PRIME FACTOR GREATER THAN 7.

178 1, 2, 3, 4, 5, 6, 7, 8, 9, 11, 22, 33, 44, 55, 66, 77, 88, 99, 101, 111, 121, 131, 141, 151, 161, 171, 181, 191, 202, 212, 222, 232, 242, 252, 262, 272, 282, 292, 303, 313, 323
PALINDROMES.

179 1, 2, 3, 4, 5, 6, 7, 9, 10, 11, 12, 14, 16, 17, 20, 21, 22, 25, 27, 29, 31, 32, 36, 39, 40, 42, 45, 46, 47, 49, 51, 54, 55, 56, 57, 60, 61, 65, 66, 67, 69, 71, 77, 84, 86, 87, 90, 94
N(N − 1) − 1 IS PRIME. REF PO1 249. LE1 46.

180 1, 2, 3, 4, 5, 6, 8, 9, 10, 11, 12, 14, 15, 16, 18, 20, 21, 22, 23, 24, 26, 27, 28, 29, 30, 32, 33, 35, 36, 39, 40, 41, 42, 44, 46, 48, 50, 51, 52, 53, 54, 55, 56, 58, 60, 63, 64, 65
VALUES OF EULER TOTIENT FUNCTION. REF BA2 64 (DIVIDED BY 2).

181 1, 2, 3, 4, 5, 6, 8, 9, 14, 15, 16, 22, 28, 29, 36, 37, 54, 59, 85, 93, 117, 119, 161, 189, 193, 256, 308, 322, 327, 411, 466, 577, 591, 902, 928, 946
45.2N − 1 IS PRIME. REF MTAC 23 874 69.**

182 1, 2, 3, 4, 5, 6, 8, 10, 11, 13, 16, 18, 20, 23, 26, 29, 32, 35, 39, 43, 46, 50, 55, 59, 63, 68, 73, 78, 83, 88, 94, 100, 105, 111, 118, 124, 130, 137, 144, 151, 158, 165, 173, 181
GENUS OF COMPLETE GRAPH. REF PNAS 60 438 68.

183 1, 2, 3, 4, 5, 7, 8, 9, 10, 11, 12, 14, 15, 16, 18, 19, 21, 22, 23, 24, 25, 26, 29, 30, 31, 32, 33, 35, 37, 38, 40, 42, 43, 44, 45, 46, 47, 49, 51, 52, 53, 54, 56, 57, 58, 60, 63, 64
A NUMBER-THEORETIC FUNCTION. REF IAS 5 382 37.

184 1, 2, 3, 4, 5, 7, 8, 9, 10, 12, 13, 14, 15
WYTHOFF GAME. REF CMB 2 189 59.

185 1, 2, 3, 4, 5, 7, 8, 9, 11, 13, 16, 17, 19, 23, 25, 27, 29, 31, 32, 37, 41, 43, 47, 49, 53, 59, 61, 64, 67, 71, 73, 79, 81, 83, 89, 97, 101, 103, 107, 109, 113, 121, 125, 127, 128
PRIME POWERS. REF AS1 870.

186 1, 2, 3, 4, 5, 7, 8, 10, 12, 14, 16, 19, 21, 24, 27, 30, 33, 37, 40, 44, 48, 52, 56, 61, 65, 70, 75, 80, 85, 91, 96, 102, 108, 114, 120, 127, 133, 140, 147, 154, 161, 169, 176, 184
PARTITIONS INTO AT MOST 3 PARTS. REF RS2 2.

187 1, 2, 3, 4, 5, 7, 9, 11, 13, 16, 17, 19, 23, 24, 25, 29, 30
A TWO-WAY CLASSIFICATION OF INTEGERS. REF CMB 2 89 59.

188 1, 2, 3, 4, 5, 7, 10, 11, 12, 14, 15, 18, 24, 25, 26, 28, 29, 31, 33, 35, 38, 39, 42, 43, 46, 49, 50, 53, 56, 59, 63, 64, 67, 68, 75, 81, 82, 87, 89, 91, 92, 94, 96, 106, 109, 120, 124
FROM CUBAN PRIMES. REF MES 41 144 12.

189 1, 2, 3, 4, 6, 6, 10, 19, 27, 33, 39, 157, 183, 386, 664, 687, 969, 1281, 1332, 2917, 2993, 3376, 6002
SIZE OF MERSENNE PRIMES. REF BE3 19. NAMS 18 608 71.

190 1, 2, 3, 4, 6, 6, 12, 15, 20, 30, 30, 60, 60, 84, 105, 140, 210, 210, 420, 420, 420, 420, 840, 840, 1260, 1260, 1540, 2310, 2520, 4620, 4620, 5460, 5460, 9240, 9240, 13860
LARGEST ORDER OF PERMUTATION OF N SYMBOLS. REF BSMF 97 187 69.

191 1, 2, 3, 4, 6, 6, 13, 10, 24, 22, 45, 30, 158, 74, 245, 368, 693, 522, 2637, 1610, 7341
NECKLACES. REF IJM 8 269 64.

192 1, 2, 3, 4, 6, 7, 8, 9, 11, 12, 13, 14, 16, 17, 18, 19, 21, 22, 23, 24, 25, 27, 28, 29, 30, 32, 33, 34, 35, 37, 38, 39, 40, 42, 43, 44, 45, 46, 48, 49, 50, 51, 53, 54, 55, 56, 58, 59
A BEATTY SEQUENCE. REF CMB 2 189 59.

193 1, 2, 3, 4, 6, 7, 9, 10, 12, 13, 14, 16, 17
WYTHOFF GAME. REF CMB 2 188 59.

194 1, 2, 3, 4, 6, 7, 9, 15, 22, 28, 30, 46, 60, 63, 127, 153, 172, 303, 471, 532, 865, 900, 1366, 2380, 3310, 4495, 6321, 7447, 10198, 11425, 21846, 24369, 27286, 28713
XN + X + 1 IS IRREDUCIBLE OVER GF(2). REF IC 16 502 70.**

195 1, 2, 3, 4, 6, 7, 11, 18, 34, 38, 43, 55, 64, 76, 94, 103, 143, 206, 216, 306, 324, 391, 458, 470, 827
3.2N − 1 IS PRIME. REF MTAC 23 874 69.**

196 1, 2, 3, 4, 6, 8, 9, 10, 12, 16, 18, 20, 24, 30, 32, 36, 40, 48, 60, 64, 72, 80, 84, 90, 96, 100, 108, 120, 128, 144, 160, 168, 180, 192, 200, 216, 224, 240, 256, 288, 320, 336
RELATED TO HIGHLY COMPOSITE NUMBERS. REF RAM 87.

197 1, 2, 3, 4, 6, 8, 9, 11, 12, 16, 17, 18, 19, 22, 24, 25, 27, 32, 33, 34, 36, 38, 41, 43, 44, 48, 49, 50
OF THE FORM X2 + 2Y**2. REF EUL (1) 1 421 11. LE1 59.**

198 1, 2, 3, 4, 6, 8, 9, 12, 15, 16, 21, 24, 24, 32, 36, 36, 45, 48, 48, 60, 66, 64, 75, 84, 81, 96, 105, 96, 120, 128, 120, 144, 144, 144, 171, 180, 168, 192, 210, 192, 231, 240, 216
DEGREES OF RATIONAL PORISMS. REF BU2 39 103 47.

199 1, 2, 3, 4, 6, 8, 10, 12, 15, 18, 21, 24, 28, 32, 36, 40, 45, 50
RESTRICTED PARTITIONS. REF CAY 2 277.

200 1, 2, 3, 4, 6, 8, 10, 12, 16, 18, 20, 24, 30, 36, 42, 48, 60, 72, 84, 90, 96, 108, 120, 144, 168, 180, 210, 216, 240, 288, 300, 336, 360, 420, 480, 504, 540, 600, 630, 660, 720
HIGHLY ABUNDANT NUMBERS. REF TAMS 56 467 44.

201 1, 2, 3, 4, 6, 8, 11, 13, 16, 18, 26, 28, 36, 38, 47, 48, 53, 57, 62, 69, 72, 77, 82, 87, 97, 99, 102, 106, 114, 126, 131, 138, 145, 148, 155, 175, 177, 180, 182, 189, 197, 206
A SELF-GENERATING SEQUENCE. REF UL1 IX. AT1 249.

202 1, 2, 3, 4, 6, 8, 13, 18, 30, 46, 78, 126, 224, 380, 687, 1224, 2250, 4112, 7685, 14310, 27012
NECKLACES. REF IJM 5 662 61.

203 1, 2, 3, 4, 6, 8, 14, 20, 36, 60, 108, 188, 352, 632, 1182, 2192, 4116, 7712, 14602, 27596, 52488, 99880, 190746, 364724, 699252, 1342184, 2581428, 4971068, 9587580
NECKLACES OF 2 COLORS. REF IJM 5 662 61. GO1 172.

204 1, 2, 3, 4, 6, 9, 12, 16, 22, 29, 38, 50, 64, 82, 105, 132, 166, 208, 258, 320, 395, 484, 592, 722, 876, 1060
COEFFICIENTS OF AN ELLIPTIC FUNCTION. REF CAY 9 128.

205 1, 2, 3, 4, 6, 9, 12, 18, 27, 36, 54, 81, 108, 162, 243, 324, 486, 729, 972, 1458, 2187, 2916, 4374, 6561, 8748, 13122, 19683, 26244, 39366, 59049, 78732, 118098
SUBGROUPS OF SYMMETRIC GROUP. REF CMB 8 627 65. JRM 4 168 71.

206 1, 2, 3, 4, 6, 9, 13, 19, 27, 38, 54, 77, 109, 154, 218, 309, 437, 618, 874, 1236, 1748, 2472, 3496, 4944, 6992, 9888, 13984, 19777, 27969, 39554, 55938, 79108, 111876
A NONLINEAR RECURRENCE. REF MMAG 43 143 70.

207 1, 2, 3, 4, 6, 9, 13, 19, 28, 41, 60, 88, 129, 189, 277, 406, 595, 872, 1278, 1873, 2745, 4023, 5896, 8641, 12664, 18560, 27201, 39865, 58425, 85626, 125491, 183916
A(N) = A(N − 1) + A(N − 3). REF LA3 13. FQ 2 225 64. JA2 91. MMAG 41 15 68.

208 1, 2, 3, 4, 6, 9, 14, 22, 35, 56, 90, 145, 234, 378, 611, 988, 1598, 2585, 4182, 6766, 10947, 17712, 28658, 46369, 75026, 121394, 196419, 317812, 514230, 832041
RESTRICTED PERMUTATIONS. REF CMB 4 32 61 (DIVIDED BY 3).

209 1, 2, 3, 4, 6, 9, 14, 23, 38
PAIRWISE PRIME POLYNOMIALS. REF IC 13 615 68.

210 1, 2, 3, 4, 6, 12, 15, 20, 30, 60, 84, 105, 140, 210, 420, 840, 1260, 1540, 2310, 2520, 4620, 5460, 9240, 13860, 16380, 27720, 30030, 32760, 60060, 120120, 180180
LARGEST ORDER OF PERMUTATION OF N SYMBOLS. REF BSMF 97 187 69.

211 1, 2, 3, 4, 6, 16, 16, 30
POINT-SYMMETRIC TOURNAMENTS. REF CMB 13 322 70.

212 1, 2, 3, 4, 7, 13, 24, 44, 83, 157, 297, 567, 1085, 2086, 4019, 7766, 15039, 29181, 56717, 110408, 215225, 420076, 820836, 1605587, 3143562, 6160098, 12080946
LANDAUS APPROXIMATION. REF MTAC 18 79 64.

213 1, 2, 3, 4, 8, 10, 12, 19, 37, 54, 65, 77, 314, 366, 770, 1327, 1373, 1937, 2561, 2663, 5834, 5985, 6751, 12003
SIZE OF EVEN PERFECT NUMBERS. REF BE3 19. NAMS 18 608 71.

214 1, 2, 3, 4, 9, 27, 512, 134217728, (NEXT TERM HAS 155 DIGITS)
AN EXPONENTIAL FUNCTION ON PARTITIONS. REF AMM 76 830 69.

215 1, 2, 3, 4, 11, 17, 29, 49, 85, 144
A PARTITION FUNCTION. REF JNSM 9 103 69.

216 1, 2, 3, 4, 40, 210, 1477, 11672, 104256, 1036050
FROM MENAGE POLYNOMIALS. REF R1 197.

217 1, 2, 3, 5, 1, 13, 7, 17, 11, 89, 1, 233, 29, 61, 47, 1597, 19, 4181, 41
PRIMITIVE DIVISORS OF FIBONACCI NUMBERS. REF FQ 1(3) 15 63.

218 1, 2, 3, 5, 4, 7, 6, 9, 13, 8, 10, 19, 14, 12, 29, 16, 21, 22, 37, 18, 27, 20, 43, 33, 34, 28, 49, 24, 61, 32, 67, 30, 73, 45, 57, 44, 40, 36, 50, 42, 52, 101, 63, 85, 109, 91, 74, 54
INVERSE OF A DIVISOR FUNCTION. REF BA2 85.

219 1, 2, 3, 5, 5, 7, 7, 7, 11, 9, 9, 11, 13, 11, 11, 15, 13, 13, 13, 17, 15, 19, 15, 19, 17, 21, 17, 19, 17, 17, 19, 21, 25, 19, 19, 23, 25, 23, 21, 23, 21, 21, 29, 23, 25, 23, 27, 29, 23
QUADRATIC PARTITIONS OF PRIMES. REF CU2 1. LE1 55.

220 1, 2, 3, 5, 6, 5, 8, 9, 11, 10, 7, 15, 15, 14, 17, 24, 24, 21, 13, 19, 27, 25, 29, 26, 44, 44, 29, 46, 39, 46, 27, 42, 47, 47, 54, 35, 41, 60, 51, 37, 48, 45, 49, 50, 49, 53
NUMBERS WITH INTEGRAL HARMONIC MEAN. REF AMM 61 95 54.

221 1, 2, 3, 5, 6, 6, 6, 7, 8, 10, 13, 13, 13, 14, 17, 17, 17, 18, 19, 20, 22, 23, 27, 29, 29, 29, 31, 32, 35, 36, 37, 40, 43, 46, 48, 50, 53, 55, 57, 60, 60, 61, 63, 66, 66, 68, 71, 74, 77
LATTICE POINTS IN CIRCLES. REF MTAC 20 306 66.

222 1, 2, 3, 5, 6, 7, 2, 10, 11, 3, 13, 14, 15, 17, 2, 19, 5, 21, 22, 23, 6, 26, 3, 7, 29, 30, 31, 2, 33, 34, 35, 37, 38, 39, 10, 41, 42, 43, 11, 5, 46, 47, 3, 2, 51, 13, 53, 6, 55, 14, 57, 58
REMOVE SQUARES FROM N. REF NCM 4 168 1878.

223 1, 2, 3, 5, 6, 7, 8, 10, 11, 12, 13, 14, 15, 17, 18, 19, 20, 21, 22, 23, 24, 26, 27, 28, 29, 30, 31, 32, 33, 34, 35, 37, 38, 39, 40, 41, 42, 43, 44, 45, 46, 47, 48, 50, 51, 52, 53, 54
NO SQUARES. REF HO2 97.

224 1, 2, 3, 5, 6, 7, 8, 10, 11, 13, 14, 15, 17, 19, 21, 22, 23, 24, 26, 27, 29, 30, 31, 32, 33, 34, 35, 37, 38, 39, 40, 41, 42, 43, 46, 47, 51, 53, 54, 55, 56, 57, 58, 59, 61, 62, 65, 66
CONTAIN ODD POWERS ONLY. REF AMM 73 139 66.

225 1, 2, 3, 5, 6, 7, 9, 10, 11, 13, 14, 15, 16, 18, 19, 20, 22, 23, 24, 26, 27, 28, 29, 31, 32, 33, 35, 36, 37, 39, 40, 41, 42, 44, 45, 46, 48, 49, 50, 52, 53, 54, 56, 57, 58, 59, 61, 62
A BEATTY SEQUENCE. REF CMB 2 188 59.

226 1, 2, 3, 5, 6, 7, 19, 21, 23, 31, 37, 38, 44, 69, 73
LEAST POSITIVE PRIMITIVE ROOTS. REF RS5 XLIV.

227 1, 2, 3, 5, 6, 8, 10, 13, 15, 18, 21, 25, 28, 32, 36, 41, 45, 50
RESTRICTED PARTITIONS. REF CAY 2 277.

228 1, 2, 3, 5, 6, 8, 12, 14, 15, 17, 20, 21, 24, 27, 33, 38, 41, 50, 54, 57, 59, 62, 66, 69, 71, 75, 77, 78, 80, 89, 90, 99, 101, 105, 110, 111, 117, 119, 131, 138, 141, 143, 147, 150
N(N + 1) + 1 IS PRIME. REF CU1 1 245. LIN 3 209 29. LE1 46.

229 1, 2, 3, 5, 6, 9, 11, 15, 18, 23, 27, 34, 39, 47, 54, 64, 72, 84, 94, 108, 120, 136, 150, 169, 185, 206, 225, 249, 270, 297, 321, 351, 378, 411, 441, 478, 511, 551, 588, 632, 672
PARTITIONS INTO AT MOST 4 PARTS. REF RS2 2.

230 1, 2, 3, 5, 6, 10, 11, 17, 21, 27, 33, 46, 53, 68, 82, 104, 123, 154, 179, 221, 262, 314, 369, 446, 515, 614, 715, 845, 977, 1148, 1321, 1544, 1778, 2060, 2361, 2736, 3121
MOCK THETA NUMBERS. REF TAMS 72 495 52.

231 1, 2, 3, 5, 7, 8, 9, 13, 14, 18, 19, 24, 25, 29, 30, 35, 36, 40, 41, 46, 51, 56, 63, 68, 72, 73, 78, 79, 83, 84, 89, 94, 115, 117, 126, 153, 160, 165, 169, 170, 175, 176, 181, 186
A SELF-GENERATING SEQUENCE. REF UL1 IX.

232 1, 2, 3, 5, 7, 8, 10, 12, 13, 18, 20, 27, 28, 33, 37, 42, 45, 47, 55, 58, 60, 62, 63, 65, 67, 73, 75, 78, 80, 85, 88, 90, 92, 102, 103, 105, 112, 115, 118, 120, 125, 128, 130, 132
(2N)2 + 1 IS PRIME. REF KR1 1 11.**

233 1, 2, 3, 5, 7, 9, 12, 15, 18, 22, 26, 30, 35, 40, 45, 51, 57, 63, 70, 77, 84, 92, 100, 108, 117, 126, 135, 145, 155, 165, 176, 187, 198, 210, 222, 234, 247, 260, 273, 287, 301
RELATED TO ZARANKIEWICZS PROBLEM. REF TI1 126 (DIVIDED BY 2).

234 1, 2, 3, 5, 7, 10, 11, 13, 14, 18, 21, 22, 31, 42, 67, 70, 71, 73, 251, 370, 375, 389, 407
39.2N + 1 IS PRIME. REF PAMS 9 674 58.**

235 1, 2, 3, 5, 7, 10, 12, 17, 18, 23, 25, 30, 32, 33, 38, 40, 45, 47, 52, 58, 70, 72, 77, 87, 95, 100, 103, 107, 110, 135, 137, 138, 143, 147, 170, 172, 175, 177, 182, 192, 205, 213
6A - 1, 6A + 1 ARE TWIN PRIMES. REF LE3 69.

236 1, 2, 3, 5, 7, 10, 13, 18, 23, 30, 37, 47, 57, 70, 83, 101, 119, 142, 165, 195, 225, 262, 299, 346, 393, 450, 507, 577, 647, 730, 813, 914, 1015, 1134, 1253, 1395, 1537
A LINEAR RECURRENCE. REF FQ 9 135 71.

237 1, 2, 3, 5, 7, 10, 13, 18, 23, 30, 37, 47, 57, 70, 84, 101, 119, 141, 164, 192, 221, 255, 291, 333, 377, 427, 480, 540, 603, 674, 748, 831, 918, 1014, 1115, 1226, 1342, 1469
PARTITIONS INTO AT MOST 5 PARTS. REF RS2 2.

238 1, 2, 3, 5, 7, 10, 14, 19, 26, 35, 47, 62, 82, 107, 139, 179, 230, 293
PLANAR PARTITIONS. REF PCPS 47 686 51.

239 1, 2, 3, 5, 7, 10, 14, 20, 27, 37, 49, 66, 86, 113, 146, 190, 242, 310, 392, 497, 623, 782, 973, 1212, 1498, 1851, 2274, 2793, 3411, 4163, 5059, 6142, 7427, 8972, 10801
REPRESENTATIONS OF THE SYMMETRIC GROUP. REF CJM 4 383 52.

240 1, 2, 3, 5, 7, 10, 14, 20, 29, 43, 65, 100, 156, 246, 391, 625, 1003, 1614, 2602, 4200, 6785, 10967, 17733, 28680, 46392, 75050, 121419, 196445, 317839, 514258
NTH FIBONACCI NUMBER + N. REF HO2 96.

241 1, 2, 3, 5, 7, 11, 13, 17, 19, 23, 29, 31, 37, 41, 43, 47, 53, 59, 61, 67, 71, 73, 79, 83, 89, 97, 101, 103, 107, 109, 113, 127, 131, 137, 139, 149, 151, 157, 163, 167, 173, 179
PRIMES. REF AS1 870.

242 1, 2, 3, 5, 7, 11, 13, 17, 19, 31, 37, 41, 61, 73, 97, 101, 109, 151, 163, 181, 193, 241, 251, 257, 271, 401, 433, 487, 541, 577, 601, 641, 751, 769, 811, 1153, 1201, 1297
A RESTRICTED CLASS OF PRIMES. REF KR1 1 53.

243 1, 2, 3, 5, 7, 11, 14, 20, 26, 35, 44, 58, 71, 90, 110, 136, 163, 199, 235, 282, 331, 391, 454, 532, 612, 709, 811, 931, 1057, 1206, 1360, 1540, 1729, 1945, 2172, 2432
PARTITIONS INTO AT MOST 6 PARTS. REF CAY 10 415. RS2 2.

244 1, 2, 3, 5, 7, 11, 15, 22, 30, 42, 56, 77, 101, 135, 176, 231, 297, 385, 490, 627, 792, 1002, 1255, 1575, 1958, 2436, 3010, 3718, 4565, 5604, 6842, 8349, 10143, 12310, 14883
NUMBER OF PARTITIONS OF N. REF RS2 90. R1 122. AS1 836.

245 1, 2, 3, 5, 7, 11, 17, 25, 38, 57, 86, 129, 194, 291, 437, 656, 985, 1477, 2216, 3325, 4987, 7481, 11222, 16834, 25251, 37876, 56815, 85222, 127834, 191751, 287626
QUOTIENT OF 3N / 2**N. REF JIMS 2 40 36. LE1 82.**

246 1, 2, 3, 5, 7, 11, 19, 43, 53, 79, 107, 149
LEAST POSITIVE PRIME PRIMITIVE ROOTS. REF RS5 XLV.

247 1, 2, 3, 5, 7, 11, 101, 131, 151, 181, 191, 313, 353, 373, 383, 727, 757, 787, 797, 919, 929, 10301, 10501, 10601, 11311, 11411, 12421, 12721, 12821, 13331, 13831
PALINDROMIC PRIMES. REF BE3 228.

248 1, 2, 3, 5, 7, 13, 17, 19, 31, 61, 89, 107, 127, 521, 607, 1279, 2203, 2281, 3217, 4253, 4423, 9689, 9941, 11213, 19937
MERSENNE PRIMES. REF MTAC 18 93 64. NAMS 18 608 71.

249 1, 2, 3, 5, 7, 13, 20, 35, 55, 96, 156, 267, 433, 747, 1239, 2089, 3498, 5912
PARAFFINS. REF JACS 54 1544 32.

250 1, 2, 3, 5, 7, 17, 31, 89, 127, 521, 607, 1279, 2281, 3217, 4423, 9689
IRREDUCIBLE MERSENNE TRINOMIALS. REF IC 15 68 69.

251 1, 2, 3, 5, 8, 9, 10, 11, 12, 18, 19, 22, 26, 28, 30, 31, 33, 35, 36, 38, 39, 40, 41, 44, 46, 47, 48, 50, 52, 54, 55, 56, 58, 61, 62, 66, 67, 68, 69, 71, 72, 74, 76, 77, 80, 82, 83, 91
ELLIPTIC CURVES. REF JRAM 212 23 63.

252 1, 2, 3, 5, 8, 11, 12, 14, 18, 20, 21, 27, 29, 30, 32, 35, 44, 45, 48, 50
OF THE FORM 2X2 + 3Y**2. REF EUL (1) 1 425 11.**

253 1, 2, 3, 5, 8, 12, 18, 26, 38, 53, 75, 103, 142, 192, 260, 346, 461, 605, 796
PLANAR PARTITIONS. REF PCPS 47 686 51.

254 1, 2, 3, 5, 8, 13, 17, 26, 34, 45, 54, 67, 81, 97, 115, 132, 153, 171, 198, 228, 256, 288, 323, 357, 400, 439, 488, 530, 581, 627, 681, 732, 790, 843, 908, 963, 1029, 1085
A SELF-GENERATING SEQUENCE. REF AMM 75 80 68. RLG.

255 1, 2, 3, 5, 8, 13, 21, 34, 55, 89, 144, 232, 375, 606, 979, 1582, 2556, 4130, 6673, 10782, 17421, 28148, 45480, 73484, 118732, 191841, 309967, 500829, 809214, 1307487
DYING RABBITS. REF FQ 2 108 64.

256 1, 2, 3, 5, 8, 13, 21, 34, 55, 89, 144, 233, 377, 610, 987, 1597, 2584, 4181, 6765, 10946, 17711, 28657, 46368, 75025, 121393, 196418, 317811, 514229, 832040, 1346269
FIBONACCI NUMBERS A(N) = A(N − 1) + A(N − 2). REF HW1 148. REC 11 20 62. HO1.

257 1, 2, 3, 5, 8, 14, 21, 39, 62, 112, 189, 352, 607, 1144, 2055, 3883, 7154, 13602
NECKLACES. REF IJM 5 663 61.

258 1, 2, 3, 5, 8, 14, 23, 39, 65, 110, 184, 310, 520, 876, 1471, 2475, 4159, 6996, 11759
PARAFFINS. REF JACS 54 1105 32.

259 1, 2, 3, 5, 8, 15, 26, 48, 87, 161, 299, 563, 1066, 2030, 3885, 7464, 14384, 27779, 53782, 104359, 202838, 394860, 769777, 1502603, 2936519, 5744932
POPULATION OF U2 − 2V**2. REF MTAC 20 560 66.**

260 1, 2, 3, 5, 8, 21, 29, 79, 661, 740, 19161, 19901, 118666, 138567, 3167140, 3305707, 29612796, 32918503, 62531299, 595700194, 658231493, 1253931687
CONVERGENTS TO FIFTH ROOT OF 5. REF AMP 46 116 1866. LE1 67. HPR.

261 1, 2, 3, 5, 9, 16, 28, 50, 89, 159, 285, 510, 914, 1639, 2938, 5269, 9451, 16952, 30410, 54555, 97871, 175586, 315016, 565168, 1013976, 1819198, 3263875, 5855833
PARTITIONS INTO POWERS OF 1/2. REF EMS 11 224 59. ST3.

262 1, 2, 3, 5, 9, 16, 28, 51, 93, 170, 315, 585, 1091, 2048, 3855, 7280, 13797, 26214
NECKLACES. REF IJM 5 663 61. ME1.

263 1, 2, 3, 5, 9, 16, 29, 52, 94, 175, 327, 616, 1169, 2231, 4273, 8215, 15832, 30628, 59345, 115208, 224040, 436343, 850981, 1661663, 3248231, 6356076, 12448925
RAMANUJANS APPROXIMATION. REF MTAC 18 79 64.

264 1, 2, 3, 5, 9, 16, 29, 53, 98, 181, 341, 640, 1218, 2321, 4449, 8546, 16482, 31845, 61707, 119760, 232865, 453511, 884493, 1727125, 3376376, 6607207
POPULATION OF 3U2 + 4V**2. REF MTAC 20 567 66.**

265 1, 2, 3, 5, 9, 16, 29, 54, 97, 180, 337, 633, 1197, 2280, 4357, 8363, 16096, 31064, 60108, 116555, 226419, 440616, 858696, 1675603, 3273643, 6402706, 12534812
POPULATION OF U2 + V**2. REF MTAC 20 560 66.**

266 1, 2, 3, 5, 9, 17, 33, 65, 129, 257, 513, 1025, 2049, 4097, 8193, 16385, 32769, 65537, 131073, 262145, 524289, 1048577, 2097153, 4194305, 8388609, 16777217
2N + 1. REF BA1.**

267 1, 2, 3, 5, 9, 18, 35, 75, 159, 355, 802, 1858, 4347, 10359, 24894, 60523, 148284, 366319, 910726, 2278658, 5731580, 14490245, 93839412, 240215803, 617105614
PARAFFINS. REF JACS 54 2919 32.

268 1, 2, 3, 5, 9, 18, 35, 75, 159, 357, 799
HYDROCARBONS. REF BS1 201.

269 1, 2, 3, 5, 10, 11, 26, 32, 39, 92, 116, 134, 170, 224
25.4N + 1 IS PRIME. REF PAMS 9 674 58.**

270 1, 2, 3, 5, 10, 18, 35, 63, 126, 231
FROM RADONS THEOREM. REF MFM 73 12 69.

271 1, 2, 3, 5, 10, 24, 69, 384
LINEAR SPACES. REF BSM 19 424 67.

272 1, 2, 3, 5, 10, 27, 119, 1113, 29375, 2730166
THRESHOLD FUNCTIONS. REF PGEC 19 821 70.

273 1, 2, 3, 5, 11, 16, 38, 54, 130, 184, 444, 628, 1516, 2144, 5176, 7320, 17672, 24992, 60336, 85328, 206000, 291328, 703328, 994656, 2401312, 3395968, 8198592
A LINEAR RECURRENCE. REF AMM 72 1024 65.

274 1, 2, 3, 5, 11, 24, 55, 136, 345, 900, 2412, 6563, 18122, 50699, 143255, 408419, 1172854, 3395964
PARAFFINS. REF JACS 54 1544 32.

275 1, 2, 3, 5, 11, 47, 923, 409619, 83763206255, 3508125906290858798171, 6153473687096578758448522809275077520433167
HAMILTON NUMBERS. REF RS3 178 288 1887. LU1 496.

276 1, 2, 3, 5, 12, 14, 11, 13, 20, 72, 19, 42, 132, 84, 114, 29, 30, 110, 156, 37, 156, 420, 210, 156, 552, 462, 72, 53, 420, 342, 59
SHUFFLING CARDS. REF SIAMR 3 296 61.

277 1, 2, 3, 5, 12, 36, 110, 326, 963, 2964, 9797, 34818, 130585, 506996, 2018454, 8238737, 34627390, 150485325, 677033911, 3147372610, 15066340824, 74025698886
FROM A DIFFERENTIAL EQUATION. REF AMM 67 766 60.

278 1, 2, 3, 5, 13, 83, 2503, 976253, 31601312113, 2560404986164794683, 20252311318903795247872230479800З
FROM A CONTINUED FRACTION. REF AMM 63 711 56.

279 1, 2, 3, 5, 16, 231, 53105, 2820087664, 7952894429824835871, 63248529811938901240357985099443351745
A(N) = A(N − 1)2 − A(N − 2)**2. REF EUR 27 6 64.**

280 1, 2, 3, 6, 2, 0, 1, 10, 0, 2, 10, 6, 7, 14, 0, 10, 12, 0, 6, 0, 9, 4, 10, 0, 18, 2, 0, 6, 14, 18, 11, 12, 0, 0, 22, 0, 20, 14, 6, 22, 0, 0, 23, 26, 0, 18, 4, 0, 14, 2, 0, 20, 0, 0, 0, 12, 3, 30
GLAISHERS CHI FUNCTION. REF QJM 20 151 1884.

281 1, 2, 3, 6, 5, 11, 14, 22, 30, 47, 66, 99, 143, 212, 308, 454, 663, 974, 1425, 2091, 3062, 4490, 6578, 9643, 14130, 20711, 30351, 44484, 65192, 95546, 140027, 205222
A(N) = A(N − 2) + A(N − 3) + A(N − 4). REF IDM 8 64 01. FQ 6(3) 68 68.

282 1, 2, 3, 6, 7, 10, 14, 15, 21, 30, 35, 42, 70, 105, 210, 221, 230, 231, 238, 247, 253, 255, 266, 273, 285, 286, 299, 322, 323, 330, 345, 357, 374, 385, 390, 391, 399, 418, 429
A SPECIALLY CONSTRUCTED SEQUENCE. REF AMM 74 874 67.

283 1, 2, 3, 6, 7, 11, 14, 17, 33, 42, 43, 63, 65, 67, 81, 134, 162, 206, 211, 366
9.2N + 1 IS PRIME. REF PAMS 9 674 58.**

284 1, 2, 3, 6, 8, 10, 22, 35, 42, 43, 46, 56, 91, 102, 106, 142, 190, 208, 266, 330, 360, 382, 462, 503, 815
33.2N − 1 IS PRIME. REF MTAC 23 874 69.**

285 1, 2, 3, 6, 8, 16, 24, 42, 69, 124, 208, 378, 668, 1214, 2220, 4110, 7630, 14308, 26931
NECKLACES. REF IJM 5 663 61.

286 1, 2, 3, 6, 9, 14, 20, 29, 42, 58, 79, 108, 145, 191, 252, 329, 427, 549, 704, 894, 1136, 1427, 1793, 2237, 2789, 3450, 4268, 5248, 6447, 7880, 9619, 11691, 14199
MIXED PARTITIONS. REF JNSM 9 91 69.

287 1, 2, 3, 6, 9, 18, 30, 56, 99, 186, 335, 630, 1161, 2182, 4080, 7710, 14532, 27594, 52377, 99858, 190557, 364722, 698870, 1342176, 2580795, 4971008, 9586395
IRREDUCIBLE POLYNOMIALS, OR NECKLACES. REF IJM 5 663 61. JSIAM 12 288 64.

288 1, 2, 3, 6, 9, 26, 53, 146, 369, 1002
NECKLACES. REF IJM 2 302 58.

289 1, 2, 3, 6, 10, 11, 21, 30, 48, 72, 110, 171, 260, 401, 613, 942, 1445, 2216, 3401, 5216, 8004, 12278, 18837, 28899, 44335, 68018, 104349, 160089, 245601, 376791
A FIELDER SEQUENCE. REF FQ 6(3) 68 68.

290 1, 2, 3, 6, 10, 17, 21, 38, 57, 92, 143, 225, 351, 555, 868, 1366, 2142, 3365, 5282, 8296, 13023, 20451, 32108, 50417, 79160, 124295, 195159, 306431, 481139, 755462
A FIELDER SEQUENCE. REF FQ 6(3) 68 68.

291 1, 2, 3, 6, 10, 17, 28, 46, 75, 122, 198, 321, 520, 842, 1363, 2206, 3570, 5777, 9348, 15126, 24475, 39602, 64078, 103681, 167760, 271442, 439203, 710646, 1149850
A(N) = A(N - 1) + A(N - 2) + 1. REF JA2 96.

292 1, 2, 3, 6, 10, 19, 35, 62, 118, 219, 414, 783, 1497, 2860, 5503, 10593, 20471, 39637, 76918, 149501, 291115, 567581, 1108022, 2165621, 4237085, 8297727
POPULATION OF U2 + 4V**2. REF MTAC 18 84 64.**

293 1, 2, 3, 6, 10, 19, 35, 67, 127, 248, 482, 952, 1885, 3765, 7546, 15221, 30802, 62620, 127702, 261335, 536278, 1103600, 2276499, 4706985, 9752585, 20247033
SERIES-REDUCED PLANTED TREES. AM1 101 150 59. CA3.

294 1, 2, 3, 6, 10, 20, 35, 70, 126, 252, 462, 924, 1716, 3432, 6435, 12870, 24310, 48620, 92378, 184756, 352716, 705432, 1352078, 2704156, 5200300, 10400600
CENTRAL BINOMIAL COEFFICIENTS C(N, N/2). REF RS1. AS1 828.

295 1, 2, 3, 6, 10, 20, 36, 72, 137, 274, 543
RESTRICTED HEXAGONAL POLYOMINOES. REF EMS 17 11 70.

296 1, 2, 3, 6, 11, 20, 37, 68, 125, 230, 423, 778, 1431, 2632, 4841, 8904, 16377, 30122, 55403, 101902, 187427, 344732, 634061, 1166220, 2145013, 3945294, 7256527
TRIBONACCI NUMBERS A(N) = A(N - 1) + A(N - 2) + A(N - 3). REF FQ 5 211 67.

297 1, 2, 3, 6, 11, 22, 42, 84, 165
RANDOM TOURNAMENTS. REF CMB 13 108 70.

298 1, 2, 3, 6, 11, 23, 46, 98, 207, 451, 983, 2179, 4850, 10905, 24631, 56011, 127912, 293547, 676157, 1563372, 3626149, 8436379, 19680277, 46026618
WEDDERBURN-ETHERINGTON NUMBERS. REF CO1 1 68.

299 1, 2, 3, 6, 11, 23, 47, 106, 235, 551, 1301, 3159, 7741, 19320, 48629, 123867, 317955, 823065, 2144505, 5623756, 14828074, 39299897, 104636890, 279793450
UNLABELED TREES. REF R1 138. HA5 232.

300 1, 2, 3, 6, 11, 24, 47, 103, 214, 481, 1030, 2337, 5131, 11813, 26329, 60958, 137821, 321690, 734428, 1721998, 3966556, 9352353, 21683445, 51296030, 119663812
STRUCTURE OF RAYLEIGH POLYNOMIAL. REF DMJ 31 517 64.

301 1, 2, 3, 6, 12, 23, 44, 85, 164, 316, 609, 1174, 2263, 4362, 8408, 16207, 31240, 60217, 116072, 223736, 431265, 831290, 1602363, 3088654, 5953572, 11475879
TETRANACCI NUMBERS. REF FQ 8 7 70.

302 1, 2, 3, 6, 12, 26, 59, 146, 368
SERIES-PARALLEL NUMBERS. REF ICM 1 646 50.

303 1, 2, 3, 6, 13, 28, 62
ALKYLS. REF ZFK 93 437 36.

304 1, 2, 3, 6, 13, 35, 116
CONNECTED WEIGHTED LINEAR SPACES. REF BSM 22 234 70.

305 1, 2, 3, 6, 14, 36, 94, 250, 675, 1838, 5053, 14016, 39169, 110194, 311751, 886160, 2529260
FIXED TRIANGULAR POLYOMINOES. REF LU5.

306 1, 2, 3, 6, 15, 63, 567, 14755, 1366318
THRESHOLD FUNCTIONS. REF PGEC 19 821 70.

307 1, 2, 3, 6, 18, 206, 7888299
BOOLEAN FUNCTIONS. REF JSIAM 12 294 64.

308 1, 2, 3, 6, 20, 150, 3287, 244158, 66291591, 68863243522
THRESHOLD FUNCTIONS. REF PGEC 19 821 70.

309 1, 2, 3, 6, 20, 168, 7581, 7828354, 2414682040998
MONOTONE BOOLEAN FUNCTIONS, OR DEDEKINDS PROBLEM. REF HA2 188. BI1 63. CO1 2 116. WE1 181.

310 1, 2, 3, 6, 22, 402, 1228158, 400507806843728
BOOLEAN FUNCTIONS. REF JSIAM 11 827 63.

311 1, 2, 3, 6, 30, 75, 81
N.2N - 1 IS PRIME. REF MTAC 23 875 69.**

312 1, 2, 3, 7, 5, 11, 103, 71, 661, 269, 329891, 39916801, 2834329, 75024347, 3790360487, 46271341, 1059511, 1000357, 123610951, 1713311273363831
LARGEST FACTOR OF FACTORIAL (N) + 1. REF SMA 14 25 48.

313 1, 2, 3, 7, 8, 10, 16, 18, 19, 40, 48, 55, 90, 96, 98, 190, 398, 456, 502
57.2N + 1 IS PRIME. REF PAMS 9 675 58.**

314 1, 2, 3, 7, 10, 13, 18, 27, 37, 51, 74, 157, 271, 458, 530, 891
21.2N - 1 IS PRIME. REF MTAC 23 874 69.**

315 1, 2, 3, 7, 10, 13, 25, 26, 46, 60, 87, 90, 95, 145, 160, 195, 216, 308, 415
7.4N + 1 IS PRIME. REF PAMS 9 674 58.**

316 1, 2, 3, 7, 12, 27, 55, 127, 284, 682
CENTERED TREES. REF CAY 9 438.

317 1, 2, 3, 7, 13, 31, 65, 154, 347, 824, 1905, 4512, 10546, 24935, 58476, 138002, 323894, 763172, 1790585, 4213061, 9878541
SQUARE FILAMENTS. REF PL2 1 337 70.

318 1, 2, 3, 7, 14, 32, 72, 171, 405, 989, 2426, 6045, 15167, 38422, 97925, 251275, 648061, 1679869, 4372872, 11428365, 29972078, 78859809, 208094977, 550603722
HYDROCARBONS. REF JACS 55 253 33.

319 1, 2, 3, 7, 15, 34, 78, 182, 429, 1019, 2433, 5830, 14004, 33694, 81159, 195635, 471819, 1138286, 2746794, 6629290, 16001193, 38624911, 93240069, 225087338
SUM OF FIBONACCI AND PELL NUMBERS.

320 1, 2, 3, 7, 15, 43, 131, 468, 1776, 7559, 34022, 166749, 853823, 4682358
REFINEMENTS OF PARTITIONS. REF GU5.

321 1, 2, 3, 7, 16, 54
DIFFERENT GRAPHS, ALLOWING COMPLEMENTATION. REF KNAW 69 339 66.

322 1, 2, 3, 7, 18, 41, 123, 367
ALTERNATING KNOTS. REF TA1 1 345. JL2 343.

323 1, 2, 3, 7, 21, 49, 166, 549
ALTERNATING AND NONALTERNATING KNOTS. REF TA1 1 345. JL2 343.

324 1, 2, 3, 7, 21, 135, 2470, 175428
THRESHOLD FUNCTIONS. REF PGEC 19 821 70.

325 1, 2, 3, 7, 23, 41, 71, 191, 409, 2161, 5881, 36721, 55441, 71761, 110881, 760321
LEAST POSITIVE PRIMITIVE ROOTS. REF RS5 XLIV.

326 1, 2, 3, 7, 23, 43, 67, 83, 103, 127, 163, 167, 223, 227, 283, 367, 383, 443, 463, 467, 487, 503, 523, 547, 587, 607, 643, 647, 683, 727, 787, 823, 827, 863, 883, 887, 907
PRIMES DIVIDING ALL FIBONACCI SEQUENCES. REF FQ 2 38 64.

327 1, 2, 3, 7, 23, 89, 113, 523, 887, 1129, 1327, 9551, 15683, 19609, 31397, 155921, 360653, 370261, 492113, 1349533, 1357201, 2010733, 4652353, 17051707
INCREASING GAPS BETWEEN PRIMES. REF KR1 1 14. MTAC 18 649 64.

328 1, 2, 3, 7, 23, 164, 3779, 619779, 2342145005, 1451612289057674, 3399886472013047316638149, 4935316984175079105557291745555191750431
A(N) = A(N − 1)A(N − 2) + A(N − 3). REF GU5.

329 1, 2, 3, 7, 43, 13, 53, 5, 6221671, 38709183810571
FROM EUCLIDS PROOF. REF BAMS 69 737 63.

330 1, 2, 3, 7, 43, 139, 50207, 340999, 3202139, 410353
FROM EUCLIDS PROOF. REF NAMS 11 376 64.

331 1, 2, 3, 7, 43, 1807, 3263443, 10650056950807, 113423713055421844361000443, 12864938683278671740537145998360961546653259485195807
A(N + 1) = A(N)2 − A(N) + 1. REF CJM 15 475 63. AMM 70 403 63.**

332 1, 2, 3, 8, 10, 12, 14, 17, 23, 24, 27, 28, 37, 40, 41, 44, 45, 53, 59, 66, 70, 71, 77, 80, 82, 87, 90, 97, 99, 102, 105, 110, 114, 119, 121, 124, 127, 133, 136, 138, 139, 144
(2N)4 + 1 IS PRIME. REF MTAC 21 246 67.**

333 1, 2, 3, 8, 13, 20, 31, 32, 53, 76, 79, 80, 117, 176, 181, 182, 193, 200, 283, 284, 285, 286, 293, 440, 443, 468, 661, 678, 683, 684, 1075, 1076, 1087, 1088, 1091, 1092
RELATED TO LIOUVILLES FUNCTION. REF IAS 12 408 40.

334 1, 2, 3, 8, 18, 44, 115, 294, 783
RECTANGULAR POLYOMINOES. REF SPH 7 203 37.

335 1, 2, 3, 8, 19, 27, 100, 227, 781, 1008, 3805, 4813, 148195, 153008, 760227, 913235, 2586697, 24193508, 147747745, 615184488, 762932233, 1378116721
CONVERGENTS TO CUBE ROOT OF 4. REF AMP 46 106 1866. LE1 67. HPR.

336 1, 2, 3, 8, 24, 108, 640, 4492, 36336, 329900, 3326788
PATTERNS. REF MES 37 61 07.

337 1, 2, 3, 8, 30, 144, 840, 5760, 45360, 403200, 3991680, 43545600, 518918400, 6706022400, 93405312000, 1394852659200, 22230464256000, 376610217984000
SUMS OF FACTORIAL NUMBERS. REF CJM 22 26 70.

338 1, 2, 3, 8, 51, 1538, 599871, 19417825808, 1573273218577214751, 124442887685693556895657990772138
FROM A CONTINUED FRACTION. REF AMM 63 711 56.

339 1, 2, 3, 9, 20, 73
PARTITIONS OF A POLYGON. REF BAMS 54 359 48.

340 1, 2, 3, 10, 27, 98
SIGNED TREES. REF AM1 101 154 59.

341 1, 2, 3, 10, 1382, 420, 10851, 438670, 7333662, 51270780, 7090922730, 2155381956, 94997844116, 68926730208040
NUMERATORS OF BERNOULLI NUMBERS. REF DA2 2 208.

342 1, 2, 3, 11, 22, 26, 101, 111, 121, 202, 212, 264, 307, 836, 1001, 1111, 2002, 2285, 2636, 10001, 10101, 10201, 11011, 11111, 11211, 20002, 20102, 22865, 24846, 30693
SQUARE IS A PALINDROME. REF JRM 3 94 70.

343 1, 2, 3, 11, 69, 701, 10584, 222965, 6253604, 225352709, 10147125509, 558317255704, 36859086001973, 2875567025409598, 261713458398275391
A(N) = N(N − 1)A(N − 1)/2 + A(N − 2).

344 1, 2, 3, 12, 10, 60, 105, 280, 252, 2520, 2310, 27720, 25740, 24024, 45045, 720720, 680680, 12252240, 11639628, 11085360, 10581480, 232792560, 223092870
L. C. M. OF BINOMIAL COEFFICIENTS C(N, 1), C(N, 2), ..., C(N, N).

345 1, 2, 3, 12, 52, 456, 6873, 191532, 9733032, 903753248, 154108311046
NONTRANSITIVE PRIME TOURNAMENTS. REF DMJ 37 332 70.

346 1, 2, 3, 24, 5, 720, 105, 2240, 189, 3628800, 385, 479001600, 19305, 896896, 2027025, 20922789888000, 85085, 6402373705728000, 8729721, 47297536000
N-PHI-TORIAL. REF AMM 60 422 53.

347 1, 2, 3, 26, 13, 1074, 1457, 61802, 7929, 4218722
SUMS OF LOGARITHMIC NUMBERS. REF MST 31 78 63.

348 1, 2, 3, 56, 43265728
INVERTIBLE BOOLEAN FUNCTIONS. REF JSIAM 12 297 64.

SEQUENCES BEGINNING 1, 2, 4

349 1, 2, 4, 1, 3, 6, 5, 2, 8, 4, 10, 9, 1, 8, 5, 11, 12, 10, 2, 4, 9, 13, 6, 11, 8, 16, 5, 13, 17, 18, 15, 2, 4, 11, 6, 19, 17, 13, 16, 10, 1, 3, 20, 12, 22, 18, 17, 22, 23, 11, 2, 16, 19, 13, 8
QUADRATIC PARTITIONS OF PRIMES. REF CU2 1. LE1 55.

350 1, 2, 4, 3, 6, 10, 12, 4, 8, 18, 6, 11, 20, 18, 28, 5, 10, 12, 36, 12, 20, 14, 12, 23, 21, 8, 52, 20, 18, 58, 60, 6, 12, 66, 22, 35, 9, 20, 30, 39, 54, 82, 8, 28, 11, 12, 10, 36, 48, 30
EXPONENTS OF 2. REF MAG 4 266 08. MOD 10 226 61. SIAMR 3 296 61.

351 1, 2, 4, 4, 6, 8, 8, 8, 13, 12, 12, 16, 14, 16, 24, 16
GENERALIZED DIVISOR FUNCTION. REF PLMS 19 111 19.

352 1, 2, 4, 4, 6, 8, 8, 12, 14
GENERALIZED TANGENT NUMBERS. REF MTAC 21 690 67.

353 1, 2, 4, 4, 6, 16, 16, 30, 88
POINT-SYMMETRIC TOURNAMENTS. REF CMB 13 322 70.

354 1, 2, 4, 5, 6, 7, 8, 10, 11, 12, 13, 15
WYTHOFF GAME. REF CMB 2 189 59.

355 1, 2, 4, 5, 6, 8, 9, 11, 12, 13, 15, 16, 18, 19, 20, 22, 23, 25, 26, 27, 29, 30, 32, 33, 34, 36, 37, 38, 40, 41, 43, 44, 45, 47, 48, 50, 51, 52, 54, 55, 57, 58, 59, 61, 62, 64, 65, 66
A BEATTY SEQUENCE. REF CMB 3 21 60.

356 1, 2, 4, 5, 7, 8, 9, 11, 12, 14, 15, 16
A BEATTY SEQUENCE. REF CMB 2 188 59.

357 1, 2, 4, 5, 7, 8, 10, 11, 13, 14, 16, 17, 19, 20, 22, 23, 25, 26, 28, 29, 31, 32, 34, 35, 37, 38, 40, 41, 43, 44, 46, 47, 49, 50, 52, 53, 55, 56, 58, 59, 61, 62, 64, 65, 67, 68, 70, 71
SEMI-TRIBONACCI NUMBERS. REF FQ 6(3) 261 68.

358 1, 2, 4, 5, 7, 8, 11, 13, 16, 17, 19, 31, 37, 41, 47, 53, 61, 71, 79, 113, 313, 353
PRIME LUCAS NUMBERS. REF JA2 36. MTAC 23 213 69.

359 1, 2, 4, 5, 7, 9, 10, 12, 14, 16, 17, 19, 21, 23, 25, 26, 28, 30, 32, 34, 36, 37, 39, 41, 43, 45, 47, 49, 50, 52, 54, 56, 58, 60, 62, 64, 65, 67, 69, 71, 73, 75, 77, 79, 81, 82, 84, 86
1 ODD, 2 EVEN, 3 ODD, ... REF AMM 67 380 60.

360 1, 2, 4, 5, 7, 9, 12, 13, 15, 17, 20, 22, 25, 28, 32, 33, 35, 37, 40, 42, 45, 48, 52, 54, 57, 60, 64, 67, 71, 75, 80, 81, 83, 85, 88, 90, 93, 96, 100, 102, 105, 108, 112, 115, 119
NUMBER OF ONES IN BINARY EXPANSION OF FIRST N NUMBERS. REF SIAMR 4 21 62. CMB 8 481 65. ANY 175 177 70.

361 1, 2, 4, 5, 8, 9, 10, 13, 16, 17, 18, 20, 25, 26, 29, 32, 34, 36, 37, 40, 41, 45, 49, 50, 52, 53, 58, 61, 64, 65, 68, 72, 73, 74, 80, 81, 82, 85, 89, 90, 97, 98, 100, 101, 104, 106
THE SUM OF 2 SQUARES. REF EUL (1) 1 417 11. KNAW 53 872 50.

362 1, 2, 4, 5, 8, 9, 12, 14, 17, 18, 23, 24, 27, 30, 34, 35, 40, 41, 46, 49, 52, 53, 60, 62, 65, 68, 73, 74, 81, 82, 87, 90, 93, 96, 104, 105, 108, 111, 118, 119, 126, 127, 132, 137
A NUMBER-THEORETIC FUNCTION. REF DVSS 2 281 1884.

363 1, 2, 4, 5, 8, 10, 14, 15, 16, 21, 22, 25, 26, 28, 33, 34, 35, 36, 38, 40, 42, 46, 48, 49, 50, 53, 57, 60, 62, 64, 65, 70, 77, 80, 81, 83, 85, 86, 90, 91, 92, 100
PRIME NUMBERS OF MEASUREMENT. REF PCPS 21 654 23.

364 1, 2, 4, 5, 8, 12, 19, 30, 48, 77, 124, 200, 323, 522, 844, 1365, 2208, 3572, 5779, 9350, 15128, 24477, 39604, 64080, 103683, 167762, 271444, 439205, 710648, 1149852
A(N) = A(N - 1) + A(N - 2) - 1. REF JA2 97.

365 1, 2, 4, 5, 10, 14, 17, 31, 41, 73, 80, 82, 116, 125, 145, 157, 172, 202, 224, 266, 289, 293, 463
15.2N - 1 IS PRIME. REF MTAC 23 874 69.**

366 1, 2, 4, 5, 10, 19, 36, 68, 138
BORON TREES. REF CAY 9 451.

367 1, 2, 4, 5, 14, 14, 39, 42, 132, 132, 424, 429, 1428, 1430, 4848, 4862, 16796, 16796, 58739, 58786, 208012, 208012, 742768, 742900, 2674426, 2674440, 9694416
DISSECTIONS OF A POLYGON. REF GU1.

368 1, 2, 4, 6, 3, 10, 25, 12, 42, 8, 40, 202, 21
FROM SEDLACEKS PROBLEM ON SOLUTIONS OF X + Y = Z. REF GU8.

369 1, 2, 4, 6, 7, 10, 11, 12, 22, 23, 25, 26, 27, 30, 36, 38, 42, 43, 44, 45, 50, 52, 54, 58, 59, 70, 71, 72, 74, 75, 76, 78, 86, 87, 91, 102, 103, 106, 107, 108, 110, 116, 118, 119
ELLIPTIC CURVES. REF JRAM 212 25 63.

370 1, 2, 4, 6, 8, 10, 12, 16, 18, 20, 22, 24, 28, 30, 32, 36, 40, 42, 44, 46, 48, 52, 54, 56, 58, 60, 64, 66, 70, 72, 78, 80, 82, 84, 88, 90, 92, 96, 100
VALUES OF REDUCED TOTIENT FUNCTION. REF NAM 17 305 1898. LE1 7.

371 1, 2, 4, 6, 8, 10, 12, 16, 18, 20, 22, 24, 28, 30, 32, 36, 40, 42, 44, 46, 48, 52, 54, 56, 58, 60, 64, 66, 70, 72, 78, 80, 82, 84, 88, 92, 96, 100, 102, 104, 106, 108, 110, 112, 116
VALUES OF EULER TOTIENT FUNCTION. REF BA2 64.

372 1, 2, 4, 6, 8, 12, 16, 24, 32, 36, 48, 64, 72, 96, 120, 128, 144, 192, 216, 240, 256, 288, 384, 432, 480, 512, 576, 720, 768, 864, 960, 1024, 1152, 1296, 1440, 1536, 1728
JORDAN-POLYA NUMBERS. REF JCT 5 25 68.

373 1, 2, 4, 6, 8, 14, 26
SNAKE-IN-THE-BOX PROBLEM. REF AMM 77 63 70.

374 1, 2, 4, 6, 9, 12, 16, 20, 25, 30, 36, 42, 49, 56, 64, 72, 81, 90, 100, 110, 121, 132, 144, 156, 169, 182, 196, 210, 225, 240, 256, 272, 289, 306, 324, 342, 361, 380, 400, 420
A PARTITION FUNCTION. REF AMS 26 304 55.

375 1, 2, 4, 6, 9, 13, 18, 24, 31, 39, 50, 62, 77, 93, 112, 134, 159, 187, 252, 292
DENUMERANTS. REF R1 152.

376 1, 2, 4, 6, 10, 12, 18, 22, 28, 32, 42, 46, 58, 64, 72, 80, 96, 102, 120, 128, 140, 150, 172, 180, 200, 212, 230, 242, 270, 278, 308, 324, 344, 360, 384, 396, 432, 450, 474, 490
SUM OF TOTIENT FUNCTION. REF SY1 4 103. LE1 7.

377 1, 2, 4, 6, 10, 12, 18, 22, 30, 34, 42, 48, 58, 60, 78, 82, 102, 108, 118, 132, 150, 154, 174, 192, 210, 214, 240, 258, 274, 282, 322, 330, 360, 372, 402, 418, 442, 454, 498
GENERATED BY A SIEVE. REF RLM 11 27 57.

378 1, 2, 4, 6, 10, 14, 20, 26, 36, 46, 60, 74, 94, 114, 140, 166, 202, 238, 284, 330, 390, 450, 524, 598, 692, 786, 900, 1014, 1154, 1294, 1460, 1626, 1828, 2030, 2268, 2506
THE BINARY PARTITION FUNCTION. REF FQ 4 117 66. PCPS 66 376 69. AT1 400.

379 1, 2, 4, 6, 10, 14, 21, 29, 41, 55, 76, 100, 134, 175, 230, 296, 384, 489, 626, 791, 1001, 1254, 1574, 1957, 2435, 3009, 3717, 4564, 5603, 6841, 8348, 10142, 12309
TREES OF HEIGHT 2. REF IBMJ 4 475 60. KU1.

380 1, 2, 4, 6, 10, 14, 24, 30
SIZE OF MINIMAL GRAPHS. REF SA1 94.

381 1, 2, 4, 6, 10, 16, 26, 44, 76, 132, 234, 420, 761, 1391, 2561, 4745, 8841, 16551, 31114, 58708, 111136, 211000, 401650, 766372, 1465422, 2807599, 5388709, 10359735
SUM OF (2N)/N**

382 1, 2, 4, 6, 10, 18, 33, 60, 111, 205, 385, 725, 1374, 2610, 4993, 9578, 18426, 35568, 68806, 133411, 259145, 504222, 982538, 1917190, 3745385, 7324822
POPULATION OF U2 + 2V**2. REF MTAC 20 560 66.**

383 1, 2, 4, 6, 11, 19, 33, 55, 95, 158, 267, 442, 731, 1193, 1947
PLANAR PARTITIONS. REF MA2 2 332.

384 1, 2, 4, 6, 11, 19, 34, 63, 117, 218, 411, 780, 1487, 2849, 5477, 10555, 20419, 39563, 76805, 149360, 290896, 567321, 1107775, 2165487, 4237384, 8299283
RAMANUJANS APPROXIMATION. REF MTAC 18 84 64.

385 1, 2, 4, 6, 12, 24, 36, 48, 60, 120, 180, 240, 360, 720, 840, 1260, 1680, 2520, 5040, 7560, 10080, 15120, 20160, 25200, 27720, 45360, 50400, 55440, 83160, 110880, 166320
HIGHLY COMPOSITE NUMBERS. REF RAM 87.

386 1, 2, 4, 6, 16, 20, 24, 28, 34, 46, 48, 54, 56, 74, 80, 82, 88, 90, 106, 118, 132, 140, 142, 154, 160, 164, 174, 180, 194, 198, 204, 210, 220, 228, 238, 242, 248, 254, 266, 272
N4 + 1 IS PRIME. REF MTAC 21 246 67.**

387 1, 2, 4, 6, 16, 20, 36, 54, 60, 96, 124, 150, 252, 356, 460, 612, 654, 664, 698, 702, 972
17.2N - 1 IS PRIME. REF MTAC 22 421 68.**

388 1, 2, 4, 7, 8, 11, 13, 14, 16, 19, 21, 22, 25, 26, 28, 31, 32, 35, 37, 38, 41, 42, 44, 47, 49, 50, 52, 55, 56, 59, 61, 62, 64, 67, 69, 70, 73, 74, 76, 79, 81, 82, 84, 87, 88, 91, 93
ODD NUMBER OF ONES IN BINARY EXPANSION. REF CMB 2 86 59.

389 1, 2, 4, 7, 8, 12, 13, 17, 20, 26, 28, 35, 37, 44, 48, 57, 60, 70, 73, 83, 88, 100, 104, 117, 121, 134, 140, 155, 160, 176, 181, 197, 204, 222, 228 247, 253, 272, 280, 301, 308
RELATED TO ZARANKIEWICZS PROBLEM. REF TI1 126.

390 1, 2, 4, 7, 11, 16, 21, 28, 35
RATIONAL POINTS IN A QUADRILATERAL. REF JRAM 226 22 67.

391 1, 2, 4, 7, 11, 16, 22, 29, 37, 46, 56, 67, 79, 92, 106, 121, 137, 154, 172, 191, 211, 232, 254, 277, 301, 326, 352, 379, 407, 436, 466, 497, 529, 562, 596, 631, 667, 704, 742
CENTRAL POLYGONAL NUMBERS N(N - 1)/2 + 1, OR SLICING A PANCAKE, REF MAG 30 150 46. HO3 22. FQ 3 296 65.

392 1, 2, 4, 7, 11, 16, 23, 31, 41, 53, 67, 83, 102, 123, 147, 174, 204, 237, 274, 314, 358, 406, 458, 514, 575, 640, 710, 785, 865, 950, 1041, 1137, 1239, 1347, 1461, 1581
A PARTITION FUNCTION. REF CAY 2 278. JACS 53 3084 31. AMS 26 304 55.

393 1, 2, 4, 7, 12, 8, 80, 84, 820
CYCLIC STEINER TRIPLE SYSTEMS. REF CSA 504.

394 1, 2, 4, 7, 12, 18, 27, 38, 53, 71, 94, 121, 155, 194, 241, 295, 359, 431, 515, 609, 717, 837, 973, 1123, 1292, 1477, 1683, 1908, 2157, 2427, 2724, 3045, 3396, 3774, 4185
A PARTITION FUNCTION. REF AMS 26 304 55.

395 1, 2, 4, 7, 12, 19, 29, 42, 60, 83, 113, 150, 197, 254, 324, 408, 509, 628, 769, 933, 1125, 1346, 1601, 1892, 2225, 2602, 3029, 3509, 4049, 4652, 5326, 6074, 6905, 7823
A PARTITION FUNCTION. REF AMS 26 304 55.

396 1, 2, 4, 7, 12, 19, 30, 45, 67, 97, 139, 195, 272, 373, 508, 684, 915, 1212, 1597, 2087, 2714, 3506, 4508, 5763, 7338, 9296, 11732, 14742, 18460, 23025, 28629, 35471
PARTITIONS INTO PARTS OF 2 KINDS. REF RS2 90. RCI 199. FQ 9 332 71.

397 1, 2, 4, 7, 12, 20, 33, 54, 88, 143, 232, 376, 609, 986, 1596, 2583, 4180, 6764, 10945, 17710, 28656, 46367, 75024, 121392, 196417, 317810, 514228, 832039, 1346268
FIBONACCI NUMBERS - 1. REF R1 155. AENS 79 203 62. FQ 3 295 65.

398 1, 2, 4, 7, 12, 21, 38, 68, 124, 229, 428, 806, 1530, 2919, 5591, 10750, 20717, 40077, 77653, 150752, 293161, 570963, 1113524, 2174315, 4250367, 8317036
RAMANUJANS INTEGRAL. REF MTAC 18 85 64.

399 1, 2, 4, 7, 12, 22, 39, 70, 126, 225, 404, 725
RESTRICTED PARTITIONS. REF EMS 11 224 59.

400 1, 2, 4, 7, 12, 22, 41, 72, 137, 254, 476, 901, 1716, 3274, 6286, 12090, 23331, 45140, 87511, 169972, 330752, 644499, 1257523, 2456736, 4804666, 9405749
POPULATION OF U2 + 4V**2. REF MTAC 20 560 66.**

401 1, 2, 4, 7, 13, 15, 18, 19, 20, 21, 22, 23, 25, 28, 29, 30, 35, 38, 40, 43, 44, 45, 48, 49, 50, 51, 54, 55, 56, 57, 58, 59, 60, 63, 65, 66, 71, 72, 74, 75, 79, 81, 84, 85, 87, 91, 93
ELLIPTIC CURVES. REF JRAM 212 23 63.

402 1, 2, 4, 7, 13, 17, 30, 60, 107, 197, 257, 454, 908, 1619
A JUMPING PROBLEM. REF DO1 259.

403 1, 2, 4, 7, 13, 22, 40, 70, 126, 225, 411, 746, 1376, 2537, 4719, 8799, 16509, 31041, 58635, 111012, 210870, 401427, 766149, 1465019, 2807195, 5387990, 10358998
IRREDUCIBLE POLYNOMIALS, OR NECKLACES. REF JSIAM 12 288 64.

404 1, 2, 4, 7, 13, 24, 42, 76, 137, 245, 441
RESTRICTED PARTITIONS. REF EMS 11 224 59.

405 1, 2, 4, 7, 13, 24, 43, 78, 141, 253, 456
RESTRICTED PARTITIONS. REF EMS 11 224 59.

406 1, 2, 4, 7, 13, 24, 44, 81, 149, 274, 504, 927, 1705, 3136, 5768, 10609, 19513, 35890, 66012, 121415, 223317, 410744, 755476, 1389537, 2555757, 4700770, 8646064
TRIBONACCI NUMBERS A(N) = A(N - 1) + A(N - 2) + A(N - 3). REF FQ 1(3) 71 63, 5 211 67.

407 1, 2, 4, 7, 13, 25, 43, 83, 157, 296, 564, 1083, 2077, 4006, 7733, 14968, 29044, 56447, 109864, 214197, 418080, 816907, 1598040, 3129063, 6132106
ODD POPULATION OF U2 + V**2. REF MTAC 18 84 64.**

408 1, 2, 4, 7, 14, 23, 42, 76, 139, 258, 482, 907, 1717, 3269, 6257, 12020, 23171, 44762, 86683, 168233, 327053, 636837, 1241723, 2424228, 4738426
POPULATION OF 2U2 + 3V**2. REF MTAC 20 563 66.**

409 1, 2, 4, 7, 14, 24, 43, 82, 149, 284, 534, 1015, 1937, 3713, 7136, 13759, 26597, 51537, 100045, 194586, 378987, 739161, 1443465, 2821923, 5522689
POPULATION OF 4U2 + 4UV + 5V**2. REF MTAC 20 567 66.**

410 1, 2, 4, 7, 14, 27, 52, 100, 193, 372, 717, 1382, 2664, 5135, 9898, 19079, 36776, 70888, 136641, 263384, 507689, 978602, 1886316, 3635991, 7008598, 13509507
TETRANACCI NUMBERS. REF FQ 8 7 70.

411 1, 2, 4, 7, 14, 29, 60, 127, 275, 598, 1320, 2936
BORON TREES. REF CAY 9 450.

412 1, 2, 4, 7, 39, 202, 1219, 9468, 83425, 80017
GRAPHS COMPOSED OF TWO CIRCUITS. REF RE4.

413 1, 2, 4, 8, 1, 6, 3, 2, 6, 4, 1, 2, 8, 2, 5, 6, 5, 1, 2, 1, 0, 2, 4, 2, 0, 4, 8, 4, 0, 9, 6, 8, 1, 9, 2, 1, 6, 3, 8, 4, 3, 2, 7, 6, 8, 6, 5, 5, 3, 6, 1, 3, 1, 0, 7, 2, 2, 6, 2, 1, 4, 4, 5, 2, 4, 2, 8, 8, 1
POWERS OF TWO. REF EUR 11 10 49.

414 1, 2, 4, 8, 7, 5, 10, 11, 13, 8, 7, 14, 19, 20, 22, 26, 25, 14, 19, 29, 31, 26, 25, 41, 37, 29, 40, 35, 43, 41, 37, 47, 58, 62, 61, 59, 64, 56, 67, 71, 61, 50, 46, 56, 58, 62, 70, 68
SUMS OF DIGITS OF POWERS OF 2. REF EUR 26 12 63.

415 1, 2, 4, 8, 10, 12, 14, 18, 32, 48, 54, 72, 148, 184, 248, 270, 274, 420
5.2N - 1 IS PRIME. REF MTAC 22 421 68.**

416 1, 2, 4, 8, 13, 21, 31, 45, 60, 76, 97, 119, 144, 170, 198, 231, 265, 300, 336, 374, 414, 456, 502, 550, 599, 649, 702, 759, 819, 881, 945, 1010, 1080, 1157, 1237, 1318
A SELF-GENERATING SEQUENCE. REF AMM 75 80 68. RLG.

417 1, 2, 4, 8, 13, 24, 42, 76, 140, 257, 483, 907, 1717, 3272, 6261, 12027, 23172, 44769, 86708, 168245, 327073, 636849, 1241720, 2424290, 4738450
POPULATION OF U2 + 6V**2. REF MTAC 20 563 66.**

418 1, 2, 4, 8, 14, 18, 28, 40, 52, 70, 88, 104, 140
GENERALIZED DIVISOR FUNCTION. REF PLMS 19 111 19.

419 1, 2, 4, 8, 15, 26, 42, 64, 93, 130, 176, 232, 299, 378, 470, 576, 697, 834, 988, 1160, 1351, 1562, 1794, 2048, 2325, 2626, 2952, 3304, 3683, 4090, 4526, 4992, 5489
SLICING A CAKE. REF MAG 30 150 46. FQ 3 296 65.

420 1, 2, 4, 8, 15, 27, 47, 79, 130, 209, 330, 512, 784, 1183, 1765, 2604, 3804, 5504, 7898, 11240
PLANAR PARTITIONS. REF PCPS 47 686 51.

421 1, 2, 4, 8, 15, 27, 47, 80, 134, 222, 365, 597, 973, 1582, 2568, 4164, 6747, 10927, 17691, 28636, 46346, 75002, 121369, 196393, 317785, 514202, 832012, 1346240
A NONLINEAR BINOMIAL SUM. REF FQ 3 295 65.

422 1, 2, 4, 8, 15, 29, 53, 98, 177, 319, 565, 1001, 1749, 3047, 5264, 9054, 15467, 26320, 44532, 75054, 125904, 210413, 350215, 580901, 960035, 1581534, 2596913
TREES OF HEIGHT AT MOST 3. REF IBMJ 4 475 60. KU1.

423 1, 2, 4, 8, 15, 29, 56, 108, 208, 401, 773, 1490, 2872, 5536, 10671, 20569, 39648, 76424, 147312, 283953, 547337, 1055026, 2033628, 3919944, 7555935, 14564533
TETRANACCI NUMBERS. REF AMM 33 232 26. FQ 1(3) 74 63.

424 1, 2, 4, 8, 15, 240, 15120, 672, 8400, 100800, 69300, 4950, 17199000, 22422400, 33633600, 201801600, 467812800, 102918816000
COEFFICIENTS FOR NUMERICAL DIFFERENTIATION. REF PHM 33 11 42. BAMS 48 922 42.

425 1, 2, 4, 8, 16, 21, 42, 51, 102, 112, 224, 235, 470, 486, 972, 990, 1980, 2002, 4004, 4027, 8054, 8078, 16156, 16181, 32362, 32389, 64778, 64806, 129612, 129641, 259282
A SELF-GENERATING SEQUENCE. REF AMM 75 80 68.

426 1, 2, 4, 8, 16, 22, 24, 28, 36, 42, 44, 48, 56, 62, 64, 68, 76, 82
PERIODIC DIFFERENCES. REF TCPS 2 219 1827.

427 1, 2, 4, 8, 16, 31, 57, 99, 163, 256, 386, 562, 794, 1093, 1471, 1941, 2517, 3214, 4048, 5036, 6196, 7547, 9109, 10903, 12951, 15276, 17902, 20854, 24158, 27841, 31931
BINOMIAL COEFFICIENT SUMS. REF MAG 30 150 46. FQ 3 296 65.

428 1, 2, 4, 8, 16, 31, 58, 105, 185, 319, 541, 906, 1503, 2476, 4058, 6626, 10790, 17537, 28464, 46155, 74791, 121137, 196139, 317508, 513901, 831686, 1345888
A NONLINEAR BINOMIAL SUM. REF FQ 3 295 65.

429 1, 2, 4, 8, 16, 31, 61, 120, 236, 464, 912, 1793, 3525, 6930, 13624, 26784, 52656, 103519, 203513, 400096, 786568, 1546352, 3040048, 5976577, 11749641
PENTANACCI NUMBERS. REF FQ 5 260 67.

430 1, 2, 4, 8, 16, 32, 63, 124, 244, 480, 944, 1856, 3649, 7174, 14104, 27728, 54512, 107168, 210687, 414200, 814296, 1600864, 3147216, 6187264, 12163841
A PROBABILITY DIFFERENCE EQUATION. REF AMM 32 369 25.

431 1, 2, 4, 8, 16, 32, 63, 125, 248, 492, 976, 1936, 3840, 7617, 15109, 29970, 59448, 117920, 233904, 463968, 920319, 1825529, 3621088, 7182728, 14247536
HEXANACCI NUMBERS. REF FQ 5 260 67.

432 1, 2, 4, 8, 16, 32, 64, 128, 256, 512, 1024, 2048, 4096, 8192, 16384, 32768, 65536, 131072, 262144, 524288, 1048576, 2097152, 4194304, 8388608, 16777216
POWERS OF TWO. REF BA1. MTAC 23 456 69.

433 1, 2, 4, 8, 16, 36, 85, 239
WEIGHTED LINEAR SPACES. REF BSM 22 234 70.

434 1, 2, 4, 8, 17, 35, 71, 152, 314, 628, 1357, 2725, 5551, 12212, 24424, 48848, 108807, 218715, 438531, 878162, 1867334, 3845668, 7802447, 16705005
POWERS OF TWO WRITTEN IN BASE 9. REF EUR 14 13 51.

435 1, 2, 4, 8, 17, 36, 78, 171, 379
DISTINCT VALUES TAKEN BY 22** ··· **2. REF GU7.**

436 1, 2, 4, 8, 17, 39, 89, 211, 507, 1238, 3057, 7639, 19241, 48865, 124906, 321198, 830219, 2156010, 5622109, 14715813, 38649152, 101821927, 269010485
ALCOHOLS OR ROOTED TREES OF DEGREE AT MOST 4. REF JACS 54 2919 32. FI2 41.397.

437 1, 2, 4, 8, 18, 40, 91, 210, 492, 1165, 2786, 6710, 16267, 39650, 97108, 238824, 589521
SUMS OF FERMAT COEFFICIENTS. REF MMAG 27 143 54.

438 1, 2, 4, 8, 18, 44, 122, 362, 1162, 3914, 13648
NECKLACES. REF IJM 5 664 61.

439 1, 2, 4, 8, 20, 52, 152, 472, 1520, 5044, 17112, 59008, 206260, 729096, 2601640, 9358944, 33904324, 123580884, 452902072, 1667837680, 6168510256
REPRESENTATIONS OF ZERO. REF CMB 11 292 68.

440 1, 2, 4, 8, 20, 56, 180, 596, 2068, 7316, 26272
NECKLACES. REF IJM 5 664 61.

441 1, 2, 4, 8, 20, 100, 2116, 1114244, 68723671300, 1180735735906024030724, 170141183460507917357914971986913657860
THE SUM OF 2C(N, K). REF GO3.**

442 1, 2, 4, 8, 21, 52, 131, 316, 765, 1846, 4494
RELATED TO PARTITIONS OF A NUMBER. REF AMM 76 1036 69.

443 1, 2, 4, 8, 22, 52, 140, 366, 992
NECKLACES. REF IJM 2 302 58.

444 1, 2, 4, 8, 24, 84, 328, 1372, 6024
ENERGY FUNCTION FOR SQUARE LATTICE. REF PHA 28 925 62.

445 1, 2, 4, 9, 10, 12, 27, 37, 38, 44, 48, 78, 112, 168, 229, 297, 339
15.2N + 1 IS PRIME. REF PAMS 9 674 58.**

446 1, 2, 4, 9, 11, 23, 32, 39, 44, 51, 53, 60, 65, 72, 86, 93, 95, 114, 123, 156, 170, 179, 186, 200, 207, 212, 219, 228, 233, 240, 249, 261, 270, 303, 317, 333, 338, 345, 375, 389
(N(N + 1) + 1)/7 IS PRIME. REF CU1 1 250.

447 1, 2, 4, 9, 16, 29, 47, 77, 118, 181, 267, 392, 560, 797, 1111, 1541, 2106, 2863, 3846, 5142, 6808, 8973, 11733, 15275, 19753, 25443, 32582, 41569, 52770, 66757
BIPARTITE PARTITIONS. REF PCPS 49 72 53. NI1 1.

448 1, 2, 4, 9, 18, 42, 96, 229, 549, 1346, 3326, 8329
CARBON TREES. REF CAY 9 454. ZFK 93 437 36.

449 1, 2, 4, 9, 19, 42, 89, 191, 402, 847, 1763, 3667, 7564, 15564, 31851, 64987, 132031, 267471, 539949, 1087004, 2181796, 4367927, 8721533, 17372967, 34524291
TREES OF HEIGHT AT MOST 4. REF IBMJ 4 475 60. KU1.

450 1, 2, 4, 9, 19, 48, 117, 307, 821, 2277
MINIMAL TRIANGLE GRAPHS. REF MTAC 21 249 67.

451 1, 2, 4, 9, 20, 45, 105, 249, 599
ESTERS. REF JACS 56 157 34.

452 1, 2, 4, 9, 20, 46, 105, 246, 583, 1393, 3355, 8133, 19825, 48554, 119412, 294761
SUMS OF FERMAT COEFFICIENTS. REF MMAG 27 143 54.

453 1, 2, 4, 9, 20, 47, 108, 252, 582, 1345, 3086, 7072, 16121, 36667, 83099, 187885, 423610, 953033, 2139158, 4792126, 10714105, 23911794, 53273599, 118497834
TREES OF HEIGHT AT MOST 5. REF IBMJ 4 475 60. KU1.

454 1, 2, 4, 9, 20, 48, 115, 286, 719, 1842, 4766, 12486, 32973, 87811, 235381, 634847, 1721159, 4688676, 12826228, 35221832, 97055181, 268282855, 743724984
ROOTED UNLABELED TREES. REF R1 138. HA5 232.

455 1, 2, 4, 9, 20, 51, 125, 329, 862, 2311, 6217, 16949, 46350, 127714, 353272, 981753
CONNECTED GRAPHS WITH AT MOST ONE CYCLE. REF FI2 41.399.

456 1, 2, 4, 9, 21, 51, 127, 323, 835, 2188, 5798, 15511, 41835, 113634, 310572, 853467, 2356779, 6536382, 18199284, 50852019, 142547559, 400763223, 1129760415
GENERALIZED BALLOT NUMBERS. REF BAMS 54 359 48. JSIAM 17 254 69.

457 1, 2, 4, 9, 21, 52, 129, 332, 859, 2261
PARAFFINS. REF JACS 56 157 34.

458 1, 2, 4, 9, 21, 56, 148, 428, 1305, 4191, 14140, 50159, 185987, 720298, 2905512, 12180208
GRAPHS BY POINTS AND LINES. REF R1 146. ST1.

459 1, 2, 4, 9, 22, 59, 167, 490, 1486, 4639, 14805, 48107, 158808, 531469, 1799659, 6157068, 21258104, 73996100, 259451116, 951695102, 3251073303
TOURNAMENT SCORES. REF CMB 7 135 64. MO1 68.

460 1, 2, 4, 9, 23, 63, 177, 514, 1527, 4625, 14230, 44357, 139779, 444558, 1425151, 4600339, 14939849, 48778197, 160019885, 527200711
RELATED TO SERIES-PARALLEL NUMBERS. REF JM2 21 92 42.

461 1, 2, 4, 9, 23, 63, 188
MIXED HUSIMI TREES. REF PNAS 42 535 56.

462 1, 2, 4, 9, 26, 101, 950
GEOMETRIES. REF JM2 49 127 70.

463 1, 2, 4, 10, 24, 55, 128, 300, 700, 1632, 3809, 8890, 20744, 48406
RESTRICTED PERMUTATIONS. REF AENS 79 207 62.

464 1, 2, 4, 10, 24, 66, 174, 504, 1406, 4210, 12198, 37378, 111278, 346846, 1053874
FOLDING A STRIP OF STAMPS. REF CJM 2 397 50. JCT 5 151 68.

465 1, 2, 4, 10, 24, 66, 176, 493, 1361
FOLDING A LINE. REF AMM 44 51 37.

466 1, 2, 4, 10, 24, 66, 180, 522, 1532, 4624, 14136, 43930, 137908, 437502, 1399068, 4507352, 14611576, 47633486, 156047204, 513477502, 1696305720, 5623993944
SERIES-PARALLEL NETWORKS. REF JM2 21 87 42. R1 142.

467 1, 2, 4, 10, 25, 64, 166
ALKYLS. REF ZFK 93 437 36.

468 1, 2, 4, 10, 25, 70, 196, 574, 1681
FOLDING A LINE. REF AMM 44 51 37.

469 1, 2, 4, 10, 26, 76, 232, 764, 2620, 9496, 35696, 140152, 568504, 2390480, 10349536, 46206736, 211799312, 997313824, 4809701440, 23758664096
A(N) = A(N − 1) + (N − 1)A(N − 2). REF LU1 1 221. R1 86. MU2 6. DMJ 35 659 68.

470 1, 2, 4, 10, 27, 74, 202, 548, 1490, 4052, 11013, 29937, 81377, 221207, 601302, 1634509, 4443055, 12077476, 32829985, 89241150, 242582598, 659407867
COSH(N). REF AMP 3 33 1843. MNAS 14(5) 14 25. HA4. LF1 93.

471 1, 2, 4, 10, 29, 90, 295, 1030, 3838, 15168, 63117, 275252, 1254801, 5968046, 29551768, 152005634, 810518729, 4472244574, 25497104007, 149993156234
FROM A DIFFERENTIAL EQUATION. REF AMM 67 766 60.

472 1, 2, 4, 10, 32, 122, 544, 2770, 15872, 101042, 707584, 5405530, 44736512, 398721962, 3807514624, 38783024290, 419730685952, 4809759350882
RELATED TO EULER NUMBERS. REF AMM 65 534 58. DKB 262.

473 1, 2, 4, 10, 36, 202
CHANGING MONEY. REF NMT 10 65 62.

474 1, 2, 4, 10, 37, 138
ROOTED PLANAR MAPS. REF CJM 15 542 63.

475 1, 2, 4, 10, 46, 1372, 475499108
BOOLEAN FUNCTIONS. REF JSIAM 12 294 64.

476 1, 2, 4, 11, 15, 18, 23, 37, 44, 57, 78, 88, 95, 106, 134, 156, 205, 221, 232, 249, 310, 323, 414, 429, 452, 550, 576, 639, 667, 715, 785, 816, 837, 946, 1003, 1038, 1122
OF THE FORM (P2 − 49)/120 WHERE P IS PRIME. REF IAS 5 382 37.**

477 1, 2, 4, 11, 19, 56, 96, 296, 554, 1593, 3093
PERMUTATION GROUPS. REF JPC 33 1069 29.

478 1, 2, 4, 11, 33, 116, 435, 1832, 8167, 39700, 201785, 1099449, 6237505
REFINEMENTS OF PARTITIONS. REF GU5.

479 1, 2, 4, 11, 34, 156, 1044, 12346, 274668, 12005168, 1018997864, 165091172592, 50502031367952, 29054155657235488, 31426485969804308768
GRAPHS OR REFLEXIVE SYMMETRIC RELATIONS. REF MI1 17 22 55. MAN 174 68 67. HA5 214.

480 1, 2, 4, 12, 32, 108, 336, 1036, 3120, 9540, 29244
RESTRICTED PERMUTATIONS. REF AENS 79 213 62.

481 1, 2, 4, 12, 34, 111, 360, 1226
ROOTED PLANAR 2-TREES. REF MAT 15 121 68.

482 1, 2, 4, 12, 39, 202, 1219, 9468, 83435, 836017, 9223092, 111255228, 1453132944, 20433309147, 307690667072, 4940118795869, 84241805734539
POLYGONS. REF AMM 67 349 60.

483 1, 2, 4, 12, 48, 200, 1040, 5600, 33600
SORTING NUMBERS. REF PSPM 19 173 71.

484 1, 2, 4, 12, 56, 456, 6880, 191536, 9733056, 903753248, 154108311168
TOURNAMENTS. REF MO1 87.

485 1, 2, 4, 12, 81, 1684, 123565, 33207256, 34448225389
THRESHOLD FUNCTIONS. REF PGEC 19 821 70.

486 1, 2, 4, 12, 81, 2646
SELF-DUAL MONOTONE BOOLEAN FUNCTIONS. REF WE1 181.

487 1, 2, 4, 12, 108, 10476, 108625644, 11798392680793836, 139202068568601568785946949658348
A NONLINEAR RECURRENCE. REF SA2.

488 1, 2, 4, 13, 41, 226, 1072, 9374, 60958, 723916, 5892536, 86402812, 837641884, 14512333928, 162925851376, 3252104882056, 41477207604872
TERMS IN A SKEW DETERMINANT. REF PRSE 21 354 1896.

489 1, 2, 4, 13, 42, 308
CONNECTED LINEAR SPACES. REF BSM 19 424 67.

490 1, 2, 4, 14, 34, 98, 270, 768, 2192, 6360, 18576, 54780, 162658, 486154, 1461174, 4413988, 13393816, 40807290
PARAFFINS. REF JACS 54 1105 32.

491 1, 2, 4, 14, 54, 332, 2246, 18264, 164950, 1664354, 18423144, 222406776, 2905943328, 40865005494, 615376173184, 9880209206458, 168483518571798
POLYGONS. REF AMM 67 349 60.

492 1, 2, 4, 14, 104, 1882, 94572, 15028134, 8378070864, 17561539552946
THRESHOLD FUNCTIONS. REF PGEC 19 821 70.

493 1, 2, 4, 14, 128, 3882, 412736, 151223522, 189581406208, 820064805806914, 12419746847290729472, 668590083306794321516802
BINOMIAL COEFFICIENT SUMS. REF PGEC 14 322 65.

494 1, 2, 4, 14, 222, 616126, 200253952527184
BOOLEAN FUNCTIONS. REF HA2 153.

495 1, 2, 4, 16, 56, 256, 1072, 6224, 33616, 218656, 1326656, 9893632, 70186624, 574017536, 4454046976, 40073925376, 347165733632, 3370414011904
PERMUTATIONS OF ORDER 4. REF CJM 7 159 55.

496 1, 2, 4, 16, 80, 520, 3640, 29120
PERMUTATIONS BY NUMBER OF CYCLES. REF R1 85.

497 1, 2, 4, 16, 256, 65536, 4294967296, 18446744073709551616, 340282366920938463463374607431768211456
2(2**N). REF MTAC 23 456 69.**

498 1, 2, 4, 24, 128, 880, 7440
SORTING NUMBERS. REF PSPM 19 173 71.

499 1, 2, 4, 24, 1104, 2435424, 11862575248704, 281441383062305809756861824, 158418504200047111075388369241884118003210485743490304
A SLOWLY CONVERGING SERIES. REF AMM 54 138 47.

500 1, 2, 4, 60, 1276, 41888, 1916064, 116522048, 9069595840, 878460379392
RELATED TO LATIN RECTANGLES. REF BU2 33 125 41.

501 1, 2, 4, 104, 272, 3104, 79808
EXPANSION OF SINH X / SIN X. REF MMAG 31 189 58.

SEQUENCES BEGINNING 1, 2, 5

502 1, 2, 5, 3, 15, 140, 5, 56
QUEENS PROBLEM. REF SL1 49.

503 1, 2, 5, 4, 12, 6, 9, 23, 11, 27, 34, 22, 10, 33, 15, 37, 44, 28, 80, 19, 81, 14, 107, 89, 64, 16, 82, 60, 53, 138, 25, 114, 148, 136, 42, 104, 115, 63, 20, 143, 29, 179, 67, 109
A NUMBER-THEORETIC FUNCTION. REF AMM 56 526 49.

504 1, 2, 5, 5, 16, 7, 50
TRANSITIVE GROUPS. REF BAMS 2 143 1896.

505 1, 2, 5, 6, 7, 10, 12, 14, 15, 20, 21, 22, 23, 25, 26, 30, 31, 34, 36, 37, 38, 39, 41, 42, 45, 46, 47, 49, 50, 52, 53, 54, 55, 57, 58, 60, 62, 66, 69, 70, 71, 72, 73, 74, 76, 78, 79
ELLIPTIC CURVES. REF JRAM 212 24 63.

506 1, 2, 5, 6, 8, 12, 18, 30, 36, 41, 66, 189, 201, 209, 276, 353, 408, 438, 534
3.2N + 1 IS PRIME. REF PAMS 9 674 58.**

507 1, 2, 5, 6, 11, 13, 17, 22, 27, 29, 37, 44, 44, 55
GENERALIZED DIVISOR FUNCTION. REF PLMS 19 112 19.

508 1, 2, 5, 6, 14, 21, 29, 30, 54, 90, 134, 155, 174, 230, 234, 251, 270, 342, 374, 461, 494, 550, 666, 750, 810, 990, 1890, 2070, 2486, 2757, 2966, 3150, 3566, 3630, 4554
LATTICE POINTS IN SPHERES. REF MTAC 20 306 66.

509 1, 2, 5, 7, 10, 13, 15, 18, 20, 23, 26, 28, 31, 34, 36, 39, 41, 44, 47, 49, 52, 54, 57, 60, 62, 65, 68, 70, 73, 75, 78, 81, 83, 86, 89, 91, 94, 96, 99, 102, 104, 107, 109, 112, 115
A BEATTY SEQUENCE. REF CMB 2 191 59. AMM 72 1144 65.

510 1, 2, 5, 7, 11, 14, 20, 24, 30, 35
CONSISTENT ARCS IN A TOURNAMENT. REF CMB 12 263 69. RE1.

511 1, 2, 5, 7, 12, 15, 22, 26, 35, 40, 51, 57, 70, 77, 92, 100, 117, 126, 145, 155, 176, 187, 210, 222, 247, 260, 287, 301, 330, 345, 376, 392, 425, 442, 477, 495, 532, 551, 590
GENERALIZED PENTAGONAL NUMBERS. REF AMM 76 884 69. HO2 119.

512 1, 2, 5, 7, 12, 19, 31, 50, 81, 131, 212, 343, 555, 898, 1453, 2351, 3804, 6155, 9959, 16114, 26073, 42187, 68260, 110447, 178707, 289154, 467861, 757015, 1224876
A(N) = A(N - 1) + A(N - 2). REF FQ 3 129 65.

513 1, 2, 5, 7, 19, 26, 71, 97, 265, 362, 989, 1351, 3691, 5042, 13775, 18817, 51409, 70226, 191861, 262087, 716035, 978122, 2672279, 3650401, 9973081, 13623482
A(2N) = A(2N - 1) + A(2N - 2), A(2N + 1) = 2A(2N) + A(2N - 1). REF MQET 1 10 16. NZ1 181.

514 1, 2, 5, 7, 26, 265, 1351, 5042, 13775, 18817, 70226, 716035, 3650401
RELATED TO GENOCCHI NUMBERS. REF AMM 36 645 35.

515 1, 2, 5, 8, 11, 14, 18, 22, 27, 31
PARTITIONS INTO NON-INTEGRAL POWERS. REF PCPS 47 214 51.

516 1, 2, 5, 8, 13, 16, 21, 26, 35
RAMSEY NUMBERS. REF CMB 8 579 65.

517 1, 2, 5, 8, 13, 18, 25, 32, 41, 50, 61, 72, 85, 98, 113, 128, 145, 162, 181, 200, 221, 242, 265, 288, 313, 338, 365, 392, 421, 450, 481, 512, 545, 578, 613, 648, 685, 722, 761
NEAREST INTEGER TO (N2 + 1)/2.**

518 1, 2, 5, 8, 14, 21, 32, 45, 65, 88, 121, 161, 215, 280, 367, 471, 607, 771, 980, 1232, 1551, 1933, 2410, 2983, 3690, 4536, 5574, 6811, 8317, 10110, 12276
TREES OF DIAMETER 4. REF IBMJ 4 476 60. KU1.

519 1, 2, 5, 8, 18, 29, 57, 96, 183, 318, 603, 1080, 2047, 3762
POLYTOPES. REF GR2 424.

520 1, 2, 5, 8, 21, 42, 96, 222, 495, 1177, 2717, 6435, 15288, 36374, 87516, 210494, 509694, 1237736, 3014882, 7370860, 18059899, 44379535, 109298070, 269766655
PARTITIONS OF POINTS ON A CIRCLE. REF BAMS 54 359 48.

521 1, 2, 5, 9, 9, 2, 1, 0, 4, 9, 8, 9, 4, 8, 7, 3, 1, 6, 4, 7, 6, 7, 2, 1, 0, 6, 0, 7, 2, 7, 8, 2, 2, 8, 3, 5, 0, 5, 7, 0, 2, 5, 1, 4, 6, 4, 7, 0, 1, 5, 0, 7, 9, 8, 0, 0, 8, 1, 9, 7, 5, 1, 1, 2, 1, 5, 5, 2, 9
CUBE ROOT OF 2. REF SMA 18 175 52.

522 1, 2, 5, 9, 14, 20, 27, 35, 44, 54, 65, 77, 90, 104, 119, 135, 152, 170, 189, 209, 230, 252, 275, 299, 324, 350, 377, 405, 434, 464, 495, 527, 560, 594, 629, 665, 702, 740, 779
N(N + 3)/2.

523 1, 2, 5, 9, 15, 23, 34, 47, 64, 84, 108, 136, 169, 206, 249, 297, 351, 411, 478, 551
HYDROCARBONS. REF JACS 55 684 33.

524 1, 2, 5, 9, 17, 27, 40, 55, 73, 117, 143
RATIONAL POINTS IN A QUADRILATERAL. REF JRAM 227 47 67.

525 1, 2, 5, 9, 17, 28, 47, 73, 114, 170, 253, 365, 525, 738, 1033, 1422, 1948, 2634, 3545, 4721, 6259, 8227, 10767, 13990, 18105, 23286, 29837, 38028, 48297, 61053
PARTITIONS INTO PARTS OF 2 KINDS. REF RS2 90. RCI 199.

526 1, 2, 5, 9, 18, 35, 57
POLYHEDRA. REF JRM 4 123 71.

527 1, 2, 5, 9, 21, 44, 103, 232, 571, 1368, 3441
TOTAL DIAMETER OF UNLABELED TREES. REF IBMJ 4 476 60.

528 1, 2, 5, 9, 22, 62, 177, 560, 1939
SERIES-REDUCED STAR GRAPHS. REF JMP 7 1585 66.

529 1, 2, 5, 10, 13, 17, 26, 29, 37, 41, 50, 53, 58, 61, 65, 73, 74, 82, 85, 89, 97, 101, 106, 109, 113, 122, 125, 130, 137, 145, 149, 157, 170, 173, 181, 185, 193, 197, 202, 218
SOLUBLE PELLIANS. REF AMP 52 48 1871. KR1 1 46. LE1 56.

530 1, 2, 5, 10, 15, 25, 37, 52, 67, 97, 117
GENERALIZED DIVISOR FUNCTION. REF PLMS 19 112 19.

531 1, 2, 5, 10, 16, 24, 33, 44, 56, 70, 85, 102, 120, 140, 161, 184, 208, 234, 261, 290, 320
SERIES-REDUCED PLANTED TREES. REF RI1.

532 1, 2, 5, 10, 18, 32, 55, 90, 144, 226, 346, 522, 777, 1138, 1648, 2362, 3348, 4704, 6554, 9056, 12425, 16932
COEFFICIENTS OF AN ELLIPTIC FUNCTION. REF CAY 9 128.

533 1, 2, 5, 10, 19, 33, 57, 92, 147, 227, 345, 512, 752, 1083, 1545, 2174, 3031, 4179, 5719, 7752, 10438, 13946, 18519, 24428, 32051, 41805, 54265, 70079, 90102, 115318
PARTITIONS INTO PARTS OF 2 KINDS. REF RS2 90. RCI 199.

534 1, 2, 5, 10, 20, 24, 26, 41, 53, 130, 149, 205, 234, 287, 340, 410, 425, 480, 586, 840, 850, 986, 1680, 1843, 2260, 2591, 3023, 3024, 3400, 3959, 3960, 5182, 5183, 7920
LATTICE POINTS IN CIRCLES. REF MTAC 20 306 66.

535 1, 2, 5, 10, 20, 35, 62, 102, 167, 262, 407, 614, 919, 1345, 1952, 2788, 3950, 5524, 7671, 10540, 14388, 19470, 26190, 34968, 46439, 61275, 80455, 105047, 136541
PARTITIONS INTO PARTS OF 2 KINDS. REF RS2 90. RCI 199.

536 1, 2, 5, 10, 20, 36, 65, 110, 185, 300, 481, 752, 1165, 1770, 2665, 3956, 5822, 8470, 12230, 17490, 24842, 35002, 49010, 68150, 94235, 129512, 177087, 240840
PARTITIONS INTO PARTS OF 2 KINDS. REF RS2 90. RCI 199.

537 1, 2, 5, 10, 20, 38, 71, 130, 235, 420, 744, 1308, 2285, 3970, 6865, 11822, 20284, 34690, 59155, 100610, 170711, 289032, 488400, 823800, 1387225, 2332418, 3916061
CONVOLVED FIBONACCI NUMBERS. REF RCI 101. FQ 3 51 65, 8 163 70.

538 1, 2, 5, 10, 20, 40, 86, 192, 440, 1038, 2492, 6071, 14960, 37198, 93193, 234956, 595561, 1516638, 3877904, 9950907, 25615653, 66127186, 171144671
EXPONENTIAL INTEGRAL OF N. REF RS3 160 384 1870. PHM 33 757 42. FMR 1 267.

539 1, 2, 5, 10, 22, 40, 75, 130, 230, 382, 636, 1016, 1633, 2540, 3942, 5978, 9057
COEFFICIENTS OF MODULAR FUNCTIONS. REF PLMS 9 386 59.

540 1, 2, 5, 10, 24, 63, 165, 467, 1405, 4435, 14775, 51814, 190443, 732472, 2939612
GRAPHS BY POINTS AND LINES. REF R1 146. ST1.

541 1, 2, 5, 10, 25, 56, 139, 338, 852
ALCOHOLS. REF BER 8 1545 1875.

542 1, 2, 5, 11, 21, 39, 73, 129, 226, 388, 659, 1100, 1821
PLANAR PARTITIONS. REF MA2 2 332.

543 1, 2, 5, 11, 23, 47, 94, 185
COMPOSITIONS. REF R1 155.

544 1, 2, 5, 11, 25, 66, 172, 485, 1446, 4541, 15036, 52496, 192218, 737248
GRAPHS BY POINTS AND LINES. REF R1 146. ST1.

545 1, 2, 5, 11, 26, 68, 177, 497, 1476
GRAPHS BY NUMBER OF LINES. REF R1 146. ST1. MAN 174 68 67.

546 1, 2, 5, 11, 28, 74, 199, 551, 1553, 4436, 12832, 37496, 110500, 328092, 980491, 2946889, 8901891, 27011286, 82299275
PARAFFINS. REF JACS 54 1105 32.

547 1, 2, 5, 11, 31, 77, 214, 576, 1592, 4375, 12183, 33864, 94741, 265461, 746372
CONNECTED GRAPHS WITH 1 CYCLE. REF FI2 41.399.

548 1, 2, 5, 11, 38, 174, 984, 6600, 51120, 448560, 4394880
BINOMIAL COEFFICIENT SUMS. REF CJM 22 26 70.

549 1, 2, 5, 12, 17, 63, 143, 492, 635, 2397, 3032, 93357, 96389, 478913, 575302, 1629517, 15240955, 93075247, 387541943, 480617190, 868159133, 2216935456
CONVERGENTS TO CUBE ROOT OF 4. REF AMP 46 106 1866. LE1 67. HPR.

550 1, 2, 5, 12, 24, 56, 113, 248, 503, 1043, 2080, 4169, 8145, 15897, 30545, 58402, 110461, 207802, 387561, 718875, 1324038, 2425473, 4416193, 7999516, 14411507
RELATED TO SOLID PARTITIONS. REF MTAC 24 956 70.

551 1, 2, 5, 12, 27, 59, 127
COMPOSITIONS. REF R1 155.

552 1, 2, 5, 12, 29, 70, 169, 408, 985, 2378, 5741, 13860, 33461, 80782, 195025, 470832, 1136689, 2744210, 6625109, 15994428, 38613965, 93222358, 225058681
PELL NUMBERS A(N) = 2A(N − 1) + A(N − 2). REF FQ 4 373 66.

553 1, 2, 5, 12, 30, 74, 188, 478, 1235, 3214, 8450, 22370, 59676, 160140, 432237, 1172436, 3194870, 8741442, 24007045, 66154654, 182864692, 506909562, 1408854940
POWERS OF ROOTED TREE ENUMERATOR. REF R1 150.

554 1, 2, 5, 12, 30, 76, 196
GENERALIZED BALLOT NUMBERS. REF JSIAM 17 254 69.

555 1, 2, 5, 12, 31, 80, 210, 555, 1479, 3959
PARAFFINS. REF JACS 56 157 34.

556 1, 2, 5, 12, 32, 94, 289, 910, 2934, 9686, 32540, 110780
BALANCING WEIGHTS. REF JCT 7 132 69.

557 1, 2, 5, 12, 33, 87, 252, 703, 2105, 6099, 18689, 55639, 173423, 526937, 1664094, 5137233, 16393315, 51255709, 164951529, 521138861, 1688959630, 5382512216
FOLDING A LINE. REF MTAC 22 198 68.

558 1, 2, 5, 12, 33, 90, 261, 766, 2312, 7068, 21965, 68954, 218751, 699534, 2253676, 7305788, 23816743, 78023602, 256738751
SERIES-REDUCED PLANTED TREES. REF CAY 3 246. RI1.

559 1, 2, 5, 12, 34, 130
TRIANGULATIONS OF SPHERE. REF MTAC 21 252 67.

560 1, 2, 5, 12, 35, 107, 363, 1248, 4460, 16094, 58937, 217117, 805475, 3001211
POLYOMINOES WITHOUT HOLES. REF PA1. JRM 2 182 69. LU2.

561 1, 2, 5, 12, 35, 108, 369, 1285, 4655, 17073, 63600, 238591, 901971, 3426576, 13079255, 50107911, 192622052
POLYOMINOES. REF AT1 363.

562 1, 2, 5, 12, 37, 123, 446, 1689, 6693, 27034, 111630, 467262, 1981353, 8487400, 36695369, 159918120, 701957539, 3101072051, 13779935438, 61557789660
RESTRICTED HEXAGONAL POLYOMINOES. REF EMS 17 11 70. RE3.

563 1, 2, 5, 12, 53, 171, 566, 737, 4251, 4988, 9239, 41944, 428679, 7329487, 7758166, 115943811, 123701977, 239645788, 731522646953, 731762292741
CONVERGENTS TO CUBE ROOT OF 5. REF AMP 46 107 1866. LE1 67. HPR.

564 1, 2, 5, 13, 17, 29, 37, 41, 53, 61, 73, 89, 97, 101, 109, 113, 137, 149, 157, 173, 181, 193, 197, 229, 233, 241, 257, 269, 277, 281, 293, 313, 317, 337, 349, 353, 373, 389
PRIMES WHICH ARE THE SUM OF 2 SQUARES. REF AMM 56 526 49.

565 1, 2, 5, 13, 19, 32, 53, 89, 139, 199, 293, 887, 1129, 1331, 5591, 8467, 9551, 15683, 19609, 31397, 370261, 1357201, 1561919, 2010733, 3826019, 3933599, 4652353
FROM GAPS BETWEEN PRIME-POWERS. REF DVSS 2 255 1884.

566 1, 2, 5, 13, 29, 34, 89, 169, 194, 233, 433
SOLUTIONS OF A DIOPHANTINE EQUATION. REF LEM 6 19 60.

567 1, 2, 5, 13, 33, 80, 184, 402, 840
EXPANSION OF BRACKET FUNCTION. REF FQ 2 256 64.

568 1, 2, 5, 13, 33, 89, 240, 657, 1806, 5026, 13999, 39260, 110381, 311465, 880840, 2497405
CONNECTED GRAPHS WITH ONE CYCLE. REF R1 150. ST1.

569 1, 2, 5, 13, 34, 89, 233, 610, 1597, 4181, 10946, 28657, 75025, 196418, 514229, 1346269, 3524578, 9227465, 24157817, 63245986, 165580141, 433494437
BISECTION OF FIBONACCI SEQUENCE. REF R1 39. FQ 9 283 71.

570 1, 2, 5, 13, 35, 95, 262, 727, 2033, 5714
PARTIALLY LABELED ROOTED TREES. REF R1 134.

571 1, 2, 5, 13, 36, 102, 296, 871, 2599
NONISENTROPIC BINARY TREES. REF GU5.

572 1, 2, 5, 13, 36, 109, 359, 1266, 4731, 18657, 77464, 337681, 1540381, 7330418, 36301105, 186688845, 995293580, 5491595645, 31310124067, 184199228226
FROM A DIFFERENTIAL EQUATION. REF AMM 67 766 60.

573 1, 2, 5, 13, 38, 116, 382, 1310, 4748, 17848, 70076, 284252, 1195240, 5174768, 23103368, 105899656, 498656912, 2404850720, 11879332048, 59976346448
A(N) = A(N − 1) + N.A(N − 2). REF R1 86 (DIVIDED BY 2).

574 1, 2, 5, 13, 44, 191, 1229, 13588, 288597
DISCONNECTED GRAPHS. REF TAMS 78 459 55. ST1.

575 1, 2, 5, 14, 39, 109
PARAFFINS. REF ZFK 93 437 36.

576 1, 2, 5, 14, 39, 120, 358, 1176, 3527, 11622, 36627, 121622, 389560, 1301140, 4215748
FOLDING A STRIP OF STAMPS. REF JCT 5 151 68.

577 1, 2, 5, 14, 42, 132, 429, 1430, 4862, 16796, 58786, 208012, 742900, 2674440, 9694845, 35357670, 129644790, 477638700, 1767263190, 6564120420, 24466267020
CATALAN NUMBERS OR BINOMIAL COEFFICIENTS C(2N, N)/(N + 1). REF AMM 72 973 65. GU1. RCI 101. CO1 1 67. GO4.

578 1, 2, 5, 14, 44, 152
PARTITION FUNCTION FOR SQUARE LATTICE. REF AIP 9 279 60.

579 1, 2, 5, 14, 46, 166, 652, 2780, 12644, 61136, 312676, 1680592, 9467680, 55704104, 341185496, 2170853456, 14314313872, 97620050080, 687418278544
THE PARTITION FUNCTION G(N, 3). REF CMB 1 87 58.

580 1, 2, 5, 14, 50, 233, 1249, 7595
TRIANGULATIONS OF SPHERE. REF MTAC 21 252 67. GR2 424. JCT 7 157 69.

581 1, 2, 5, 14, 51, 267
NUMBER OF GROUPS OF ORDERS 2, 4, 8, 16, 32, 64. REF HS1.

582 1, 2, 5, 15, 32, 99, 210, 650, 1379, 4268, 9055, 28025, 59458, 184021, 390420, 1208340, 2563621, 7934342, 16833545, 52099395
A TERNARY CONTINUED FRACTION. REF TOH 37 441 33.

583 1, 2, 5, 15, 49, 169, 602, 2191
PERMUTATIONS BY INVERSIONS. REF NET 96.

584 1, 2, 5, 15, 51, 196, 827, 3795, 18755, 99146, 556711, 3305017, 20655285, 135399720, 927973061, 6631556521, 49294051497, 380306658250, 3039453750685
THE PARTITION FUNCTION G(N, 4). REF CMB 1 87 58.

585 1, 2, 5, 15, 52, 203, 877, 4140, 21147, 115975, 678570, 4213597, 27644437, 190899322, 1382958545, 10480142147, 82864869804, 682076806159, 5832742205057
BELL NUMBERS. REF MTAC 16 418 62. AMM 71 498 64. PSPM 19 172 71. GO4.

586 1, 2, 5, 16, 52, 208
INVERSE SEMIGROUPS. REF PL1. MA4 2 2 67.

587 1, 2, 5, 16, 61, 272, 1385, 7936, 50521, 353792, 2702765, 22368256, 199360981, 1903757312, 19391512145, 209865342976, 2404879675441, 29088885112832
EULER NUMBERS. REF JDM 7 171 1881. JO1 238. NET 110. DKB 262. CO1 2 101.

588 1, 2, 5, 16, 63, 318, 2045
UNLABELED PARTIALLY ORDERED SETS. REF BI1 4. NAMS 17 646 70. WH1. WR1.

589 1, 2, 5, 16, 65, 326, 1957, 13700, 109601, 986410, 9864101, 108505112, 1302061345, 16926797486, 236975164805, 3554627472076, 56874039553217
PERMUTATIONS OF N THINGS. REF R1 16. MST 31 79 63.

590 1, 2, 5, 16, 67, 435
CIRCUITS BY NULLITY. REF AIEE 51 311 32.

591 1, 2, 5, 16, 73, 538
CIRCUITS BY RANK. REF AIEE 51 313 32.

592 1, 2, 5, 17, 37, 101, 197, 257, 401, 577, 677, 1297, 1601, 2917, 3137, 4357, 5477, 7057, 8101, 8837, 12101, 13457, 14401, 15377, 15877, 16901, 17957, 21317, 22501
PRIMES OF FORM N2 + 1. REF EUL (1) 3 22 17.**

593 1, 2, 5, 17, 55, 186, 635, 2199, 7691, 27101, 96061
SPHEROIDAL HARMONICS. REF MES 54 75 24.

594 1, 2, 5, 17, 73, 388, 2461, 18155, 152531, 1436714, 14986879, 171453343, 2134070335, 28708008128, 415017867707, 6416208498137, 105630583492969
A(N) = NA(N - 1) - (N - 1)(N - 2)A(N - 3)/2. REF CAY 9 190. PLMS 17 29 17. EMN 34 1 44.

595 1, 2, 5, 19, 87
CUBIC GRAPHS. REF HA5 195. KO1.

596 1, 2, 5, 19, 132, 3107
SEMIGROUPS WITH ONE IDEMPOTENT. REF MA4 2 2 67.

597 1, 2, 5, 20, 87, 616, 4843, 44128
MENAGE PERMUTATIONS. REF SMA 22 233 56. R1 195. BE5 162.

598 1, 2, 5, 20, 115, 790, 6217, 55160, 545135, 5938490, 70686805, 912660508, 12702694075, 189579135710, 3019908731105
PERMUTATIONS BY NUMBER OF PAIRS. REF DKB 263.

599 1, 2, 5, 21, 61, 214, 669, 2240, 7330, 24695, 83257, 284928, 981079, 3410990, 11937328, 42075242, 149171958, 531866972, 1905842605, 6861162880
DISSECTIONS OF A POLYGON. REF GU1.

600 1, 2, 5, 21, 106, 643, 4547, 36696, 332769, 3349507
PERMUTATIONS WITHOUT 3-SEQUENCES. REF BAMS 51 748 45.

601 1, 2, 5, 22, 138, 1579, 33366, 1348674, 105925685, 15968704512, 4520384306832, 2402814904220039, 2425664021535713098
CONNECTED GRAPHS BY POINTS AND LINES. REF ST1.

602 1, 2, 5, 24, 23, 76, 249, 168, 599, 1670, 1026, 3272, 8529, 5232
COEFFICIENTS OF MODULAR FUNCTIONS. REF PLMS 9 384 59.

603 1, 2, 5, 27, 923, 909182, 1046593950039
CONVERGENTS TO LEHMERS CONSTANT. REF DMJ 4 334 38.

604 1, 2, 5, 30, 2288, 67172352, 144115192303714304
BOOLEAN FUNCTIONS. REF HA2 153.

605 1, 2, 5, 34, 985, 1151138, 1116929202845
FROM A CONTINUED FRACTION. REF DMJ 4 334 38.

606 1, 2, 5, 34, 2136
RELATIONS WITH THREE ARGUMENTS. REF MAN 174 69 67.

SEQUENCES BEGINNING 1, 2, 6

607 1, 2, 6, 2, 10, 2, 10, 14, 10, 6, 10, 18, 2, 6, 14, 22, 14, 22, 26, 18, 14, 2, 30, 26, 30, 2, 26, 18, 10, 34, 26, 22, 18, 10, 34, 14, 34, 38, 2, 6, 30, 34, 14, 42, 38, 10, 22, 42, 38, 26
GLAISHERS CHI FUNCTION. REF QJM 20 152 1884.

608 1, 2, 6, 4, 30, 12, 84, 24, 90, 20, 132
DENOMINATORS OF GENERALIZED BERNOULLI NUMBERS. REF MT1 136.

609 1, 2, 6, 8, 5, 4, 5, 2, 0, 0, 1, 0, 6, 5, 3, 0, 6, 4, 4, 5, 3, 0, 9, 7, 1, 4, 8, 3, 5, 4, 8, 1, 7, 9, 5, 6, 9, 3, 8, 2, 0, 3, 8, 2, 2, 9, 3, 9, 9, 4, 4, 6, 2, 9, 5, 3, 0, 5, 1, 1, 5, 2, 3, 4, 5, 5, 5, 7, 2
KHINTCHINES CONSTANT. REF MTAC 14 371 60.

610 1, 2, 6, 8, 13, 29, 44, 66, 122, 184, 269, 448, 668, 972, 1505, 2205
COEFFICIENTS OF MODULAR FUNCTIONS. REF PLMS 9 386 59.

611 1, 2, 6, 8, 20, 12, 42, 32, 54, 40, 110, 48
PATTERNS. REF MES 37 61 07.

612 1, 2, 6, 8, 90, 288, 840, 17280, 28350, 89600, 598752, 87091200, 63063000, 301771008000, 5003856000, 6199345152, 976924698750, 3766102179840000
COTESIAN NUMBERS. REF QJM 46 63 14.

613 1, 2, 6, 9, 12, 15, 18, 21, 24, 27, 31, 34, 37, 40, 43, 46, 49, 53, 56, 59, 62, 65, 68, 71, 75, 78, 81, 84, 87, 90, 93, 97, 100, 103, 106, 109, 112, 115, 119, 122, 125, 128, 131
ZEROS OF BESSEL FUNCTION OF ZERO ORDER. REF BA5 171. AS1 409.

614 1, 2, 6, 9, 18, 22, 32, 46
VAN DER WAERDEN NUMBERS. REF CSA 31.

615 1, 2, 6, 10, 14, 18, 26, 30, 38
CONFERENCE MATRICES. REF AB1 82 15 68.

616 1, 2, 6, 12, 20, 30, 42, 56, 72, 90, 110, 132, 156, 182, 210, 240, 272, 306, 342, 380, 420, 462, 506, 552, 600, 650, 702, 756, 812, 870, 930, 992, 1056, 1122, 1190, 1260
THE PRONIC NUMBERS N(N + 1). REF DI2 2 232.

617 1, 2, 6, 12, 24, 40, 72, 126, 240, 272
LATTICE SPHERE PACKINGS. REF CJM 16 674 64. WA1.

618 1, 2, 6, 12, 31, 72, 178
ALKYLS. REF ZFK 93 437 36.

619 1, 2, 6, 12, 60, 20, 140, 280, 2520, 2520, 27720, 27720, 360360, 360360, 360360, 720720, 12252240, 4084080, 77597520, 15519504, 5173168, 5173168, 118982864
DENOMINATORS OF HARMONIC NUMBERS. REF KN1 1 615.

620 1, 2, 6, 12, 60, 120, 360, 2520, 5040, 55440, 720720, 1441440, 4324320, 21621600, 367567200, 6983776800, 13967553600, 321253732800, 2248776129600
SUPERIOR HIGHLY COMPOSITE NUMBERS. REF RAM 87.

621 1, 2, 6, 12, 60, 168, 360
LARGEST GROUP WITH N CONJUGATE CLASSES. REF CJM 20 456 68.

622 1, 2, 6, 13, 24, 42, 73, 125, 204, 324
PARTITIONS INTO NON-INTEGRAL POWERS. REF PCPS 47 215 51.

623 1, 2, 6, 13, 40, 100, 291, 797, 2273, 6389
FUNCTIONAL DIGRAPHS. REF MAN 143 110 61.

624 1, 2, 6, 14, 24, 46, 88, 162, 300, 562, 1056
SETS WITH A CONGRUENCE PROPERTY. REF MFC 15 58 65.

625 1, 2, 6, 14, 30, 62, 126, 254, 510, 1022, 2046, 4094, 8190, 16382, 32766, 65534, 131070, 262142, 524286, 1048574, 2097150, 4194302, 8388606, 16777214, 33554430
DIFFERENCES OF ZERO, 2N – 2. REF VO1 31. DA2 2 212. R1 33.**

626 1, 2, 6, 14, 31, 73, 172, 400, 932, 2177, 5081, 11854, 27662, 64554
RESTRICTED PERMUTATIONS. REF AENS 79 207 62.

627 1, 2, 6, 14, 38, 97, 260, 688, 1856
GLYCOLS. REF JACS 56 157 34.

628 1, 2, 6, 15, 40, 104, 273, 714, 1870, 4895, 12816, 33552, 87841
FROM FIBONACCI IDENTITIES. REF FQ 6 82 68.

629 1, 2, 6, 16, 50, 144, 448, 7472, 17676, 41600
FOLDING A STRIP OF STAMPS. REF SL1 41.

630 1, 2, 6, 16, 50, 144, 462, 1392, 4536, 14060, 46310, 146376, 485914, 1557892, 5202690, 16861984, 56579196, 184940388, 622945970, 2050228360, 6927964218
FOLDING A LINE. REF MTAC 22 198 68. JCT 5 135 68.

631 1, 2, 6, 16, 50, 165
SELF-DUAL POLYTOPES. REF JCT 7 157 69.

632 1, 2, 6, 17, 44, 112, 304, 918, 3040, 10623, 38161, 140074, 528594, 2068751, 8436893, 35813251, 157448068, 713084042, 3315414747, 15805117878, 77273097114
FROM A DIFFERENTIAL EQUATION. REF AMM 67 766 60.

633 1, 2, 6, 18, 46, 146, 460, 1436, 4352, 13252, 40532
RESTRICTED PERMUTATIONS. REF AENS 79 213 62.

634 1, 2, 6, 18, 50, 142, 390, 1086, 2958, 8134, 22050, 60146, 162466, 440750, 1187222, 3208298, 8622666
WALKS ON A SQUARE LATTICE. REF AIP 9 354 60.

635 1, 2, 6, 18, 57, 186, 622, 2120, 7338, 25724, 91144, 325878, 1174281, 4260282, 15548694, 57048048, 210295326, 778483932, 2892818244, 10786724388
A SIMPLE RECURRENCE. REF IC 16 352 70.

636 1, 2, 6, 18, 58, 186, 614, 2034, 6818, 22970
SERIES-PARALLEL NUMBERS. REF R1 142.

637 1, 2, 6, 18, 60, 184, 560, 1695, 5200, 15956, 48916
RESTRICTED PERMUTATIONS. REF AENS 79 213 62.

638 1, 2, 6, 18, 90, 540, 3780, 31500
PERMUTATIONS BY NUMBER OF CYCLES. REF R1 85.

639 1, 2, 6, 19, 61, 196, 629, 2017, 6466, 20727, 66441, 212980, 682721, 2188509, 7015418, 22488411, 72088165, 231083620, 740754589, 2374540265, 7611753682
BOARD-PILE POLYOMINOES. REF JCT 6 103 69. AT1 363.

640 1, 2, 6, 19, 63, 216, 760, 2723, 9880, 36168, 133237, 492993, 1829670, 6804267, 25336611, 94416842, 351989967, 1312471879, 4894023222, 18248301701
BOARD-PAIR-PILE POLYOMINOES. REF AT1 363.

641 1, 2, 6, 19, 63, 216, 760, 2725, 9910, 36446, 135268, 505861, 1903890, 7204874, 27394666, 104592937, 400795860, 1540820542
FIXED POLYOMINOES. REF AT1 363.

642 1, 2, 6, 20, 60, 176, 512, 1488, 4326, 12648, 37186, 109980, 327216, 979020, 2944414, 8897732, 27004290, 82287516
PARAFFINS. REF JACS 54 1105 32.

643 1, 2, 6, 20, 70, 252, 924, 3432, 12870, 48620, 184756, 705432, 2704156, 10400600, 40116600, 155117520, 601080390, 2333606220, 9075135300, 35345263800
CENTRAL BINOMIAL COEFFICIENTS C(2N, N). REF RS1. AS1 828.

644 1, 2, 6, 20, 71, 259, 961
PERMUTATIONS BY INVERSIONS. REF NET 96.

645 1, 2, 6, 20, 76, 312, 1384
SYMMETRIC PERMUTATIONS. REF LU1 1 221.

646 1, 2, 6, 20, 90, 544, 5096, 79264, 2208612, 113743760, 10926227136, 1956363435360, 652335084592096, 405402273420996800, 470568642161119963904
SYMMETRIC RELATIONS. REF MI1 17 21 55. MAN 174 70 67.

647 1, 2, 6, 21, 65, 221, 771, 2769, 10250, 39243, 154658, 628635, 2632420, 11353457, 50411413, 230341716
GRAPHS BY POINTS AND LINES. REF R1 146. ST1.

648 1, 2, 6, 21, 94, 512, 3485
CONNECTED UNLABELED TOPOLOGIES. REF WR1.

649 1, 2, 6, 21, 112, 853, 11117, 261080, 11716571
CONNECTED GRAPHS. REF TAMS 78 459 55. ST1. JCT 9 352 70.

650 1, 2, 6, 22, 67, 213, 744, 2609, 9016, 31427, 110384
PERFECT SQUARED RECTANGLES. REF GA1 207. BO4.

651 1, 2, 6, 22, 91, 408, 1938, 9614, 49335, 260130, 1402440, 7702632, 42975796, 243035536, 1390594458, 8038677054, 46892282815, 275750636070
NONSEPARABLE PLANAR GRAPHS. REF CJM 15 257 63. AT1 363.

652 1, 2, 6, 22, 92, 422, 2074, 10754, 58202, 326240
BAXTER PERMUTATIONS. REF MA4 2 25 67.

653 1, 2, 6, 22, 94, 454, 2430, 14214, 89918, 610182, 4412798
VALUES OF BELL POLYNOMIALS. REF RI1. PSPM 19 173 71.

654 1, 2, 6, 22, 101, 546, 3502, 25586, 214062, 1987516, 20599076, 232482372, 2876191276, 38228128472, 549706132536, 8408517839416, 137788390312712
TERMS IN A BORDERED SKEW DETERMINANT. REF PRSE 21 354 1896.

655 1, 2, 6, 22, 101, 573, 3836, 29228, 250749, 2409581, 25598186, 296643390, 3727542188, 50626553988, 738680521142
KENDALL-MANN NUMBERS. REF DKB 241. PGEC 19 1226 70.

656 1, 2, 6, 23, 109, 618, 4096, 31133, 267219, 2557502
MATRICES WITH 2 ROWS. REF PLMS 17 29 17.

657 1, 2, 6, 24, 78, 230, 675, 2069, 6404, 19708, 60216, 183988
RESTRICTED PERMUTATIONS. REF AENS 79 213 62.

658 1, 2, 6, 24, 80, 450, 2142, 17696, 112464, 1232370
LOGARITHMIC NUMBERS. REF MST 31 77 63.

659 1, 2, 6, 24, 120, 720, 5040, 40320, 362880, 3628800, 39916800, 479001600, 6227020800, 87178291200, 1307674368000, 20922789888000, 355687428096000
FACTORIAL NUMBERS. REF AS1 833. MTAC 24 231 70.

660 1, 2, 6, 25, 135, 892, 6937, 61886
LABELED TREES WITH UNLABELED END-POINTS. REF JCT 6 63 69.

661 1, 2, 6, 26, 135, 875
SEMIGROUPS BY NUMBER OF IDEMPOTENTS. REF MA4 2 2 67.

662 1, 2, 6, 26, 147, 892, 5876, 40490
TRIANGULATIONS OF THE DISK. REF PLMS 14 759 64.

663 1, 2, 6, 26, 159, 1347
RELATED TO NUMBER OF TOPOLOGIES. REF PURB 19 240 68.

664 1, 2, 6, 26, 166, 1626, 25510
FROM THE BINARY PARTITION FUNCTION. REF PRSE 65 190 59. PCPS 66 376 69.

665 1, 2, 6, 28, 180, 662, 7266, 24568
SECOND ORDER EULER NUMBERS. REF JRAM 79 69 1875. FMR 1 75.

666 1, 2, 6, 28, 244, 2544, 35600, 659632
ZERO-SUM ARRAYS. REF JA1 7 25 67.

667 1, 2, 6, 30, 42, 30, 66, 2730, 6, 510, 798, 330, 138, 2730, 6, 870, 14322, 510, 6, 1919190, 6, 13530, 1806, 690, 282, 46410, 66, 1590, 798, 870, 354, 56786730
DENOMINATORS OF BERNOULLI NUMBERS. REF DA2 2 230. AS1 810.

668 1, 2, 6, 30, 210, 2310, 30030, 510510, 9699690, 223092870, 6469693230, 200560490130, 7420738134810, 304250263527210, 13082761331670030
PRIME FACTORIALS. REF FMR 1 50.

669 1, 2, 6, 30, 390, 32370, 81022110, 79098077953830, 249960304895738623374279O, 6399996109983215106481566902449146981585570
FROM A CONTINUED FRACTION. REF AMM 63 711 56.

670 1, 2, 6, 32, 353, 8390, 436399, 50468754
GEOMETRIES. REF BSM 19 421 67.

671 1, 2, 6, 34, 250, 972, 15498, 766808, 5961306, 54891535, 2488870076
COEFFICIENTS FOR NUMERICAL INTEGRATION. REF MTAC 6 217 52.

672 1, 2, 6, 36, 220, 1590, 12978, 118664, 1201464, 13349610, 161530270, 2114578092, 29780308116, 448995414686, 7215997736010
3-LINE LATIN RECTANGLES. REF R1 210. DKB 263 (MULTIPLIED BY 2).

673 1, 2, 6, 36, 240, 1800, 15120, 141120, 1693440
SORTING NUMBERS. REF PSPM 19 172 71.

674 1, 2, 6, 36, 240, 1800, 16800, 191520, 2328480
SORTING NUMBERS. REF PSPM 19 172 71.

675 1, 2, 6, 36, 876, 408696, 83762796636, 3508125906207095591916, 6153473687096578758445014683368786661634996
HYPOTHENUSAL NUMBERS. REF RS3 178 288 1887. LU1 1 496.

676 1, 2, 6, 38, 390, 6062, 134526
COLORED GRAPHS. REF CJM 22 596 70.

677 1, 2, 6, 40, 1992, 18666624, 12813206169137152
BOOLEAN FUNCTIONS. REF HA2 153.

678 1, 2, 6, 42, 4094, 98210640
SELF-COMPLEMENTARY BOOLEAN FUNCTIONS. REF PGEC 12 561 63.

679 1, 2, 6, 46, 522, 7970, 152166, 3487246, 93241002, 2849229890, 97949265606, 3741386059246, 157201459863882, 7205584123783010, 357802951084619046
GENERALIZED EULER NUMBERS. REF MTAC 21 693 67.

680 1, 2, 6, 56, 528, 6193, 86579, 1425518, 27298230, 601580875, 15116315766, 429614643062, 13711655205087, 488332318973594, 19296579341940067
INTEGERS RELATED TO BERNOULLI NUMBERS. REF MTAC 21 678 67.

681 1, 2, 6, 60, 2880, 2246400, 135862272000, 10376834265907200000, 7754011537434823832371 2000000000
AN OPERATIONAL RECURRENCE. REF FQ 1(1) 31 63.

682 1, 2, 6, 70, 700229
SWITCHING NETWORKS. REF JFI 276 317 63.

683 1, 2, 6, 74, 169112
BOOLEAN FUNCTIONS. REF PGEC 9 265 60.

SEQUENCES BEGINNING 1, 2, 7

684 1, 2, 7, 1, 8, 2, 8, 1, 8, 2, 8, 4, 5, 9, 0, 4, 5, 2, 3, 5, 3, 6, 0, 2, 8, 7, 4, 7, 1, 3, 5, 2, 6, 6, 2, 4, 9, 7, 7, 5, 7, 2, 4, 7, 0, 9, 3, 6, 9, 9, 9, 5, 9, 5, 7, 4, 9, 6, 6, 9, 6, 7, 6, 2, 7, 7, 2, 4, 0
DIGITS OF E. REF MTAC 4 14 50, 23 679 69.

685 1, 2, 7, 8, 37, 40, 200, 258, 1039, 1500
PERMUTATION GROUPS. REF JPC 33 1069 29.

686 1, 2, 7, 9, 43, 52, 303, 355, 658, 4303, 9264, 50623, 414248, 1293367, 4294349, 18470763, 41235875, 265886013, 1104779927, 4685005721, 5789785648
CONVERGENTS TO CUBE ROOT OF 3. REF AMP 46 106 1866. LE1 67. HPR.

687 1, 2, 7, 11, 15, 20, 24, 28, 32, 37, 41, 45, 50, 54
WYTHOFF GAME. REF CMB 2 188 59.

688 1, 2, 7, 11, 101, 111, 1001, 2201, 10001, 10101, 11011
CUBE IS A PALINDROME. REF JRM 3 97 70.

689 1, 2, 7, 13, 18, 23, 28, 34, 39, 44, 49, 54, 60, 65
WYTHOFF GAME. REF CMB 2 189 59.

690 1, 2, 7, 14, 32, 58, 110, 187, 322, 519, 839, 1302, 2015, 3032, 4542, 6668, 9738, 14006, 20036, 28324, 39830, 55473, 76875, 105692, 144629, 196585, 266038, 357952
TREES OF DIAMETER 5. REF IBMJ 4 476 60. KU1.

691 1, 2, 7, 15, 28, 45, 70, 100, 138
PARTITIONS INTO NON-INTEGRAL POWERS. REF PCPS 47 214 51.

692 1, 2, 7, 18, 28, 182, 845, 904, 5235, 36028, 74713, 526624, 977572, 4709369, 9959574, 96696762, 7724076630, 35354759457, 138217852516, 642742746639
DIGITS OF E. REF MTAC 4 14 50, 23 679 69.

693 1, 2, 7, 18, 60, 196, 704, 2500, 9189, 33896, 126759, 476270, 1802312, 6849777, 26152418, 100203198, 385221143
ONE-SIDED POLYOMINOES. REF GO2 105. LU2.

694 1, 2, 7, 18, 64, 226, 856, 3306, 13248, 53794, 222717
RESTRICTED HEXAGONAL POLYOMINOES. REF EMS 17 11 70.

695 1, 2, 7, 20, 54, 148, 403, 1096, 2980, 8103, 22026, 59874, 162754, 442413, 1202604, 3269017, 8886110, 24154952, 65659969, 178482300, 485165195, 1318815734
EN. REF MNAS 14(5) 14 25. FW1. FMR 1 230.**

696 1, 2, 7, 20, 66, 212, 715, 2424, 8398, 29372, 104006, 371384, 1337220, 4847208, 17678835, 64821680, 238819350, 883629164, 3282060210, 12233125112
DISSECTIONS OF A POLYGON: REF GU1. MAT 15 121 68.

697 1, 2, 7, 23, 88, 414, 2371, 16071, 125672, 1112082
BINOMIAL COEFFICIENT SUMS. REF CJM 22 26 70.

698 1, 2, 7, 23, 115, 694, 5282, 46066, 456454, 4999004, 59916028
SYMMETRIC PERMUTATIONS. REF LU1 1 222.

699 1, 2, 7, 23, 122, 888
SYMMETRICAL ALIORELATIVE RELATIONS. REF AJM 49 453 27.

700 1, 2, 7, 26, 97, 362, 1351, 5042, 18817, 70226, 262087, 978122, 3650401, 13623482, 50843527, 189750626, 708158977, 2642885282, 9863382151, 36810643322
A(N) = 4A(N − 1) − A(N − 2). REF NCM 4 167 1878. MMAG 40 78 67. FQ 7 239 69.

701 1, 2, 7, 26, 107, 458, 2058, 9498, 44947, 216598, 1059952, 5251806, 26297238, 132856766, 676398395, 3466799104, 17873808798, 92630098886, 482292684506
ORIENTED ROOTED UNLABELED TREES. REF R1 138.

702 1, 2, 7, 26, 111, 562, 3151, 19252, 128449, 925226
FORESTS OF LEAST HEIGHT. REF JCT 5 97 68. RI1.

703 1, 2, 7, 28, 124, 588, 2938, 15268
WALKS ON A SQUARE LATTICE. REF JCP 31 1333 59.

704 1, 2, 7, 30, 157, 972, 6961, 56660, 516901, 5225670, 57999271, 701216922, 9173819257, 129134686520, 1946194117057, 31268240559432
A(N) = NA(N − 1) + A(N − 2). REF EUR 20 15 57.

705 1, 2, 7, 31, 164, 999, 6841, 51790, 428131, 3929021
SORTING NUMBERS. REF PSPM 19 173 71.

706 1, 2, 7, 32, 181, 1214, 9403, 82508, 808393, 8743994, 103459471, 1328953592, 18414450877, 273749755382, 4345634192131, 73362643649444
A(N) = NA(N − 1) + (N − 2)A(N − 2). REF R1 188.

707 1, 2, 7, 32, 184, 1268, 10186, 93356, 960646, 10959452, 137221954, 1870087808, 27548231008, 436081302248, 7380628161076, 132975267434552
STOCHASTIC MATRICES OF INTEGERS. REF DMJ 35 659 68.

708 1, 2, 7, 34, 209, 1546, 13327, 130922
RELATED TO GAMMA FUNCTION. REF SE2 78.

709 1, 2, 7, 34, 257
POLYTOPES. REF GR2 424.

710 1, 2, 7, 35, 228, 1834, 17582, 195866, 2487832, 35499576, 562356672, 9794156448, 186025364016, 3826961710272, 84775065603888, 2011929826983504
COEFFICIENTS OF ITERATED EXPONENTIALS. REF SMA 11 353 45.

711 1, 2, 7, 36, 317, 5624, 251610, 33642660, 14685630688
INCIDENCE MATRICES. REF CPM 89 217 64.

712 1, 2, 7, 37, 216, 1780
SEMIGROUPS WITH TWO IDEMPOTENTS. REF MA4 2 2 67.

713 1, 2, 7, 37, 266, 2431, 27007, 353522, 5329837, 90960751, 1733584106, 36496226977, 841146804577, 21065166341402, 569600638022431
FROM BESSEL POLYNOMIALS. REF RCI 77. RI1.

714 1, 2, 7, 38, 291, 2932, 36961, 561948, 10026505, 205608536
FORESTS OF LABELED TREES. REF JCT 5 96 68. RI1.

715 1, 2, 7, 42, 582, 21480, 2142288, 575016219, 415939243032, 816007449011040, 4374406209970747314, 64539836938720749739356
ANTISYMMETRIC RELATIONS. REF PAMS 4 494 53. MI1 17 23 55.

716 1, 2, 7, 44, 361, 3654, 44207, 622552, 10005041, 180713290
MODIFIED BESSEL FUNCTIONS. REF AS1 429.

717 1, 2, 7, 44, 447, 6749, 142176, 3987677, 143698548, 6470422337, 356016927083, 23503587609815, 1833635850492653, 166884365982441238
A(N) = N(N − 1)A(N − 1)/2 + A(N − 2).

718 1, 2, 7, 56, 2212, 2595782, 3374959180831, 5695183504489239067484387, 16217557574922386301420531277071365103168734284282
A NONLINEAR RECURRENCE. REF PRSE 59(2) 159 39. CMB 11 87 68.

719 1, 2, 7, 60, 13733
SWITCHING NETWORKS. REF JFI 276 317 63.

720 1, 2, 7, 97, 18817, 708158977, 1002978273411373057, 2011930833870518011412817828051050497
A(N) = 2A(N − 1)2 − 1. REF D12 1 399.**

721 1, 2, 7, 111, 308063, 100126976263592
BOOLEAN FUNCTIONS. REF HA2 153 (DIVIDED BY 2).

722 1, 2, 7, 124, 494298
SWITCHING NETWORKS. REF JFI 276 317 63.

723 1, 2, 7, 1172, 36325278240, 18272974787063551687986348306336
INVERTIBLE BOOLEAN FUNCTIONS. REF PGEC 13 530 64.

SEQUENCES BEGINNING 1, 2, 8 THROUGH 1, 2, 10

724 1, 2, 8, 9, 10, 11, 15, 19, 21, 22, 25, 26, 27, 28, 30, 31, 34, 40, 42, 45, 46, 47, 50, 55, 57, 58, 59, 62, 64, 65, 66, 70, 74, 75, 78, 79, 80, 84, 86, 94, 96, 97, 98, 100, 101, 103
NUMBERS WITH AN EVEN NUMBER OF PARTITIONS. REF JLMS 1 226 26. MTAC 21 470 67.

725 1, 2, 8, 10, 24, 53, 74, 153, 280, 436, 793, 1322, 2085, 3510, 5648, 8796
COEFFICIENTS OF MODULAR FUNCTIONS. REF PLMS 9 385 59.

726 1, 2, 8, 19, 41, 78, 134, 218
PARTITIONS INTO NON-INTEGRAL POWERS. REF PCPS 47 214 51.

727 1, 2, 8, 20, 80, 350, 1232, 5768, 31040, 142010, 776600, 4874012, 27027728, 168369110, 1191911840, 7678566800, 53474964992, 418199988338
PERMUTATIONS OF ORDER EXACTLY 3. REF CJM 7 159 55.

728 1, 2, 8, 20, 152, 994, 7888, 70152, 695760
FROM MENAGE POLYNOMIALS. REF R1 197.

729 1, 2, 8, 21, 48, 99, 186
PARTITIONS INTO NON-INTEGRAL POWERS. REF PCPS 47 214 51.

730 1, 2, 8, 26, 80, 268, 944, 3474, 13072, 49672, 191272, 744500
MAGNETISATION FOR DIAMOND LATTICE. REF PHA 29 382 63.

731 1, 2, 8, 29, 166, 1023
POLYOMINOES MADE FROM CUBES. REF FQ 3 19 65. BO4.

732 1, 2, 8, 34, 136, 538, 2080, 7970, 30224, 113874
SERIES-PARALLEL NUMBERS. REF R1 142.

733 1, 2, 8, 34, 152, 714, 3472, 17318, 88048
MAGNETISATION FOR SQUARE LATTICE. REF PHA 22 934 56.

734 1, 2, 8, 36, 184, 1110, 7776, 62216, 559952, 5599530, 61594840, 739138092, 9608795208, 134523132926, 2017846993904, 32285551902480
GENERATING PERMUTATIONS. REF CJ1 13 155 70.

735 1, 2, 8, 38, 192, 1002, 5336, 28814, 157184, 864146, 4780008, 26572086, 148321344, 830764794, 4666890936
BINOMIAL COEFFICIENT SUMS. REF AMM 43 29 36.

736 1, 2, 8, 38, 212, 1370, 10112, 84158, 780908, 8000882
BINOMIAL COEFFICIENT SUMS. REF CJM 22 26 70.

737 1, 2, 8, 40, 240, 1680, 13440, 120960, 1209600, 13305600
GENERALIZED TANGENT NUMBERS. REF TOH 42 152 36.

738 1, 2, 8, 42, 268, 1994, 16852
SORTING NUMBERS. REF PSPM 19 173 71.

739 1, 2, 8, 42, 296, 2635
POLYTOPES. REF JCT 7 157 69.

740 1, 2, 8, 44, 436, 7176, 222368
SELF-CONVERSE RELATIONS. REF MAT 13 157 66.

741 1, 2, 8, 44, 490, 14074, 1349228
THRESHOLD FUNCTIONS. REF PGEC 19 823 70.

742 1, 2, 8, 48, 384, 3840, 46080, 645120, 10321920, 185794560, 3715891200, 81749606400, 1961990553600, 51011754393600, 1428329123020800
DOUBLE FACTORIALS, (2N).FACTORIAL N. REF AMM 55 425 48. MTAC 24 231 70.**

743 1, 2, 8, 50, 416, 4322, 53888, 783890, 13031936, 243733442, 5064892768
FROM FIBONACCI SUMS. REF FQ 5 48 67.

744 1, 2, 8, 50, 418, 4348, 54016, 779804, 12824540, 236648024, 4841363104, 108748223128, 2660609220952, 70422722065040, 2005010410792832
A(N) = (2N - 1)A(N - 1) - (N - 1)A(N - 2). REF AJM 2 94 1879. LU1 1 223.

745 1, 2, 8, 60, 320, 1980, 10512, 60788, 320896, 1787904, 9381840, 51081844
FOLDING A MAP. REF CJ1 14 77 71.

746 1, 2, 8, 60, 672, 9953, 184557, 4142631, 109813842, 3373122370, 118280690398, 4678086540493, 206625802351035, 10107719377251109, 543762148079927802
EVEN GRAPHS. REF CJM 8 410 56. CA3.

747 1, 2, 8, 72, 1536, 86080, 14487040
MAJORITY DECISION FUNCTIONS. REF MTAC 16 471 62.

748 1, 2, 8, 75, 8949, 11964723
CONTINUED COTANGENT FOR E. REF DMJ 4 339 38.

749 1, 2, 8, 96, 1152, 7680, 18432
FROM HIGHER ORDER BERNOULLI NUMBERS. REF NO1 459.

750 1, 2, 8, 96, 4608, 798720, 361267200
FOLDING A MAP. REF CJ1 14 77 71.

751 1, 2, 8, 96, 10368, 108615168, 11798392572168192, 139202068568601556987554268864512
A NONLINEAR RECURRENCE. REF SA2.

752 1, 2, 8, 214, 10740500
SWITCHING NETWORKS. REF JFI 276 317 63.

753 1, 2, 8, 502, 547849868
SWITCHING NETWORKS. REF JFI 276 317 63.

754 1, 2, 9, 4, 28, 18, 118, 80, 504, 466, 1631, 2160, 5466, 7498
COEFFICIENTS OF MODULAR FUNCTIONS. REF PLMS 9 384 59.

755 1, 2, 9, 9, 50, 267, 413, 2180, 17731, 50533, 110176, 1966797, 9938669, 8638718, 278475061, 2540956509, 9816860358, 27172288399, 725503033401
EXPANSION OF EXP(1 - EXP(X)). REF JIA 76 153 50.

756 1, 2, 9, 20, 149, 467, 237385, 237852, 1426645, 7371077, 8797722, 16168799, 24966521, 66101841, 91068362, 157170203, 3863153234, 4020323437
CONVERGENTS TO CUBE ROOT OF 6. REF AMP 46 107 1866. LE1 67. HPR.

757 1, 2, 9, 20, 670
CUTTING NUMBERS OF GRAPHS. REF CSA 149.

758 1, 2, 9, 28, 101, 342, 1189, 4088, 14121, 48682, 167969, 579348, 1998541, 6893822, 23780349, 82029808, 282961361, 976071762, 3366950329, 11614259468
A(N) = 2A(N - 1) + 5A(N - 2). REF MQET 1 11 16.

759 1, 2, 9, 28, 185, 846, 7777, 47384, 559953
LOGARITHMIC NUMBERS. REF MST 31 78 63.

760 1, 2, 9, 31, 109, 399, 1043, 2998, 8406, 22652, 59521, 151958, 379693, 927622, 2224235, 5236586, 12130780, 27669593, 62229990, 138095696, 302673029
BIPARTITE PARTITIONS. REF PCPS 49 72 53. NI1 1.

761 1, 2, 9, 34, 119, 401, 1316, 4247, 13532, 42712
PARTIALLY LABELED ROOTED TREES. REF R1 134.

762 1, 2, 9, 35, 132, 494, 1845, 6887, 25704, 95930, 358017, 1336139, 4986540, 18610022, 69453549, 259204175, 967363152, 3610248434, 13473630585, 50284273907
FROM THE SOLUTION TO A PELLIAN. REF AMM 56 175 49.

763 1, 2, 9, 37, 183, 933, 5314
RELATIONS ON AN INFINITE SET. REF MAN 174 67 67.

764 1, 2, 9, 38, 161, 682, 2889, 12238, 51841, 219602, 930249, 3940598, 16692641, 70711162, 299537289, 1268860318, 5374978561, 22768774562, 96450076809
A(N) = 4A(N - 1) + A(N - 2). REF TH2 282.

765 1, 2, 9, 40, 355, 11490, 7758205, 549758283980
PRECOMPLETE POST FUNCTIONS. REF SMD 10 619 69. RO3.

766 1, 2, 9, 44, 265, 1854, 14833, 133496, 1334961, 14684570, 176214841, 2290792932, 32071101049, 481066515734, 7697064251745, 130850092279664
SUBFACTORIAL OR RENCONTRES NUMBERS. REF R1 65. DB1 168. RYS 23. MTAC 21 502 67. CO1 2 12.

767 1, 2, 9, 49, 306, 2188, 17810, 162482, 1642635, 18231462, 220420179, 2883693795, 40592133316, 611765693528, 9828843229764, 167702100599524
MODIFIED BESSEL FUNCTIONS. REF AS1 429. HPR.

768 1, 2, 9, 54, 378, 2916, 24057, 208494, 1876446, 17399772, 165297834, 1602117468, 15792300756, 157923007560, 1598970451545, 16365932856990
ROOTED MAPS. REF CJM 15 254 63. JCT 3 121 67.

769 1, 2, 9, 54, 450, 4500, 55125, 771750, 12502350
EXPANSION OF AN INTEGRAL. REF CO1 1 176.

770 1, 2, 9, 60, 525, 5670, 72765, 1081080, 18243225
EXPANSION OF AN INTEGRAL. REF CO1 1 176.

771 1, 2, 9, 64, 625, 7776, 117649, 2097152, 43046721, 1000000000, 25937424601, 743008370688, 23298085122481, 793714773254144, 29192926025390625
N(N - 1). REF BA1. R1 128.**

772 1, 2, 9, 88, 1802, 75598, 6421599, 1097780312, 376516036188, 258683018091900, 355735062429124915, 978786413996934006272
NUMBER OF FULL SETS. REF PAMS 13 828 62.

773 1, 2, 9, 96, 2500, 162000, 26471025, 11014635520, 11759522374656, 32406091200000000, 231627686043080250000, 4311500661703860387840000
PRODUCT OF BINOMIAL COEFFICIENTS. REF AS1 828.

774 1, 2, 9, 443, 11211435
SWITCHING NETWORKS. REF JFI 276 317 63.

775 1, 2, 10, 4, 40, 92, 352, 724, 2680, 14200, 73712, 365596
QUEENS PROBLEM. REF PSAM 10 93 60. LI2.

776 1, 2, 10, 12, 21, 102, 111, 122, 201, 212, 1002, 1011, 1101, 1112, 1121, 1202, 1222, 2012, 2021, 2111, 2122, 2201, 2221, 10002, 10022, 10121, 10202, 10211, 10222
PRIMES IN TERNARY. REF EUR 23 23 60.

777 1, 2, 10, 28, 106, 344, 1272, 4592, 17692, 69384
SYMMETRIC PERMUTATIONS. REF LU1 1 222.

778 1, 2, 10, 30, 70, 140, 252, 420, 660, 990
RELATED TO BINOMIAL MOMENTS. REF JO2 449.

779 1, 2, 10, 36, 145, 560, 2197, 8568, 33490, 130790, 510949, 1995840, 7796413, 30454814, 118965250, 464711184, 1815292333, 7091038640, 27699580729
PRODUCT OF FIBONACCI AND PELL NUMBERS. REF FQ 3 213 65.

780 1, 2, 10, 36, 720, 5600, 703760, 11220000
SELF-COMPLEMENTARY GRAPHS. REF JLMS 38 103 63.

781 1, 2, 10, 56, 346, 2252, 15184, 104960, 739162, 5280932, 38165260, 278415920, 2046924400, 15148345760, 112738423360, 843126957056, 6332299624282
CARD MATCHING. REF R1 193.

782 1, 2, 10, 74, 518, 3934, 29914
WALKS ON A DIAMOND LATTICE. REF PCPS 58 100 62.

783 1, 2, 10, 74, 706, 8162, 109960
A PROBLEM OF CONFIGURATIONS. REF CJM 4 25 52.

784 1, 2, 10, 104, 3044, 291968, 96928992, 112282908928, 458297100061728, 6666621572153927936, 349390545493499839161856
UNRESTRICTED RELATIONS. REF PAMS 4 494 53. MI1 17 19 55. MAN 174 66 67.

785 1, 2, 10, 208, 615904, 200253951911058
NONDEGENERATE BOOLEAN FUNCTIONS. REF PGEC 14 323 65.

786 1, 2, 10, 2104, 13098898366
SWITCHING NETWORKS. REF JFI 276 317 63.

SEQUENCES BEGINNING 1, 2, 11, 1, 2, 12, ...

787 1, 2, 11, 23, 24, 26, 33, 47, 49, 50, 59, 73, 74, 88, 96, 97, 107, 121, 122, 146, 169, 177, 184, 191, 193, 194, 218, 239, 241, 242, 249, 289, 297, 299, 311, 312, 313, 337, 338
FORMING PERFECT SQUARES. REF MMAG 37 218 64.

788 1, 2, 11, 32, 50, 132, 380, 368, 1135
THE NO-THREE-IN-LINE PROBLEM. REF GU3. WE1 124.

789 1, 2, 11, 35, 85, 175, 322, 546, 870, 1320, 1925, 2717, 3731, 5005, 6580, 8500, 10812, 13566, 16815, 20615, 25025, 30107, 35926, 42550, 50050, 58500, 67977, 78561
STIRLING NUMBERS OF FIRST KIND. REF AS1 833. DKB 226.

790 1, 2, 11, 38, 946, 4580, 202738, 3786092, 261868876, 1992367192, 2381255244240
RELATED TO ZEROS OF BESSEL FUNCTION. REF MTAC 1 406 45.

791 1, 2, 11, 46, 128, 272, 522, 904, 1408, 2160, 3154
GENERALIZED TANGENT NUMBERS. REF MTAC 21 690 67.

792 1, 2, 11, 62, 406, 3046, 25737, 242094
PERMUTATIONS WITH 1 3-SEQUENCE. REF BAMS 51 748 45.

793 1, 2, 11, 64, 426, 3216, 27240, 256320, 2656080, 30078720, 369774720, 4906137600, 69894316800, 1064341555200, 17255074636800, 296754903244800
DIFFERENCES OF FACTORIAL NUMBERS. REF JRAM 198 61 57.

794 1, 2, 11, 123, 1364, 15127, 167761, 1860498, 20633239, 228826127, 2537720636, 28143753123, 312119004989, 3461452808002, 38388099893011
RELATED TO BERNOULLI NUMBERS. REF RCI 139.

795 1, 2, 11, 590, 7644658
SWITCHING NETWORKS. REF JFI 276 317 63.

796 1, 2, 12, 8, 720, 288, 60480, 17280, 3628800, 89600, 95800320, 17418240, 2615348736000, 402361344000, 4483454976000, 98402304, 32011868528640000
FROM BERNOULLI POLYNOMIALS. REF JM2 22 49 43.

797 1, 2, 12, 24, 720, 160, 60480, 24192, 3628800, 1036800, 479001600, 788480, 2615348736000, 475517952000, 31384184832000, 689762304000
DENOMINATORS OF LOGARITHMIC NUMBERS. REF JM2 22 49 43. MTAC 20 465 66.

798 1, 2, 12, 32, 110, 310, 920
ALKYLS. REF ZFK 93 437 36.

799 1, 2, 12, 48, 160, 480, 1344, 3584, 9216, 23040, 56320, 135168
COEFFICIENTS OF HERMITE POLYNOMIALS. REF AS1 801.

800 1, 2, 12, 58, 300, 1682, 10332, 69298, 505500
PERMUTATIONS BY NUMBER OF SEQUENCES. REF CO1 2 103.

801 1, 2, 12, 60, 292, 1438, 7180, 36566
COLORED SERIES-PARALLEL NETWORKS. REF R1 159.

802 1, 2, 12, 70, 408, 2378, 13860, 80782, 470832, 2744210, 15994428, 93222358, 543339720, 3166815962, 18457556052, 107578520350, 627013566048
A(N) = 6A(N − 1) − A(N − 2). REF NCM 4 166 1878. ANN 30 72 28. AMM 75 683 68.

803 1, 2, 12, 70, 442, 3108, 24216, 208586, 1972904, 20373338, 228346522, 2763259364, 35927135944
PERMUTATIONS BY LENGTH OF RUNS. REF DKB 262.

804 1, 2, 12, 71, 481, 3708, 32028
PERMUTATIONS WITH 2 3-SEQUENCES. REF BAMS 51 748 45.

805 1, 2, 12, 72, 240, 2400, 907200, 4233600, 25401600, 1371686400
RELATED TO NUMERICAL INTEGRATION FORMULAS. REF MTAC 11 198 57.

806 1, 2, 12, 72, 720, 7200, 100800, 1411200, 24501600
SORTING NUMBERS. REF PSPM 19 172 71.

807 1, 2, 12, 72, 1440, 7200, 302400, 4233600, 101606400, 914457600, 100590336000
COEFFICIENTS FOR STEP-BY-STEP INTEGRATION. REF JACM 11 231 64.

808 1, 2, 12, 120, 1680, 30240, 665280, 17297280, 518918400, 17643225600, 670442572800, 28158588057600, 1295295050649600, 64764752532480000
COEFFICIENTS OF HERMITE POLYNOMIALS. REF MTAC 3 168 48.

809 1, 2, 12, 146, 3060, 101642, 5106612
CONNECTED LABELED PARTIALLY ORDERED SETS. REF WR1.

810 1, 2, 12, 152, 3472, 126752, 6781632, 500231552, 48656756992, 6034272215552, 929227412759552
EXPANSION OF COSH X / COS X. REF MMAG 34 37 60.

811 1, 2, 12, 288, 34560, 24883200, 125411328000, 5056584744960000, 1834933472251084800000, 6658606584104736522240000000
PRODUCT OF FIRST N FACTORIALS. REF FMR 1 50. RYS 53.

812 1, 2, 12, 576, 161280, 812851200, 61479419904000, 108776032459082956800
TOTAL NUMBER OF LATIN SQUARES. REF R1 210. RYS 53. FY1 22. RMM 193. JCT 3 98 67.

813 1, 2, 12, 2828, 8747130342
SWITCHING NETWORKS. REF JFI 276 317 63.

814 1, 2, 13, 44, 205, 806, 3457, 14168, 59449, 246410, 1027861, 4273412, 17797573, 74055854, 308289865, 1283082416, 5340773617, 22229288978, 92525540509
A(N) = 2A(N - 1) + 9A(N - 2). REF MQET 1 11 16.

815 1, 2, 13, 80, 579, 4738, 43387, 439792, 4890741, 59216642, 775596313, 10927434464, 164806435783, 2649391469058, 45226435601207, 817056406224416
MENAGE NUMBERS. REF CJM 10 478 58. R1 197.

816 1, 2, 13, 116, 1393, 20894, 376093, 7897952, 189550849, 5117872922, 153536187661, 5066694192812, 182400990941233, 7113638646708086
PERMUTATIONS WITH NO CYCLES OF LENGTH 3. REF R1 83.

817 1, 2, 14, 72, 330, 1430, 6006, 24052, 100776, 396800, 1634380, 6547520
PARTITIONS OF A POLYGON BY NUMBER OF PARTS. REF CAY 13 95.

818 1, 2, 14, 90, 646, 5242, 47622, 479306, 5296790, 63779034
HERTZSPRUNGS PROBLEM. REF IDM 26 121 19. AH1 271. AMS 38 1253 67.

819 1, 2, 14, 182, 3614, 99302, 3554894, 159175382
QUADRATIC INVARIANTS. REF CJM 8 310 56.

820 1, 2, 15, 60, 469, 3660, 32958, 328920
FROM MENAGE POLYNOMIALS. REF R1 197.

821 1, 2, 15, 84, 420, 1980, 9009, 40040
COEFFICIENTS FOR EXTRAPOLATION. REF SE2 97.

822 1, 2, 15, 148, 1785, 26106, 449701, 8927192, 200847681
TOTAL HEIGHT OF LABELED TREES. REF IBMJ 4 478 60.

823 1, 2, 15, 150, 1707, 20910, 268616, 3567400, 48555069, 673458874, 9481557398, 135119529972, 1944997539623, 28235172753886
COEFFICIENTS OF JACOBI NOME. REF BAMS 48 738 42. MTAC 3 234 48. CACM 4 317 61.

824 1, 2, 16, 88, 416, 1824, 7680, 31616, 128512, 518656, 2084864, 8361984, 33497088, 134094848
PERMUTATIONS BY LENGTH OF RUNS. REF DKB 261.

825 1, 2, 16, 96, 512, 2560, 12288, 57344, 262144, 1179648, 5242880, 23068672, 100663296, 436207616
COEFFICIENTS OF CHEBYSHEV POLYNOMIALS. REF LA4 518.

826 1, 2, 16, 130, 1424, 23682
COEFFICIENTS OF BELLS FORMULA. REF NMT 10 65 62.

827 1, 2, 16, 134, 1164, 10982, 112354, 1245676, 14909340, 191916532, 2646066034, 38932027996
PERMUTATIONS BY LENGTH OF RUNS. REF DKB 262.

828 1, 2, 16, 136, 1232, 12096, 129024, 1491840, 18627840
GENERALIZED TANGENT NUMBERS. REF TOH 42 152 36.

829 1, 2, 16, 272, 7936, 353792, 22368256, 1903757312, 209865342976, 29088885112832, 4951498053124096, 1015423886506852352
TANGENT NUMBERS. REF MTAC 21 672 67.

830 1, 2, 16, 980, 9332768
COMPLETE POST FUNCTIONS. REF ZML 7 198 61. PLMS 16 191 66.

831 1, 2, 17, 40, 5126, 211888, 134691268742, 28539643139633848, 244353369161294832262756363893 2102
A SIMPLE RECURRENCE. REF MMAG 37 167 64.

832 1, 2, 17, 62, 1382, 21844, 929569, 6404582, 443861162, 18888466084, 113927491862, 58870668456604, 8374643517010684, 689005380505609448
MULTIPLES OF BERNOULLI NUMBERS. REF RO2 329. FMR 1 74.

833 1, 2, 17, 5777, 192900153617, 717790523757994658974359292468 4177
A(N) = A(N - 1)3 + 3A(N - 1)**2 - 3. REF CR 83 1287 1876. DI2 1 397.**

834 1, 2, 18, 144, 1200, 10800, 105840, 1128960, 13063680, 163296000, 2195424000, 31614105600
COEFFICIENTS OF LAGUERRE POLYNOMIALS. REF AS1 799.

835 1, 2, 20, 104, 775, 6140, 55427
HIT POLYNOMIALS. REF RI3.

836 1, 2, 20, 110, 2600, 16150, 208012, 1376550, 74437200, 511755750, 7134913500, 50315410002, 1433226830360
COEFFICIENTS OF LEGENDRE POLYNOMIALS. REF MTAC 3 17 48.

837 1, 2, 20, 120, 560, 2240, 8064, 26880, 84480, 253440, 732160, 2050048, 5591040, 14909440, 38993920, 100270080, 254017536, 635043840, 1568931840
PRODUCT OF BINOMIAL COEFFICIENTS. REF MFM 74 62 70.

838 1, 2, 20, 144, 1265, 12072, 126565, 1445100, 17875140, 238282730, 3407118041, 52034548064, 845569542593, 14570246018686, 265397214435860
DISCORDANT PERMUTATIONS. REF SMA 20 23 54. KYU 10 13 56.

839 1, 2, 20, 198, 1960, 19402, 192060, 1901198, 18819920, 186298002, 1844160100, 18255302998, 180708869880, 1788833395802, 17707625088140
A(N) = 10A(N - 1) - A(N - 2). REF TH2 281.

840 1, 2, 20, 198, 2048, 22468, 264538, 3340962, 45173518, 652197968, 10024549190
PERMUTATIONS BY LENGTH OF RUNS. REF DKB 262.

841 1, 2, 20, 210, 2520, 34650, 540540, 9459450, 183783600, 3928374450, 91662070500, 2319050383650, 63246828645000, 1849969737866250
ASSOCIATED STIRLING NUMBERS. REF TOH 37 259 33. JO2 152. DB1 296. CO1 2 98.

842 1, 2, 20, 402, 14440, 825502, 69055260, 7960285802, 1209873973712
SOME SPECIAL NUMBERS. REF FMR 1 77.

843 1, 2, 22, 164, 1030, 5868, 31388, 160648, 795846, 3845020
ROOTED PLANAR MAPS. REF BAMS 74 74 68.

844 1, 2, 23, 44, 563, 3254, 88069, 11384, 1593269, 15518938, 31730711, 186088972, 3788707301, 5776016314
SUMS OF RECIPROCALS. REF RO2 313. FMR 1 89.

845 1, 2, 24, 11, 1085, 2542, 64344, 56415, 4275137
SUMS OF LOGARITHMIC NUMBERS. REF MST 31 77 63.

846 1, 2, 24, 48, 5760, 11520, 35840, 215040, 51609600, 103219200, 13624934400
COEFFICIENTS FOR NUMERICAL DIFFERENTIATION. REF PHM 33 13 42.

847 1, 2, 24, 140, 1232, 11268, 115056
FROM MENAGE POLYNOMIALS. REF R1 197.

848 1, 2, 24, 180, 1120, 6300, 33264, 168168
COEFFICIENTS FOR EXTRAPOLATION. REF SE2 93.

849 1, 2, 24, 272, 3424, 46720, 676608, 10251520, 160900608
ALMOST CUBIC MAPS. REF PL2 1 292 70.

850 1, 2, 24, 312, 4720, 82800, 1662024, 37665152, 952401888, 26602156800, 813815035000, 27069937855488, 972940216546896, 37581134047987712
TREES BY TOTAL HEIGHT. REF JA1 10 281 69.

851 1, 2, 24, 552, 21280, 1073760, 70299264, 5792853248, 587159944704, 71822743499520, 10435273503677440, 1776780700509416448
3-LINE LATIN RECTANGLES. REF PLMS 31 336 28. BU2 33 124 41. JMSJ 1(4) 241 50. R1 210.

852 1, 2, 24, 912, 87360, 19226880, 9405930240
COLORED GRAPHS. REF CJM 22 596 70.

853 1, 2, 24, 40320, 20922789888000, 263130836933693530167218012160000000
INVERTIBLE BOOLEAN FUNCTIONS. REF PGEC 13 530 64.

854 1, 2, 26, 50, 54, 126, 134, 246, 354, 362, 950
11.2N − 1 IS PRIME. REF MTAC 22 421 68.**

855 1, 2, 26, 938, 42800, 2130458
SETS WITH A CONGRUENCE PROPERTY. REF MFC 15 316 65.

856 1, 2, 28, 182, 4760, 31654, 428260, 2941470, 163761840, 1152562950, 16381761396, 117402623338
COEFFICIENTS OF LEGENDRE POLYNOMIALS. REF MTAC 3 17 48.

857 1, 2, 28, 236, 1852, 14622, 119964, 1034992
PERMUTATIONS BY NUMBER OF SEQUENCES. REF CO1 2 103.

858 1, 2, 30, 3522, 1066590, 604935042, 551609685150, 737740947722562, 1360427147514751710, 3308161927353377294082
GENERALIZED EULER NUMBERS. REF MTAC 21 689 67.

859 1, 2, 34, 5678, 91011121314151617181920212223242526272829303132333435 36
EACH TERM DIVIDES THE NEXT. REF JRM 3 40 70.

860 1, 2, 36, 840, 29680, 1429920, 90318144, 7237943552, 717442928640
3-LINE LATIN RECTANGLES. REF PLMS 31 336 28. BU2 33 125 41.

861 1, 2, 37, 329, 1501, 31354, 1451967, 39284461, 737652869
RELATED TO MENAGE NUMBERS. REF BU2 39 83 47.

862 1, 2, 44, 1008, 34432, 1629280, 101401344, 8030787968, 788377273856
RELATED TO LATIN RECTANGLES. REF BU2 33 125 41.

863 1, 2, 46, 406, 718, 832, 950, 1148, 1648, 1698, 3990, 39880, 59012, 65300, 89478, 317722
CLASS NUMBERS OF QUADRATIC FIELDS. REF MTAC 24 447 70.

864 1, 2, 46, 3362, 515086, 135274562, 54276473326, 30884386347362, 23657073914466766, 23471059057478981762, 2927935785185659513540 6
GENERALIZED TANGENT NUMBERS. REF MTAC 21 690 67.

865 1, 2, 46, 7970, 3487246, 2849229890, 3741386059246, 7205584123783010, 19133892392367261646, 67000387673723462963330
GENERALIZED EULER NUMBERS. REF MTAC 21 689 67.

866 1, 2, 48, 5824, 2887680, 5821595648
COUNTING BINARY MATRICES. REF JSIAM 20 377 71.

867 1, 2, 49, 629, 6961, 38366, 1899687, 133065253, 6482111309
RELATED TO 3-LINE LATIN RECTANGLES. REF BU2 39 72 47.

868 1, 2, 52, 142090700, 17844701940501123640681816160
INVERTIBLE BOOLEAN FUNCTIONS. REF PGEC 13 350 64.

869 1, 2, 56, 16256, 1073709056
COMPLETE POST FUNCTIONS. REF PLMS 16 191 66.

870 1, 2, 60, 836, 9576, 103326, 1106820
PERMUTATIONS BY NUMBER OF SEQUENCES. REF CO1 2 103.

871 1, 2, 88, 3056, 319616, 18940160, 283936226304
FROM HIGHER ORDER BERNOULLI NUMBERS. REF NO1 462.

872 1, 2, 136, 22377984, 768614354122719232, 354460798875983863749270670915141632
RELATIONS WITH THREE ARGUMENTS. REF OB1.

873 1, 2, 154, 2270394624
INVERTIBLE BOOLEAN FUNCTIONS. REF JACM 10 27 63.

874 1, 2, 720, 620448401733239439360000
FACTORIAL (FACTORIAL N). REF MTAC 24 231 70.

875 1, 2, 32896, 4029752732059759479357 44
RELATIONS WITH FOUR ARGUMENTS. REF OB1.

SEQUENCES BEGINNING 1, 3

876 1, 3, 0, 5, 3, 7, 8, 3, 15, 22, 15, 39, 35, 38, 72, 85, 111, 152, 175, 241, 308, 414, 551, 655, 897, 1164, 1463, 2001, 2538, 3286, 4296, 5503, 7259, 9357, 12147, 15910
FROM SYMMETRIC FUNCTIONS. REF PLMS 23 310 23.

877 1, 3, 0, 9, 5, 7, 12, 6, 15, 13, 3, 9, 17, 4, 21, 3, 23, 16, 21, 25, 15, 20, 1, 5, 27, 18, 30, 12, 19, 27, 35, 9, 37, 25, 39, 15, 2, 30, 24, 10, 29, 21, 39, 31, 3, 43, 40, 45, 15, 47, 48
QUADRATIC PARTITIONS OF PRIMES. REF CU2 1. LE1 55.

878 1, 3, 1, 2, 2, 4, 2, 6, 1, 8, 2, 10, 2, 5, 4, 14, 3, 16, 2, 7, 4, 20, 4, 10, 5, 18, 4, 26, 2, 28, 8, 16, 7, 8, 6, 34, 8, 20, 4, 38, 3, 40, 8, 12, 10, 44, 8, 28, 5, 30, 10, 50, 9, 16, 8, 33, 13
SOLUTIONS TO A SYSTEM OF CONGRUENCES. REF AMM 41 585 34.

879 1, 3, 1, 3, 11, 9, 8, 27, 37, 33, 67, 117, 131, 192, 341, 459, 613, 999, 1483, 2013, 3032, 4623, 6533, 9477, 14311, 20829, 30007, 44544, 65657, 95139, 139625, 206091
ASSOCIATED MERSENNE NUMBERS. REF EUR 11 22 49.

880 1, 3, 1, 4, 1, 5, 9, 2, 6, 5, 3, 5, 8, 9, 7, 9, 3, 2, 3, 8, 4, 6, 2, 6, 4, 3, 3, 8, 3, 2, 7, 9, 5, 0, 2, 8, 8, 4, 1, 9, 7, 1, 6, 9, 3, 9, 9, 3, 7, 5, 1, 0, 5, 8, 2, 0, 9, 7, 4, 9, 4, 4, 5, 9, 2, 3, 0, 7, 8
DIGITS OF PI. REF MTAC 16 80 62.

881 1, 3, 1, 5, 3, 7, 1, 9, 5, 11, 3, 13, 7, 15, 1, 17, 9, 19, 5, 21, 11, 23, 3, 25, 13, 27, 7, 29, 15, 31, 1, 33, 17, 35, 9, 37, 19, 39, 5, 41, 21, 43, 11, 45, 23, 47, 3, 49, 25, 51, 13, 53
REMOVE TWOS FROM N. REF FQ 6 52 68.

882 1, 3, 1, 5, 3, 15, 3, 20, 1, 1, 1, 32, 37, 22, 36, 8, 36, 10, 1, 7, 49, 48, 23, 77, 92, 81, 13, 95, 49, 1, 17, 95, 30, 96, 66, 132, 67, 107, 3, 50, 148, 25, 52, 175, 167, 109, 143, 201
FERMAT QUOTIENTS. REF BE4 35 666 13. LE1 10.

883 1, 3, 1, 5, 7, 3, 17, 11, 23, 45, 1, 91, 89, 93, 271, 85, 457, 627, 287, 1541, 967, 2115, 4049, 181, 8279, 7917, 8641, 24475, 7193, 41757, 56143, 27371, 139657, 84915
A(N) = -A(N - 1) - 2A(N - 2). REF JA2 82.

884 1, 3, 1, 6, 1, 7, 4, 8, 1, 16, 1, 10, 9, 15, 1, 21, 1, 22, 11, 14, 1, 36, 6, 16, 13, 28, 1, 42, 1, 31, 15, 20, 13, 55, 1, 22, 17, 50, 1, 54, 1, 40, 33, 26, 1, 76, 8, 43, 21, 46, 1, 66, 17
SUM OF THE DIVISORS OF N. REF AS1 840.

885 1, 3, 1, 11, 43, 19, 683, 2731, 331, 43691, 174763, 5419, 2796203, 251, 87211, 59, 715827883, 67, 281, 1777, 22366891, 83, 2932031007403, 18837001, 283
SMALLEST PRIMITIVE FACTOR OF 2(2N + 1) + 1. REF KR1 2 85.**

886 1, 3, 1, 11, 43, 19, 683, 2731, 331, 43691, 174763, 5419, 2796203, 4051, 87211, 3033169, 715827883, 20857, 86171, 25781083, 22366891, 8831418697
LARGEST FACTOR OF 2(2N + 1) + 1. REF KR1 2 85.**

887 1, 3, 1, 19, 25, 11, 161, 227, 681, 1019, 3057, 5075, 15225, 29291, 55105, 34243, 233801, 439259, 269201, 1856179, 3471385, 6219851, 1882337, 5647011
REMAINDER OF 3N / 2**N. REF JIMS 2 40 36. LE1 82.**

888 1, 3, 2, 1, 5, 23, 25, 27, 49, 74, 62, 85
GENERALIZED DIVISOR FUNCTION. REF PLMS 19 111 19.

889 1, 3, 2, 1, 7, 4, 1, 1, 8, 5, 2, 9, 8, 2, 1, 6, 8, 5, 2, 3, 8, 5, 4, 8, 5, 9, 9, 7, 0, 9, 4, 3, 5, 2, 2, 3, 3, 8, 5, 4, 3, 6, 6, 2, 0, 6, 2, 4, 8, 3, 7, 3, 4, 8, 7, 3, 1, 2, 3, 7, 5, 9, 2, 5, 6, 0, 6, 2, 2
MIX PI AND E. REF EUR 13 11 50.

890 1, 3, 2, 1, 9, 5, 8, 3, 1, 19, 10, 7, 649, 15, 4, 1, 33, 17, 170, 9, 55, 197, 24, 5, 1, 51, 26, 127, 9801, 11, 1520, 17, 23, 35, 6, 1, 73, 37, 25, 19, 2049, 13, 3482, 199, 161
SOLUTIONS OF PELLIANS. REF DE1. CAY 13 430. LE1 55.

891 1, 3, 2, 5, 5, 4, 2, 9, 5, 8, 5, 13, 12, 8, 5, 17, 8, 6, 11, 14, 11, 23, 7, 23, 26, 11, 16, 14, 15, 31, 10, 28, 16, 24, 15, 37, 9, 39, 16, 20, 27, 20, 31, 14, 43, 47, 23, 32, 20, 51, 17
RELATED TO PERFECT POWERS. REF FQ 8 268 70.

892 1, 3, 2, 45, 72, 105, 6480, 42525, 22400, 56133, 32659200, 7882875
RELATED TO CHEBYSHEV INTEGRATION FORMULA. REF JM2 26 192 47.

893 1, 3, 2, 115, 11, 5887, 151, 259723, 15619, 381773117, 655177, 20646903199, 27085381, 467168310097, 2330931341
VALUES OF AN INTEGRAL. REF PHM 36 295 45. MTAC 19 114 65.

894 1, 3, 3, 1, 3, 5, 3, 7, 1, 9, 9, 5, 3, 9, 9, 3, 11, 1, 9, 11, 7, 15, 15, 13, 3, 15, 9, 11, 17, 5, 13, 7, 3, 15, 19, 3, 11, 9, 19, 21, 21, 13, 15, 21, 7, 3, 19, 23, 15, 21, 11, 17, 3, 9, 23, 15
QUADRATIC PARTITIONS OF PRIMES. REF CU2 1. LE1 55. MTAC 23 459 69.

895 1, 3, 3, 1, 3, 6, 3, 0, 3, 6, 6, 3, 1, 6, 6, 0, 3, 9, 6, 3, 6, 6, 3, 0, 3, 9, 12, 4, 0, 12, 6, 0, 3, 6, 9, 6, 6, 6, 9, 0, 6, 15, 6, 3, 3, 12, 6, 0, 1, 9, 15, 6, 6, 12, 12, 0, 6, 6, 6, 9, 0, 12, 12, 0, 3
POPULATION OF U2 + V**2 + W**2. REF PNISI 13 39 47.**

896 1, 3, 3, 1, 5, 3, 1, 7, 5, 3, 5, 3, 5, 5, 3, 7, 1, 11, 5, 13, 9, 3, 7, 5, 15, 7, 13, 11, 3, 3, 19, 3, 5, 19, 9, 3, 17, 9, 21, 15, 5, 7, 7, 25, 7, 9, 3, 21, 5, 3, 9, 5, 7, 25, 13, 5, 13, 3, 23, 11
CLASS NUMBERS H(-P) FOR PRIMES P = 4N - 1. REF MTAC 23 458 69.

897 1, 3, 3, 2, 48, 362, 49711, 13952
FROM A HYPERGEOMETRIC FUNCTION. REF JACM 3 14 56.

898 1, 3, 3, 3, 4, 5, 5, 6, 6, 7, 9, 9, 10, 11, 11, 12, 13, 14, 15, 16, 17, 17, 18, 19, 20, 21, 22, 22, 23, 23, 24, 26, 27, 28, 28, 30, 30, 31, 32, 33, 34, 35, 35, 36, 37, 37, 38, 39, 41, 42
COMPRESSED PRIMES. REF AMM 74 43 67.

899 1, 3, 3, 3, 5, 3, 3, 5, 3, 3, 5, 3, 5, 7, 3, 3, 5, 7, 3, 5, 3, 3, 5, 3, 5, 7, 3, 5, 7, 3, 3, 5, 7, 3, 5, 3, 3, 5, 7, 3, 5, 3, 5, 7, 3, 5, 7, 19, 3, 5, 3, 3, 5, 3, 3, 5, 3, 5, 7, 13, 11, 13, 19, 3, 5, 3, 5
RELATED TO GOLDBACH CONJECTURE. REF FVS 4(4) 7 27. LE1 80.

900 1, 3, 3, 5, 5, 7, 5, 7, 7, 11, 11, 13, 11, 13, 13, 17, 17, 19, 17, 19, 13, 23, 19, 19, 23, 23, 19, 29, 29, 31, 23, 29, 31, 29, 31, 37, 29, 37, 37, 41, 41, 43, 41, 43, 31, 47, 43, 37, 47
RELATED TO GOLDBACH CONJECTURE. REF FVS 4(4) 7 27. LE1 80.

901 1, 3, 3, 5, 7, 11, 17, 27, 43, 69, 111, 179, 289, 467, 755, 1221, 1975, 3195, 5169, 8363, 13531, 21893, 35423, 57315, 92737, 150051, 242787, 392837, 635623, 1028459
A(N) = A(N - 1) + A(N - 2) - 1. REF FQ 5 288 67.

902 1, 3, 3, 5, 9, 21, 21, 81, 81, 81, 243, 243, 441, 1215, 1701, 1701, 6561, 6561, 6561, 45927, 45927, 45927, 137781, 137781, 229635, 1594323
LARGEST GROUP OF A TOURNAMENT. REF MO1 81.

903 1, 3, 3, 7, 4, 2, 30, 1, 8, 3, 1, 1, 1, 9, 2, 2, 1, 3, 22986, 2, 1, 32, 8, 2, 1, 8, 55, 1, 5, 2, 28, 1, 5, 1, 1501790
AN EXOTIC CONTINUED FRACTION. REF AT1 21.

904 1, 3, 3, 7, 6, 12, 13, 20, 21, 34, 36, 51, 58, 78, 89, 118, 131, 171, 197, 245, 279, 349, 398, 486, 557, 671, 767, 920, 1046, 1244, 1421, 1667, 1898, 2225, 2525, 2937, 3333
MOCK THETA NUMBERS. REF TAMS 72 495 52.

905 1, 3, 3, 9, 15, 38, 73, 174, 380
HYDROCARBONS. REF BS1 201.

906 1, 3, 3, 15, 30, 101, 261, 807, 2308, 7065, 21171
PARTITION FUNCTION FOR CUBIC LATTICE. REF PCPS 47 425 51.

907 1, 3, 4, 5, 6, 8, 9, 10, 11, 12, 14, 15
WYTHOFF GAME. REF CMB 2 189 59.

908 1, 3, 4, 5, 6, 8, 9, 10, 12, 13, 14, 16
WYTHOFF GAME. REF CMB 2 188 59.

909 1, 3, 4, 5, 6, 8, 10, 12, 17, 21, 23, 28, 32, 34, 39, 43, 48, 52, 54, 59, 63, 68, 72, 74, 79, 83, 98, 99, 101, 110, 114, 121, 125, 132, 136, 139, 143, 145, 152, 161, 165, 172, 176
A SELF-GENERATING SEQUENCE. REF UL1 IX.

910 1, 3, 4, 5, 7, 9, 14, 18, 24, 31, 43, 55, 72, 94, 123, 156, 200, 254, 324, 408, 513, 641, 804, 997, 1236, 1526, 1883, 2308, 2829, 3451, 4209, 5109, 6194, 7485, 9038, 10871
REPRESENTATIONS OF THE ALTERNATING GROUP. REF CJM 4 383 52.

911 1, 3, 4, 5, 7, 11, 13, 17, 23, 29, 43, 47, 83, 131, 137, 359, 431, 433, 449, 509, 569, 571
PRIME FIBONACCI NUMBERS. REF MTAC 23 213 69.

912 1, 3, 4, 5, 8, 10, 7, 9, 18, 24, 14, 30, 19, 20, 44, 16, 27, 58, 15, 68, 70, 37, 78, 84, 11, 49, 50, 104, 36, 27, 19, 128, 130, 69, 46, 37, 50, 79, 164, 168, 87, 178, 90, 190, 97, 99
FIBONACCI ENTRY POINTS. REF JA2 7. MTAC 20 618 66.

913 1, 3, 4, 5, 9, 15, 27, 50, 92, 171, 322, 610, 1161, 2220, 4260, 8201, 15828, 30622, 59362, 115287, 224260, 436871, 852161, 1664196, 3253531, 6366973, 12471056
POPULATION OF U2 + V**2. REF MTAC 18 79 64.**

914 1, 3, 4, 6, 5, 12, 8, 6, 12, 15, 10, 12, 7, 24, 20, 12, 9, 12, 18, 30, 8, 30, 24, 12, 25, 21, 36, 24, 14, 60, 30, 24, 20, 9, 40, 12, 19, 18, 28, 30, 20, 24, 44, 30, 60, 24, 16, 12, 56
FIBONACCI ENTRY POINTS. REF HM1. MTAC 23 459 69. ACA 16 109 69.

915 1, 3, 4, 6, 6, 12, 8, 12, 12, 18, 12, 24, 14, 24, 24, 24, 18, 36, 20, 36, 32, 36, 24, 48, 30, 42, 36, 48, 30, 72, 32, 48, 48, 54, 48, 72, 38, 60, 56, 72, 42, 96, 44, 72, 72, 72, 48, 96
RELATED TO A MODULAR GROUP. REF NBS 67B 62 63.

916 1, 3, 4, 6, 7, 8, 12, 13, 14, 15, 18, 20, 24, 28, 30, 31, 32, 36, 38, 39, 40, 42, 44, 48, 54, 56, 57, 60, 62, 63, 68, 72, 74, 78, 80, 84, 90, 91, 93, 96, 98, 102, 104, 108, 110, 112
VALUES OF A DIVISOR FUNCTION. REF BA2 85.

917 1, 3, 4, 6, 8, 9, 11, 12, 14, 16, 17, 19, 21, 22, 24, 25, 27, 29, 30, 32, 33, 35, 37, 38, 40, 42, 43, 45, 46, 48, 50, 51, 53, 55, 56, 58, 59, 61, 63, 64, 66, 67, 69, 71, 72, 74, 76, 77
A BEATTY SEQUENCE. REF CMB 2 191 59. AMM 72 1144 65.

918 1, 3, 4, 6, 8, 9, 11, 12, 14, 16, 17, 19, 21, 22, 24, 25, 27, 29, 30, 32, 34, 35, 37, 38, 40, 42, 43, 45, 47, 48, 50, 51, 53, 55, 56, 58, 60, 61, 63, 64, 66, 68, 69, 71, 73, 74, 76, 77
A CURIOUS SEQUENCE. REF FQ 1(4) 50 63.

919 1, 3, 4, 6, 8, 9, 11, 13, 15, 17, 19, 20, 22, 26, 28, 30, 31, 33, 35, 37, 39, 41, 43, 45, 48, 50, 52, 54, 56, 58, 62, 64, 65, 67, 69, 71, 73, 75, 79, 81, 83, 85, 86, 90, 92, 94, 96, 98
POPULATION OF U2 + V**2. REF PURB 20 14 52.**

920 1, 3, 4, 6, 11, 45, 906, 409182, 83762797735
RELATED TO HAMILTON NUMBERS. REF SY1 4 551.

921 1, 3, 4, 7, 6, 12, 8, 15, 13, 18, 12, 28, 14, 24, 24, 31, 18, 39, 20, 42, 32, 36, 24, 60, 31, 42, 40, 56, 30, 72, 32, 63, 48, 54, 48, 91, 38, 60, 56, 90, 42, 96, 44, 84, 78, 72, 48, 124
SUM OF THE DIVISORS OF N. REF AS1 840.

922 1, 3, 4, 7, 8, 11, 15, 19, 20, 24, 35, 40, 43, 51, 52, 67, 84, 88, 91, 115, 120, 123, 132, 148, 163, 168, 187, 195, 228, 232, 235, 267, 280, 312, 340, 372, 403, 408, 420, 427
DISCRIMINANTS. REF BO1 426.

923 1, 3, 4, 7, 10, 50
GRAPHS BY CUTTING CENTER. REF CSA 149.

924 1, 3, 4, 7, 11, 18, 29, 47, 76, 123, 199, 322, 521, 843, 1364, 2207, 3571, 5778, 9349, 15127, 24476, 39603, 64079, 103682, 167761, 271443, 439204, 710647, 1149851
LUCAS NUMBERS A(N) = A(N - 1) + A(N - 2). REF HW1 148. HO1.

925 1, 3, 4, 7, 11, 29, 40, 109, 912, 1021, 26437, 27458, 163727, 191185, 4369797, 4560982, 40857653, 45418635, 86276288, 821905227, 908181515, 1730086742
CONVERGENTS TO FIFTH ROOT OF 5. REF AMP 46 116 1866. LE1 67. HPR.

926 1, 3, 4, 8, 9, 11, 13, 18, 19, 24, 27, 28, 29, 33, 35, 40, 43, 44, 51, 59, 61, 63, 67, 68, 75, 83, 88, 91, 92, 93, 98, 100, 104, 107, 108, 109, 115, 120, 121, 123, 125, 126, 129
ELLIPTIC CURVES. REF JRAM 212 24 63.

927 1, 3, 4, 8, 11, 18, 24, 36, 47, 66, 84, 113, 141, 183, 225, 284, 344, 425, 508, 617, 729, 872, 1020, 1205, 1397, 1632, 1877, 2172, 2480, 2846, 3228, 3677
EXPANSION OF A GENERATING FUNCTION. REF CAY 10 415.

928 1, 3, 4, 8, 14, 25, 45, 82, 151, 282, 531, 1003, 1907, 3645, 6993, 13456, 25978, 50248, 97446, 189291, 368338, 717804, 1400699, 2736534, 5352182, 10478044
POPULATION OF U2 + 3V**2. REF MTAC 20 560 66.**

929 1, 3, 4, 8, 65536, (THERE IS NO ROOM TO DESCRIBE THE NEXT TERM)
ACKERMANNS SEQUENCE. REF AMM 70 133 63.

930 1, 3, 4, 9, 12, 23, 31, 54, 73, 118, 159, 246, 329, 489, 651, 940, 1242, 1751, 2298, 3177, 4142, 5630, 7293, 9776, 12584, 16659, 21320, 27922, 35532, 46092, 58342, 75039
TYPES OF ROOTS OF AN EQUATION. REF AMM 76 194 69.

931 1, 3, 4, 9, 12, 23, 31, 54, 73, 118, 159, 246, 340, 500, 684, 984, 1341, 1883
COEFFICIENTS OF MODULAR FUNCTIONS. REF PLMS 9 387 59.

932 1, 3, 4, 9, 13, 26, 40, 74, 118, 210, 342, 595, 981, 1684, 2798, 4763
PARAFFINS. REF JACS 54 1105 32.

933 1, 3, 4, 9, 14, 27, 48, 93, 163, 315, 576, 1085
SEQUENCES RELATED TO TRANSFORMATIONS ON THE UNIT INTERVAL. REF ME1.

934 1, 3, 4, 11, 15, 41, 56, 153, 209, 571, 780, 2131, 2911, 7953, 10864, 29681, 40545, 110771, 151316, 413403, 564719, 1542841, 2107560, 5757961, 7865521
A(2N) = A(2N − 1) + A(2N − 2), A(2N + 1) = 2A(2N) + A(2N − 1). REF MQET 1 10 16. NZ1 181.

935 1, 3, 4, 11, 16, 30, 50, 91, 157, 278, 485, 854, 1496, 2628, 4609, 8091, 14196, 24915, 43720, 76726, 134642, 236283, 414645, 727654, 1276941, 2240878, 3932464
A FIELDER SEQUENCE. REF FQ 6(3) 69 68.

936 1, 3, 4, 11, 20, 51, 108, 267, 619
BI-CENTERED TREES. REF CAY 9 438.

937 1, 3, 4, 11, 21, 36, 64, 115, 211, 383, 694, 1256, 2276, 4126, 7479, 13555, 24566, 44523, 80694, 146251, 265066, 480406, 870689, 1578040, 2860046, 5183558, 9394699
A FIELDER SEQUENCE. REF FQ 6(3) 69 68.

938 1, 3, 4, 11, 21, 42, 71, 131, 238, 443, 815, 1502, 2757, 5071, 9324, 17155, 31553, 58038, 106743, 196331, 361106, 664183, 1221623, 2246918, 4132721, 7601259
A FIELDER SEQUENCE. REF FQ 6(3) 69 68.

939 1, 3, 4, 11, 136, 283, 419, 1121, 1540, 38081, 39621, 117323, 156944, 431211, 5331476, 11094163, 16425639, 43945441, 60371080, 1492851361, 1553222441
A CONTINUED FRACTION. REF IC 13 623 68.

940 1, 3, 4, 12, 22, 71, 181, 618, 1957, 6966
NECKLACES. REF IJM 5 664 61.

941 1, 3, 4, 12, 24, 66, 160, 448, 1186, 3334, 9235, 26166, 73983, 211297
TRIANGULAR POLYOMINOES. REF HA1 37. JRM 2 216 69. LU5.

942 1, 3, 4, 12, 27, 82, 228, 733, 2282, 7528, 24834, 83898, 285357, 983244, 3412420, 11944614, 42080170, 149197152, 531883768, 1905930975, 6861221666
DISSECTIONS OF A POLYGON. REF BAMS 54 355 48. CMB 6 175 63. GU1. MAT 15 121 68.

943 1, 3, 4, 12, 28, 94, 298, 1044, 3658, 13164
NECKLACES. REF IJM 5 664 61.

944 1, 3, 4, 13, 53, 690, 36571, 25233991, 922832284862, 23286741570717144243, 21489756930695820973683319349467
A(N) = A(N − 1)A(N − 2) + 1. REF EUR 19 13 57.

945 1, 3, 4, 23, 27, 50, 227, 277, 504, 4309, 4813, 71691, 76504, 836731, 1749966, 2586697, 12096754, 147747745, 307592244, 1070524477, 2448641198, 3519165675
CONVERGENTS TO CUBE ROOT OF 2. REF AMP 46 105 1866. LE1 67. HPR.

946 1, 3, 5, 3, 9, 3, 51, 675, 5871
FROM DISCORDANT PERMUTATIONS. REF KYU 10 11 56.

947 1, 3, 5, 3, 17, 3, 5, 3, 257, 3, 5, 3, 17, 3, 5, 3, 65537, 3, 5, 3, 17, 3, 5, 3, 97, 3, 5, 3, 17, 3, 5, 3, 641, 3, 5, 3, 17, 3, 5, 3, 257, 3, 5, 3, 17, 3, 5, 3, 193, 3, 5, 3, 17, 3, 5, 3, 257, 3
SMALLEST FACTOR OF 2N + 1. REF AJM 1 239 1878.**

948 1, 3, 5, 3, 17, 11, 13, 43, 257, 19, 41, 683, 241, 2731, 113, 331, 65537, 43691, 109, 174763, 61681, 5419, 2113, 2796203, 673, 4051, 1613, 87211, 15790321, 3033169
LARGEST FACTOR OF 2N + 1. REF AJM 1 239 1878.**

949 1, 3, 5, 6, 8, 9, 10, 12, 14, 16, 17, 24, 27, 31, 32, 33, 34, 36, 37, 41, 42, 46, 52, 62, 68, 69, 70, 73, 77, 78, 80, 82, 86, 88, 90, 92, 96, 97, 98, 99, 103, 108, 111, 114, 117, 119
ELLIPTIC CURVES. REF JRAM 212 23 63.

950 1, 3, 5, 6, 8, 10, 12, 13, 15, 17, 18, 20, 22, 24, 25, 27, 29, 30, 32, 34, 36, 37, 39, 41, 42, 44, 46, 48, 49, 51, 53, 54, 56, 58, 60, 61, 63, 65, 67, 68, 70, 72, 73, 75, 77, 79, 80
A BEATTY SEQUENCE. REF CMB 3 21 60.

951 1, 3, 5, 6, 8, 11, 12, 14, 17, 20, 29, 41, 44, 59, 62, 71, 92, 101, 107, 116, 137, 149, 164, 179, 191, 197, 212, 227, 239, 254, 269, 281, 311, 332, 347, 356, 419, 431, 452, 461
RELATED TO EULERS TOTIENT FUNCTION. REF AMM 56 22 49.

952 1, 3, 5, 6, 9, 10, 12, 15, 17, 18, 20, 23, 24, 27, 29, 30, 33, 34, 36, 39, 40, 43, 45, 46, 48, 51, 53, 54, 57, 58, 60, 63, 65, 66, 68, 71, 72, 75, 77, 78, 80, 83, 85, 86, 89, 90, 92
EVEN NUMBER OF ONES IN BINARY EXPANSION. REF CMB 2 86 59.

953 1, 3, 5, 6, 9, 13, 20, 31, 49, 78, 125, 201, 324, 523, 845, 1366, 2209, 3573, 5780, 9351, 15129, 24478, 39605, 64081, 103684, 167763, 271445, 439206, 710649, 1149853
A(N) = A(N - 1) + A(N - 2) - 2. REF SMA 20 23 54. R1 233. JCT 7 292 69.

954 1, 3, 5, 7, 10, 13, 16, 19, 22, 26, 30
STEINHAUS SORTING PROBLEM. REF AMM 66 389 59. WE2 207.

955 1, 3, 5, 7, 11, 13, 17, 19, 23, 29, 31, 105, 165, 195, 231, 255, 273, 285, 345, 357, 385, 399, 429, 435, 455, 465, 483, 561, 595, 609, 627, 651, 663, 665, 715, 741, 759, 805
RELATED TO LIOUVILLES FUNCTION. REF JIMS 7 71 43.

956 1, 3, 5, 7, 11, 13, 17, 19, 23, 31, 43, 61, 79
(2P + 1)/3 IS PRIME. REF MMAG 27 157 54.**

957 1, 3, 5, 7, 15, 11, 13, 17, 19, 25, 23, 35, 29, 31, 51, 37, 41, 43, 69, 47, 65, 53, 81, 87, 59, 61, 85, 67, 71, 73, 79, 123, 83, 129, 89, 141, 97, 101, 103, 159, 107, 109, 121, 113
INVERSE OF EULER TOTIENT FUNCTION. REF BA2 64.

958 1, 3, 5, 7, 17, 19, 37, 97, 113, 257, 401, 487, 631, 971, 1297, 1801, 19457, 22051, 28817, 65537, 157303, 160001
A SPECIAL SEQUENCE OF PRIMES. REF ACA 5 425 59.

959 1, 3, 5, 7, 17, 29, 47, 61, 73, 83, 277, 317, 349, 419, 503, 601, 709, 829
FROM A GOLDBACH CONJECTURE. REF BIT 6 49 66.

960 1, 3, 5, 7, 19, 21, 43, 81, 125, 127, 209, 211
11.2N + 1 IS PRIME. REF PAMS 9 674 58.**

961 1, 3, 5, 7, 32, 11, 13, 17, 19, 25, 23, 224, 29, 31, 128, 37, 41, 43, 115, 47, 119, 53, 81, 928, 59, 61, 256, 67, 71, 73, 79, 187, 83, 203, 89, 209, 235, 97, 101
INVERSE OF REDUCED TOTIENT FUNCTION. REF NAM 17 305 1898. LE1 7.

962 1, 3, 5, 8, 9, 13, 15, 18, 19, 20, 21, 24, 28, 29, 31, 35, 37, 40, 47, 49, 51, 53, 56, 60, 61, 67, 69, 77, 79, 83, 84, 85, 88, 90, 92, 93, 95, 98, 100, 101, 104, 109, 111, 115, 120
ELLIPTIC CURVES. REF JRAM 212 25 63.

963 1, 3, 5, 8, 11, 14, 17, 21, 25, 29, 33, 37, 41, 45, 49, 54, 59, 64, 69, 74, 79, 84, 89, 94, 99, 104, 109, 114, 119, 124, 129, 135, 141, 147, 153, 159, 165, 171, 177, 183, 189
RELATED TO PERMUTATION NETWORKS. REF AFI 32 519 68.

964 1, 3, 5, 8, 11, 15, 19, 23, 27, 32, 36, 42, 47, 52, 58, 64, 70, 76, 83, 89, 96, 103, 110, 118, 125, 133, 140, 148, 156, 164, 173, 181, 190, 198, 207, 216, 225, 234, 244, 253, 263
N(3/2). REF BO3 46. LF1 17. AT1 177.**

965 1, 3, 5, 8, 11, 15, 19, 23, 28, 33, 38, 44, 50, 56, 62, 69, 76, 83, 90, 98, 106, 114, 122, 131, 140, 149, 158, 167, 177, 187, 197, 207, 217, 228, 239, 250, 261, 272, 284, 296
FROM A SELF-GENERATING SEQUENCE. REF AMM 74 740 67.

966 1, 3, 5, 8, 12, 16, 21, 27, 33, 40, 48, 56, 65, 75, 85, 96, 108, 120, 133, 147, 161, 176, 192, 208, 225, 243, 261, 280, 300, 320, 341, 363, 385, 408, 432, 456, 481, 507, 533
(N2)/3**

967 1, 3, 5, 8, 12, 17, 23, 30, 37, 45, 54
RATIONAL POINTS IN A QUADRILATERAL. REF CR 265 161 67.

968 1, 3, 5, 8, 12, 18, 24, 30, 36, 42, 52, 60, 68, 78, 84, 90, 100, 112, 120, 128, 138, 144, 152, 162, 172, 186, 198, 204, 210, 216, 222, 240, 258, 268, 276, 288, 300, 308, 320
SUMS OF SUCCESSIVE PRIMES. REF EUR 26 12 63.

969 1, 3, 5, 8, 12, 18, 24, 33, 43, 55, 69, 86, 104, 126, 150, 177, 207, 241
RESTRICTED PARTITIONS. REF CAY 2 278.

970 1, 3, 5, 8, 20, 12, 9, 28, 11, 48, 39, 65, 20, 60, 15, 88, 51, 85, 52, 19, 95, 28, 60, 105, 120, 32, 69, 115, 160, 68, 25, 75, 175, 180, 225, 252, 189, 228, 40, 120, 29, 145, 280
QUADRATIC PARTITIONS OF PRIME-SQUARES. REF CU3 77. LE1 60.

971 1, 3, 5, 9, 11, 15, 19, 25, 29, 35, 39, 45, 49, 51, 59, 61, 65, 69, 71, 79, 85, 95, 101, 121, 131, 139, 141, 145, 159, 165, 169, 171, 175, 181, 195, 199, 201, 205, 209, 219, 221
(N2 + 1)/2 IS PRIME. REF EUL (1) 3 24 17.**

972 1, 3, 5, 9, 13, 17, 21, 27, 33, 41, 47, 55, 65
POSTAGE STAMP PROBLEM. REF CJ1 12 379 69.

973 1, 3, 5, 9, 13, 22, 30, 45, 61, 85, 111
EXPANSION OF A GENERATING FUNCTION. REF CAY 10 415.

974 1, 3, 5, 9, 15, 25, 41, 67, 109, 177, 287, 465, 753, 1219, 1973, 3193, 5167, 8361, 13529, 21891, 35421, 57313, 92735, 150049, 242785, 392835, 635621, 1028457
A(N) = A(N − 1) + A(N − 2) + 1. REF FQ 8 267 70.

975 1, 3, 5, 9, 17, 31, 57, 105, 193, 355, 653, 1201, 2209, 4063, 7473, 13745, 25281, 46499, 85525, 157305, 289329, 532159, 978793, 1800281, 3311233, 6090307, 11201821
A(N) = A(N − 1) + A(N − 2) + A(N − 3). REF FQ 1(3) 72 63, 2 260 64.

976 1, 3, 5, 10, 13, 26, 25, 50, 49, 73, 81, 133, 109, 196, 169, 241, 241, 375, 289, 476, 421, 568, 529, 806, 577, 1001, 833, 1081, 1009, 1393, 1081, 1768, 1441, 1849, 1633
GENUS OF MODULAR GROUPS. REF GU6 15.

977 1, 3, 5, 10, 14, 21, 26, 36, 43, 55, 64, 78, 88, 105, 117, 136, 150, 171, 186, 210, 227, 253, 272, 300, 320, 351, 373, 406, 430, 465, 490, 528, 555, 595, 624, 666, 696, 741
RELATED TO ZARANKIEWICZS PROBLEM. REF TI1 126.

978 1, 3, 5, 10, 16, 29, 45, 75, 115, 181, 271, 413, 605, 895, 1291, 1866, 2648, 3760, 5260, 7352, 10160, 14008, 19140, 26085, 35277, 47575, 63753, 85175, 113175, 149938
2-LINE PARTITIONS. REF DMJ 31 272 64.

979 1, 3, 5, 10, 25, 64, 160, 390, 940, 2270, 5515, 13440, 32735, 79610, 193480, 470306
RELATED TO PARTITIONS OF A NUMBER. REF AMM 76 1034 69.

980 1, 3, 5, 10, 32, 382, 15768919
BOOLEAN FUNCTIONS. REF JACM 13 154 66.

981 1, 3, 5, 11, 13, 19, 29, 37, 53, 59, 61, 67, 83, 101, 107, 131, 139, 149, 163, 173, 179, 181, 197, 211, 227, 269, 293, 317, 347, 349, 373, 379, 389, 419, 421, 443, 461, 467
PRIMES WITH 2 AS PRIMITIVE ROOT. REF KR1 1 56. AS1 864.

982 1, 3, 5, 11, 17, 29, 41, 59, 71, 101, 107, 137, 149, 179, 191, 197, 227, 239, 269, 281, 311, 347, 419, 431, 461, 521, 569, 599, 617, 641, 659, 809, 821, 827, 857, 881, 1019
PRIME PAIRS. REF EUR 18 17 55. AS1 870.

983 1, 3, 5, 11, 21, 43, 85, 171, 341, 683, 1365, 2731, 5461, 10923, 21845, 43691, 87381, 174763, 349525, 699051, 1398101, 2796203, 5592405, 11184811, 22369621
A(N) = A(N - 1) + 2A(N - 2). REF NCM 6 146 1880. EUR 26 12 63.

984 1, 3, 5, 11, 29, 97, 127, 541, 907, 1151, 1361, 9587, 15727, 19661, 31469, 156007, 360749, 370373, 492227, 1349651, 1357333, 2010881, 4652507, 17051887
INCREASING GAPS BETWEEN PRIMES. REF MTAC 18 649 64.

985 1, 3, 5, 12, 30, 79, 227, 710, 2322, 8071, 29503, 112822, 450141
CONNECTED GRAPHS BY LINES. REF PRV 164 801 67. ST1.

986 1, 3, 5, 13, 17, 241, 257, 65281, 65537
AN INFINITE COPRIME SEQUENCE. REF MAG 48 420 64.

987 1, 3, 5, 13, 27, 66, 153, 377, 914, 2281, 5690, 14397, 36564, 93650, 240916, 623338, 1619346, 4224993
HYDROCARBONS. REF JACS 55 685 33, 56 157 34.

988 1, 3, 5, 15, 17, 51, 85, 255, 257, 771, 1285, 3855, 4369, 13107, 21845, 65535, 65537
RELATED TO PARITY OF BINOMIAL COEFFICIENTS. REF GO3.

989 1, 3, 5, 17, 41, 127, 365, 1119, 3413, 10685, 33561, 106827, 342129, 1104347, 3584649, 11701369, 38374065, 126395259
RELATED TO SERIES-PARALLEL NUMBERS. REF JM2 21 87 42.

990 1, 3, 5, 17, 257, 65537, 4294967297, 18446744073709551617, 340282366920938463463374607431768211457
FERMAT NUMBERS 2(2**N) + 1. REF HW1 14.**

991 1, 3, 5, 21, 41, 49, 89, 133, 141, 165, 189, 293, 305, 395, 651, 665, 771, 801, 923, 953
19.2N - 1 IS PRIME. REF MTAC 22 421 68.**

992 1, 3, 5, 35, 63, 231, 429, 6435, 12155, 46189, 88179, 676039, 1300075, 5014575, 9694845, 300540195, 583401555, 2268783825, 4418157975, 34461632205
COEFFICIENTS OF LEGENDRE POLYNOMIALS. REF PHM 33 13 42. MTAC 3 17 48. RG1 414.

993 1, 3, 5, 691, 35, 3617, 43867, 1222277, 854513, 1181820455, 76977927, 23749461029, 8615841276005, 84802531453387, 90219075042845
RELATED TO BERNOULLI NUMBERS. REF EUL (1) 15 93 27. FMR 1 73.

994 1, 3, 6, 6, 10, 16, 28, 28, 28, 28, 28, 28, 28
EQUIANGULAR LINES. REF KNAW 69 336 66. SE3.

995 1, 3, 6, 9, 9, 0, 27, 81, 162, 243, 243
EXPANSION OF BRACKET FUNCTION. REF FQ 2 254 64.

996 1, 3, 6, 9, 13, 17, 22, 27, 32, 37, 43, 49, 56, 63, 70, 77, 85, 93, 102
PRIMES IN A SEQUENCE OF DIFFERENCES. REF IDM 7 136 1900.

997 1, 3, 6, 9, 13, 18, 24, 31
RATIONAL POINTS IN A QUADRILATERAL. REF CR 265 161 67.

998 1, 3, 6, 9, 14, 18, 23
RAMSEY NUMBERS. REF RYS 42. CO1 2 134.

999 1, 3, 6, 9, 15, 18, 27, 30, 45, 42, 66
COMPOSITIONS INTO RELATIVELY PRIME PARTS. REF FQ 2 250 64.

1000 1, 3, 6, 9, 15, 25, 34, 51, 73, 97, 132, 178, 226, 294, 376, 466, 582, 722, 872, 1062, 1282, 1522, 1812, 2147, 2507, 2937, 3422, 3947, 4557, 5243, 5978, 6825, 7763, 8771
A GENERALIZED PARTITION FUNCTION. REF PNISI 17 237 51.

1001 1, 3, 6, 10, 13, 17, 20, 23, 27, 30, 34, 37, 40
A BEATTY SEQUENCE. REF CMB 2 188 59.

1002 1, 3, 6, 10, 15, 21, 28, 36, 45, 55, 66, 78, 91, 105, 120, 136, 153, 171, 190, 210, 231, 253, 276, 300, 325, 351, 378, 406, 435, 465, 496, 528, 561, 595, 630, 666, 703, 741
TRIANGULAR NUMBERS OR BINOMIAL COEFFICIENTS N(N + 1)/2. REF DI2 2 1. RS1. BE3 189. AS1 828.

1003 1, 3, 6, 10, 30, 126, 448, 1296, 4140, 17380, 76296, 296088, 1126216, 4940040, 23904000, 110455936, 489602448, 2313783216, 11960299360, 61878663840
PERMUTATIONS OF ORDER TWO. REF CJM 7 167 55.

1004 1, 3, 6, 11, 17, 26, 35, 45, 58, 73, 90, 106, 123, 146, 168, 193, 216, 243, 271, 302, 335, 365, 402, 437, 473, 516, 557, 600, 642, 687, 736, 782, 835, 886, 941, 999, 1050
POPULATION OF U2 + V**2. REF PNISI 13 37 47.**

1005 1, 3, 6, 11, 18, 27, 39, 54, 72, 94, 120, 150, 185, 225, 270, 321, 378, 441, 511, 588
HYDROCARBONS. REF JACS 55 684 33.

1006 1, 3, 6, 11, 19, 32, 48, 71, 101, 141, 188, 249, 322, 414, 518, 645, 791, 966
RESTRICTED PARTITIONS. REF CAY 2 278.

1007 1, 3, 6, 11, 19, 32, 53, 87, 142, 231, 375, 608, 985, 1595, 2582, 4179, 6763, 10944, 17709, 28655, 46366, 75023, 121391, 196416, 317809, 514227, 832038, 1346267
A SIMPLE RECURRENCE. REF R1 233.

1008 1, 3, 6, 11, 24, 51, 130, 315, 834, 2195, 5934, 16107, 44368, 122643, 341802, 956635, 2690844, 7596483, 21524542, 61171659, 174342216, 498112275, 1426419858
NECKLACES OF 3 COLORS. REF R1 162. IJM 5 658 61.

1009 1, 3, 6, 11, 24, 69, 227, 753, 2451, 8004, 27138, 97806, 375313, 1511868, 6292884, 26826701, 116994453, 523646202, 2414394601, 11487130362, 56341183365
FROM A DIFFERENTIAL EQUATION. REF AMM 67 766 60.

1010 1, 3, 6, 12, 20, 32, 49, 73, 102, 141, 190, 252, 325, 414, 521, 649, 795, 967
RESTRICTED PARTITIONS. REF CAY 2 278.

1011 1, 3, 6, 12, 21, 40, 67, 117, 193, 319, 510, 818, 1274, 1983, 3032, 4610, 6915, 10324, 15235, 22371, 32554, 47119, 67689, 96763, 137404, 194211, 272939, 381872
3-LINE PARTITIONS. REF DMJ 31 272 64.

1012 1, 3, 6, 12, 24, 48, 90, 168, 318, 600, 1098, 2004, 3696, 6792, 12270, 22140, 40224, 72888, 130650, 234012, 421176, 756624, 1348998, 2403840, 4299018
SUSCEPTIBILITY FOR HONEYCOMB. REF PHA 28 931 62.

1013 1, 3, 6, 12, 24, 48, 90, 174, 336, 648, 1218, 2328, 4416, 8388, 15780, 29892, 56268, 106200, 199350, 375504, 704304, 1323996, 2479692, 4654464, 8710212
WALKS ON A HONEYCOMB. REF JMP 2 61 61.

1014 1, 3, 6, 13, 23, 45, 78, 141, 239, 409
4-LINE PARTITIONS. REF MES 54 115 24. DMJ 31 272 64. CH1.

1015 1, 3, 6, 13, 24, 47, 83, 152, 263, 457, 768, 1292, 2118, 3462, 5564, 8888, 14016, 21973, 34081, 52552, 80331, 122078, 184161, 276303, 411870, 610818, 900721
5-LINE PARTITIONS. REF MES 54 115 24. DMJ 31 272 64. CH1.

1016 1, 3, 6, 13, 24, 48, 86, 160, 282, 500, 859, 1479, 2485, 4167, 6879, 11297, 18334, 29601, 47330, 75278, 118794, 186475, 290783, 451194, 696033, 1068745, 1632658
PLANAR PARTITIONS. REF MA2 2 332. MES 54 115 24. PCPS 63 1099 67. CH1.

1017 1, 3, 6, 13, 28, 60, 129, 277, 595, 1278, 2745, 5896, 12664, 27201, 58425, 125491, 269542, 578949, 1243524, 2670964, 5736961, 12322413, 26467299, 56849086
A SIMPLE RECURRENCE. REF EUL (1) 1 322 11.

1018 1, 3, 6, 14, 25, 53, 89, 167, 278, 480, 760
RESTRICTED PARTITIONS. REF JCT 9 373 70.

1019 1, 3, 6, 14, 27, 58, 111, 223, 424, 817, 1527
FUNCTIONAL DETERMINANTS. REF CAY 2 219.

1020 1, 3, 6, 14, 36, 98, 276, 794, 2316, 6818, 20196, 60074, 179196, 535538, 1602516, 4799354, 14381676, 43112258, 129271236, 387682634, 1162785756, 3487832978
1N + 2**N + 3**N. REF AS1 813.**

1021 1, 3, 6, 15, 27, 63, 120, 252, 495, 1023, 2010, 4095
COMPOSITIONS INTO RELATIVELY PRIME PARTS. REF FQ 2 251 64.

1022 1, 3, 6, 15, 29
ASYMMETRIC TREES. REF AM1 101 156 59. HA5 232.

1023 1, 3, 6, 15, 33, 82, 194, 482, 1188, 2988, 7528, 19181, 49060, 126369, 326863, 849650, 2216862, 5806256, 15256265, 40210657, 106273050, 281593237, 747890675
ALCOHOLS. REF JACS 54 2919 32.

1024 1, 3, 6, 15, 41, 115, 345, 1103, 3664, 12763, 46415, 175652, 691001, 2821116, 11932174, 52211412
GRAPHS BY POINTS AND LINES. REF R1 146. ST1.

1025 1, 3, 6, 18, 48, 156, 492, 1740, 6168, 23568, 91416, 374232, 1562640, 6801888, 30241488, 139071696, 653176992, 3156467520, 15566830368, 78696180768
A(N) = A(N − 1) + N.A(N − 2).

1026 1, 3, 6, 22, 402, 1228158, 400507806843728
BOOLEAN FUNCTIONS. REF HA2 149.

1027 1, 3, 6, 24, 148, 1646, 34040, 1358852, 106321628, 16006173014, 4525920859198, 2404130854745735, 2426376196165902704
GRAPHS BY POINTS AND LINES. REF R1 146. ST1.

1028 1, 3, 6, 30, 360, 504, 44016, 204048, 8261760, 128422272, 1816480512, 76562054400, 124207469568
A PARTITION FUNCTION. REF PRV 135 A1275 64.

1029 1, 3, 6, 38, 213, 1479, 11692, 104364, 1036809
FROM MENAGE POLYNOMIALS. REF R1 198.

1030 1, 3, 6, 42, 618, 15990, 668526, 43558242
COLORED GRAPHS. REF CJM 22 596 70.

1031 1, 3, 6, 44, 180, 1407, 10384, 92896
HIT POLYNOMIALS. REF RI3.

1032 1, 3, 7, 5, 93, 637, 1425, 22341
RELATED TO WEBER FUNCTIONS. REF KNAW 66 751 63.

1033 1, 3, 7, 8, 13, 17, 18, 21, 30, 31, 32, 38, 41, 43, 46, 47, 50, 55, 57, 68, 70, 72, 73, 75, 76, 83, 91, 93, 98, 99, 100, 105, 111, 112, 117, 119, 122, 123, 128, 129, 132, 133, 142
REDUCIBLE NUMBERS. REF AMM 56 525 49.

1034 1, 3, 7, 8, 14, 29, 31, 42, 52, 66, 85, 99, 143, 161, 185, 190, 267, 273, 304, 330, 371, 437, 476, 484, 525, 603, 612, 658, 806, 913, 1015, 1074, 1197, 1261, 1340, 1394
OF THE FORM (P2 – 1)/120 WHERE P IS PRIME. REF IAS 5 382 37.**

1035 1, 3, 7, 9, 13, 15, 21, 25, 31, 33, 37, 43, 49, 51, 63, 67, 69, 73, 75, 79, 87, 93, 99, 105, 111, 115, 127, 129, 133, 135, 141, 151, 159, 163, 169, 171, 189, 193, 195, 201, 205
LUCKY NUMBERS. REF MMAG 29 119 55.

1036 1, 3, 7, 10, 19, 32, 34, 37, 51, 81, 119, 122, 134, 157, 160, 161, 174, 221, 252, 254, 294, 305, 309, 364, 371, 405, 580, 682, 734, 756, 763, 776, 959, 1028, 1105, 1120, 1170
LATTICE POINTS IN SPHERES. REF MTAC 20 306 66.

1037 1, 3, 7, 11, 14, 18, 22, 26, 29, 33, 37, 40, 44, 48, 52, 55, 59, 63, 66, 70, 74, 78, 81, 85, 89, 92, 96, 100, 104, 107, 111, 115, 118, 122, 126, 130, 133, 137, 141, 145, 148, 152
A BEATTY SEQUENCE. REF CMB 3 21 60.

1038 1, 3, 7, 11, 16, 22, 27, 33, 40, 46, 53, 60, 67, 74, 81, 89, 96, 104, 112, 120, 128, 136, 144, 153, 161, 169, 178, 187, 195, 204, 213, 222, 231, 240, 249, 258, 267, 276, 286
NEAREST INTEGER TO 2N LOG N. REF NBS 66B 229 62.

1039 1, 3, 7, 11, 19, 23, 31, 43, 47, 59, 67, 71, 79, 83, 103, 107, 127, 131, 139, 151, 163, 167, 179, 191, 199, 211, 223, 227, 239, 251, 263, 271, 283, 307, 311, 331, 347, 359, 367
PRIMES OF THE FORM 4N + 3. REF AS1 870.

1040 1, 3, 7, 11, 21, 39, 71, 131, 241, 443, 815, 1499, 2757, 5071, 9327, 17155, 31553, 58035, 106743, 196331, 361109, 664183, 1221623, 2246915, 4132721, 7601259
A FIELDER SEQUENCE. REF FQ 6(3) 69 68.

1041 1, 3, 7, 11, 26, 45, 85, 163, 304, 578, 1090, 2057, 3888, 7339, 13862, 26179, 49437, 93366, 176321, 332986, 628852, 1187596, 2242800, 4235569, 7998951
A FIELDER SEQUENCE. REF FQ 6(3) 69 68.

1042 1, 3, 7, 12, 18, 26, 35, 45, 57, 70, 84, 100, 117, 135, 155, 176, 198, 222, 247, 273, 301, 330, 360, 392, 425, 459, 495, 532, 570, 610, 651, 693, 737, 782, 828, 876, 925, 975
FERMAT COEFFICIENTS. REF MMAG 27 141 54.

1043 1, 3, 7, 12, 19, 30, 43, 49, 53, 70, 89, 112, 141, 172, 209, 250, 293, 301
RELATED TO A HIGHLY COMPOSITE SEQUENCE. REF BSMF 97 152 69.

1044 1, 3, 7, 12, 20, 30, 44, 59, 75, 96, 118, 143, 169, 197, 230, 264, 299, 335, 373, 413, 455, 501, 549, 598, 648, 701, 758, 818, 880, 944, 1009, 1079, 1156, 1236, 1317, 1400
PRIME NUMBERS OF MEASUREMENT. REF PCPS 21 654 23.

1045 1, 3, 7, 13, 15, 21, 43, 63, 99, 109, 159, 211, 309, 343, 415, 469, 781, 871, 939
9.2N - 1 IS PRIME. REF MTAC 23 874 69.**

1046 1, 3, 7, 13, 15, 25, 39, 55, 75, 85, 127, 1947
5.2N + 1 IS PRIME. REF PAMS 9 674 58.**

1047 1, 3, 7, 13, 17, 23, 27, 33, 37, 53, 63, 67, 77, 87, 97, 103, 113, 127, 137, 147, 153, 163, 167, 197, 223, 227, 247, 263, 267, 277, 283, 287, 297, 303, 323, 347, 363, 367, 373
(N2 + 1)/10 IS PRIME. REF EUL (1) 3 25 17.**

1048 1, 3, 7, 13, 19, 27, 39, 49, 63, 79, 91, 109, 133, 147, 181, 207, 223, 253, 289, 307, 349, 387, 399, 459, 481, 529, 567, 613, 649, 709, 763, 807, 843, 927, 949, 1009, 1093
FLAVIUS SIEVE. REF MMAG 29 117 55.

1049 1, 3, 7, 13, 21, 31, 43, 57, 73, 91, 111, 133, 157, 183, 211, 241, 273, 307, 343, 381, 421, 463, 507, 553, 601, 651, 703, 757, 813, 871, 931, 993, 1057, 1123, 1191, 1261
CENTRAL POLYGONAL NUMBERS N(N - 1) + 1. REF HO3 22. HO2 87.

1050 1, 3, 7, 13, 22, 34, 50, 70, 95, 125, 161, 203, 252, 308, 372, 444, 525, 615, 715, 825, 946, 1078, 1222, 1378, 1547, 1729, 1925, 2135, 2360, 2600, 2856, 3128, 3417, 3723
A PARTITION FUNCTION. REF AMS 26 308 55.

1051 1, 3, 7, 13, 31, 43, 73, 157, 211, 241, 307, 421, 463, 601, 757, 1123, 1483, 1723, 2551, 2971, 3307, 3541, 3907, 4423, 4831, 5113, 5701, 6007, 6163, 6481, 8011, 8191
PRIMES OF FORM N(N + 1) + 1. REF LIN 3 209 29. LE1 46.

1052 1, 3, 7, 14, 18, 30, 34, 51, 65, 91, 105, 140
RELATED TO ZARANKIEWICZS PROBLEM. REF TI1 126.

1053 1, 3, 7, 14, 26, 46, 79, 133, 221, 364, 596, 972, 1581, 2567, 4163, 6746, 10926, 17690, 28635, 46345, 75001, 121368, 196392, 317784, 514201, 832011, 1346239
FROM ROOK POLYNOMIALS. REF SMA 20 18 54.

1054 1, 3, 7, 15, 1, 292, 1, 1, 1, 2, 1, 3, 1, 14, 2, 1, 1, 2, 2, 2, 2, 1, 84, 2, 1, 1, 15, 3, 13, 1, 4, 2, 6, 6, 99, 1, 2, 2, 6, 3, 5, 1, 1, 6, 8, 1, 7, 1, 2, 3, 7, 1, 2, 1, 1, 12, 1, 1, 1, 3, 1, 1, 8, 1, 1
CONTINUED FRACTION EXPANSION OF PI. REF LE4. MFM 67 312 63. MTAC 25 403 71.

1055 1, 3, 7, 15, 26, 51, 99, 191, 367, 708, 1365, 2631, 5071, 9775, 18842, 36319, 70007, 134943, 260111, 501380, 966441, 1862875, 3590807, 6921503, 13341626
A FIELDER SEQUENCE. REF FQ 6(3) 70 68.

1056 1, 3, 7, 15, 26, 57, 106, 207, 403, 788, 1530, 2985, 5812, 11322, 22052, 42959, 83675, 162993, 317491, 618440, 1204651, 2346534, 4570791, 8903409, 17342876
A FIELDER SEQUENCE. REF FQ 6(3) 70 68.

1057 1, 3, 7, 15, 27, 41, 62, 85, 115, 150, 186, 229, 274, 323, 380, 443, 509, 577, 653, 733, 818, 912, 1010, 1114, 1222, 1331, 1448, 1572, 1704, 1845, 1994, 2138, 2289, 2445
A NUMBER-THEORETIC FUNCTION. REF ACA 6 372 61.

1058 1, 3, 7, 15, 29, 469, 29531, 1303, 16103, 190553, 128977, 9061, 30946717, 39646461, 58433327, 344499373, 784809203, 169704792667
COEFFICIENTS FOR NUMERICAL DIFFERENTIATION. REF PHM 33 11 42. BAMS 48 922 42.

1059 1, 3, 7, 15, 31, 63, 127, 255, 511, 1023, 2047, 4095, 8191, 16383, 32767, 65535, 131071, 262143, 524287, 1048575, 2097151, 4194303, 8388607, 16777215, 33554431
2N − 1. REF BA1.**

1060 1, 3, 7, 16, 31, 57, 97, 162, 257, 401, 608, 907, 1325, 1914, 2719, 3824, 5313, 7316, 9973, 13495, 18105, 24132, 31938, 42021, 54948, 71484, 92492, 119120, 152686
BIPARTITE PARTITIONS. REF PCPS 49 72 53. NI1 1.

1061 1, 3, 7, 16, 33, 71, 141, 284, 552, 1067, 2020, 3803, 7043, 12957, 23566, 42536, 76068, 135093, 238001, 416591
SOLID PARTITIONS, DISTINCT ALONG ROWS. REF AT1 404.

1062 1, 3, 7, 16, 49, 104, 322, 683, 2114, 4485, 13881, 29450, 91147, 193378, 598500, 1269781, 3929940, 8337783, 25805227, 54748516, 169445269
A TERNARY CONTINUED FRACTION. REF TOH 37 441 33.

1063 1, 3, 7, 17, 40, 102, 249, 631, 1594, 4074, 10443, 26981, 69923, 182158, 476141, 1249237, 3287448, 8677074, 22962118, 60915508, 161962845, 431536102
ALCOHOLS. REF JACS 54 2919 32.

1064 1, 3, 7, 17, 41, 99, 239, 577, 1393, 3363, 8119, 19601, 47321, 114243, 275807, 665857, 1607521, 3880899, 9369319, 22619537, 54608393, 131836323, 318281039
A(N) = 2A(N − 1) + A(N − 2). REF MQET 1 9 16. AMM 56 445 49.

1065 1, 3, 7, 18, 42, 109
AMMONIUM COMPOUNDS. REF JACS 56 157 34.

1066 1, 3, 7, 18, 44, 117, 299
CONNECTED GRAPHS WITH ONE CYCLE. REF R1 150.

1067 1, 3, 7, 18, 47, 123, 322, 843, 2207, 5778, 15127, 39603, 103682, 271443, 710647, 1860498, 4870847, 12752043, 33385282, 87403803, 228826127, 599074578
BISECTION OF LUCAS SEQUENCE. REF FQ 9 284 71.

1068 1, 3, 7, 19, 25, 51, 109, 153, 213, 289, 992, 1121, 2968, 5092, 21934, 24503, 25817, 51460, 122106, 1499699, 1870227
CLASS NUMBERS OF QUADRATIC FIELDS. REF MTAC 24 437 70.

1069 1, 3, 7, 19, 47, 130, 343, 951, 2615, 7318, 20491, 57903, 163898, 466199, 1328993, 3799624, 10884049, 31241170, 89814958, 258604642
FUNCTIONAL DIGRAPHS. REF FI2 41.401. MAN 143 110 61. ST3.

1070 1, 3, 7, 19, 51, 141, 393, 1107, 3139, 8953, 25653, 73789, 212941, 616227, 1787607, 5196627, 15134931, 44152809, 128996853, 377379369
EXPANSION OF (1 + X + X2)**N. REF EUL (1) 15 59 27. SPS 37-64-3 37 70.**

1071 1, 3, 7, 20, 55, 148, 403, 1097, 2981, 8103, 22026, 59874, 162755, 442413, 1202604, 3269017, 8886111, 24154953, 65659969, 178482301, 485165195, 1318815734
POWERS OF E. REF MNAS 14(5) 14 25. FW1. FMR 1 230.

1072 1, 3, 7, 22, 82, 333, 1448, 6572, 30490, 143552, 683101
HEXAGONAL POLYOMINOES. REF CJM 19 857 67. LU5.

1073 1, 3, 7, 23, 47, 1103, 2207, 2435423, 4870847
AN INFINITE COPRIME SEQUENCE. REF MAG 48 418 64.

1074 1, 3, 7, 23, 71, 311, 479, 1559, 5711, 10559, 18191, 31391, 422231, 701399, 366791, 3818929, 9257329
QUADRATIC NONRESIDUES. REF PCPS 61 672 65. MNR 29 114 65.

1075 1, 3, 7, 23, 89, 139, 199, 113, 1831, 523, 887, 1129, 1669, 2477, 2971, 4297, 5591, 1327, 9551, 30593, 19333, 16141, 15683, 81463, 28229, 31907, 19609, 35617, 82073
INCREASING GAPS BETWEEN PRIMES. REF MTAC 21 485 67.

1076 1, 3, 7, 23, 287, 291, 795
13.2N – 1 IS PRIME. REF MTAC 22 421 68.**

1077 1, 3, 7, 25, 90, 350, 1701, 7770, 42525, 246730, 1379400, 9321312, 63436373, 420693273, 3281882604, 25708104786, 197462483400, 1709751003480
LARGEST STIRLING NUMBERS OF SECOND KIND. REF AS1 835. PSPM 19 172 71.

1078 1, 3, 7, 31, 100, 331, 431, 2486, 2917, 5403, 24529, 250693, 4286310, 4537003, 67804352, 72341355, 140145707, 427797039119, 427937184826, 855734223945
CONVERGENTS TO CUBE ROOT OF 5. REF AMP 46 107 1866. LE1 67. HPR.

1079 1, 3, 7, 31, 127, 2047, 8191, 131071, 524287, 8388607, 536870911, 2147483647, 137438953471, 2199023255551, 8796093022207, 140737488355327
MERSENNE NUMBERS. REF HW1 16.

1080 1, 3, 7, 31, 127, 8191, 131071, 524287, 2147483647, 2305843009213693951, 618970019642690137449562111, 162259276829213363391578010288127
MERSENNE PRIMES. REF MTAC 18 93 64, 22 232 68. NAMS 18 608 71.

1081 1, 3, 7, 31, 211, 2311, 509, 277, 27953, 703763, 34231, 200560490131, 676421, 11072701, 78339888213593, 13808181181, 18564761860301
LARGEST FACTORS OF A SEQUENCE. REF SMA 14 26 48.

1082 1, 3, 7, 31, 703, 459007, 210066847231, 44127887746116242376703, 1947270476915296449559747573381594836628779007
A NONLINEAR RECURRENCE. REF SA2.

1083 1, 3, 7, 46, 4336, 134281216, 288230380379570176
BOOLEAN FUNCTIONS. REF JSIAM 11 827 63. HA2 143.

1084 1, 3, 7, 47, 2207, 4870847, 23725150497407, 562882766124611619513723647, 316837008400094222150776738483768236006420971486980607
A(N) = A(N – 1)2 – 2. REF CR 83 1286 1876. DI2 1 397. HW1 223.**

1085 1, 3, 7, 83, 109958
SELF-DUAL BOOLEAN FUNCTIONS. REF PGEC 11 284 62.

1086 1, 3, 8, 3, 56, 217, 64, 2951, 12672, 5973, 309376, 1237173, 2917888, 52635599, 163782656
EXPANSION OF EXP(SIN X). REF AMM 41 418 34. HPR.

1087 1, 3, 8, 6, 20, 24, 16, 12, 24, 60, 10, 24, 28, 48, 40, 24, 36, 24, 18, 60, 16, 30, 48, 24, 100, 84, 72, 48, 14, 120, 30, 48, 40, 36, 80, 24, 76, 18, 56, 60, 40, 48, 88, 30, 120, 48
PISANO PERIODS. REF HM1. MTAC 23 459 69. ACA 16 109 69.

1088 1, 3, 8, 9, 37, 121, 211, 695, 4889, 41241, 76301, 853513, 3882809, 11957417, 100146415, 838216959, 13379363737, 411322824001, 3547404378125
FIRST FACTOR OF PRIME CYCLOTOMIC FIELDS. REF MTAC 24 217 70.

1089 1, 3, 8, 14, 14, 25, 24, 23, 22, 25, 59, 98, 97, 98, 97, 174, 176, 176, 176, 176, 291, 290, 289, 740, 874, 873, 872, 873, 872, 871, 870, 869, 868, 867, 866, 2180, 2179, 2178
RELATED TO GAPS BETWEEN PRIMES. REF MTAC 13 122 59. SI2 35.

1090 1, 3, 8, 16, 30, 46, 64, 96, 126, 158
GENERALIZED CLASS NUMBERS. REF MTAC 21 689 67.

1091 1, 3, 8, 16, 30, 50, 80, 120, 175, 245, 336, 448, 588, 756, 960, 1200, 1485, 1815, 2200, 2640, 3146, 3718, 4368, 5096, 5915, 6825, 7840, 8960, 10200, 11560, 13056
A PARTITION FUNCTION. REF AMS 26 308 55.

1092 1, 3, 8, 16, 32, 55, 94, 147, 227, 332, 480, 668, 920, 1232, 1635
RESTRICTED PARTITIONS. REF CAY 2 279.

1093 1, 3, 8, 17, 33, 58, 97, 153, 233, 342, 489, 681, 930, 1245, 1641, 2130, 2730, 3456, 4330, 5370, 6602, 8048, 9738, 11698, 13963, 16563, 19538, 22923, 26763, 31098, 35979
A PARTITION FUNCTION. REF AMS 26 308 55.

1094 1, 3, 8, 17, 34, 61, 105, 170, 267, 403, 594, 851, 1197, 1648, 2235, 2981, 3927, 5104, 6565, 8351, 10529, 13152, 16303, 20049, 24492, 29715, 35841, 42972, 51255
A PARTITION FUNCTION. REF AMS 26 308 55.

1095 1, 3, 8, 18, 37, 72, 136, 251, 445, 770
PARTITIONS INTO NON-INTEGRAL POWERS. REF PCPS 47 215 51.

1096 1, 3, 8, 18, 38, 74, 139, 249, 434, 734, 1215, 1967, 3132, 4902, 7567, 11523, 17345, 25815, 38045, 55535, 80377, 115379, 164389, 232539, 326774, 456286, 633373
PARTITIONS INTO PARTS OF 3 KINDS. REF RS2 122.

1097 1, 3, 8, 18, 38, 76, 147, 277, 509, 924, 1648, 2912, 5088, 8823, 15170, 25935, 44042, 74427, 125112, 209411, 348960, 579326, 958077, 1579098, 2593903, 4247768
TREES OF HEIGHT 3. REF IBMJ 4 475 60. KU1.

1098 1, 3, 8, 19, 42, 88, 176, 339, 633, 1150, 2040, 3544, 6042, 10128, 16720, 27219, 43746, 69483, 109160, 169758, 261504, 399272, 604560, 908248, 1354427, 2005710
COEFFICIENTS IN AN ELLIPTIC FUNCTION. REF QJM 21 66 1885.

1099 1, 3, 8, 20, 44, 80, 343, 399
CONSECUTIVE RESIDUES. REF MTAC 24 738 70.

1100 1, 3, 8, 20, 48, 112, 256, 576, 1280, 2816, 6144, 13312, 28672, 61440, 131072, 278528, 589824, 1245184, 2621440
COEFFICIENTS OF CHEBYSHEV POLYNOMIALS. REF PRSE 62 190 46. AS1 795.

1101 1, 3, 8, 21, 55, 144, 377, 987, 2584, 6765, 17711, 46368, 121393, 317811, 832040, 2178309, 5702887, 14930352, 39088169, 102334155, 267914296, 701408733
BISECTION OF FIBONACCI SEQUENCE. REF IDM 22 23 15. R1 39. PLMS 21 729 70. FQ 9 283 71.

1102 1, 3, 8, 22, 58, 158, 425, 1161, 3175, 8751, 24192, 67239
TOTAL HEIGHT OF UNLABELED TREES. REF IBMJ 4 475 60.

1103 1, 3, 8, 23, 68, 215, 680, 2226, 7327
TRIANGULATIONS OF THE DISK. REF PLMS 14 765 64.

1104 1, 3, 8, 24, 75, 243, 808, 2742, 9458, 33062, 116868, 417022, 1500159, 5434563, 19808976, 72596742, 267343374, 988779258, 3671302176, 13679542632
A SIMPLE RECURRENCE. REF IC 16 351 70.

1105 1, 3, 8, 24, 89, 415, 2372, 16072, 125673, 1112083, 10976184, 119481296, 1421542641, 18348340127, 255323504932, 3809950977008, 60683990530225
LOGARITHMIC NUMBERS. REF MST 31 78 63. CACM 13 726 70.

1106 1, 3, 8, 25, 72, 231, 696, 2268, 7030, 23155, 73188, 242957, 778946, 2601345, 8430992, 28289598, 92470194, 311472985, 1025114180, 3463982109, 11465054942
FOLDING A LINE. REF MTAC 22 198 68. JCT 5 135 68. (DIVIDED BY 2.)

1107 1, 3, 8, 26, 84, 297, 1066
MIXED HUSIMI TREES. REF PNAS 42 535 56.

1108 1, 3, 8, 27, 91, 350, 1376, 5743, 24635, 108968, 492180, 2266502, 10598452, 50235931, 240872654, 1166732814, 5682001435, 48068787314, 139354922608
ORIENTED UNLABELED TREES. REF R1 138.

1109 1, 3, 8, 28, 143, 933, 7150, 62310, 607445, 6545935, 77232740, 989893248, 13692587323, 203271723033, 3223180454138
PERMUTATIONS BY NUMBER OF PAIRS. REF DKB 264.

1110 1, 3, 8, 33, 164, 985, 6894, 55153, 496376, 4963761, 54601370, 655216441, 8517813732, 119249392249, 1788740883734, 28619854139745
A(N) = N.A(N - 1) + (-1)N.**

1111 1, 3, 8, 45, 264, 1855, 14832, 133497, 1334960, 14684571, 176214840, 2290792933, 32071101048, 481066515735, 7697064251744, 130850092279665
RENCONTRES NUMBERS. REF R1 65.

1112 1, 3, 8, 49, 3963
SWITCHING NETWORKS. REF JFI 276 317 63.

1113 1, 3, 9, 14, 19, 24, 30, 35, 40, 45, 51, 56, 61, 66
WYTHOFF GAME. REF CMB 2 189 59.

1114 1, 3, 9, 15, 30, 45, 67, 99, 135, 175, 231, 306, 354, 465
GENERALIZED DIVISOR FUNCTION. REF PLMS 19 111 19.

1115 1, 3, 9, 18, 36, 60, 100, 150, 225, 315, 441, 588
CROSSING NUMBER OF THE COMPLETE GRAPH. REF GU2.

1116 1, 3, 9, 19, 21, 55, 115, 193, 323, 611, 1081, 1571, 10771, 13067, 16321, 44881, 57887, 93167, 189947
FROM A GOLDBACH CONJECTURE. REF BIT 6 49 66.

1117 1, 3, 9, 21, 47, 95, 186, 344, 620, 1078, 1835, 3045, 4967, 7947, 12534, 19470, 29879, 45285, 67924, 100820, 148301, 216199, 312690, 448738, 639464, 905024
PARTITIONS INTO PARTS OF 3 KINDS. REF RS2 122.

1118 1, 3, 9, 21, 81, 351, 1233, 5769, 31041, 142011, 776601, 4874013, 27027729, 168369111, 1191911841, 7678566801, 53474964993, 418199988339
PERMUTATIONS OF ORDER 3. REF CJM 7 159 55.

1119 1, 3, 9, 22, 42, 84, 140, 231, 351, 551, 783
GENERALIZED DIVISOR FUNCTION. REF PLMS 19 111 19.

1120 1, 3, 9, 22, 48, 99, 194, 363, 657, 1155, 1977, 3312, 5443, 8787, 13968, 21894, 33873, 51795, 78345, 117412, 174033, 255945
COEFFICIENTS OF AN ELLIPTIC FUNCTION. REF CAY 9 128.

1121 1, 3, 9, 22, 50, 104, 208, 394, 724, 1286, 2229, 3769, 6253, 10176, 16303, 25723, 40055, 61588, 93647, 140875, 209889, 309846, 453565, 658627, 949310, 1358589
PARTITIONS INTO PARTS OF 3 KINDS. REF RS2 122.

1122 1, 3, 9, 22, 51, 107, 217, 416, 775, 1393, 2446, 4185, 7028, 11569, 18749, 29908, 47083, 73157, 112396, 170783, 256972, 383003, 565961, 829410, 1206282, 1741592
PARTITIONS INTO PARTS OF 3 KINDS. REF RS2 122.

1123 1, 3, 9, 22, 51, 108, 221, 429, 810, 1479, 2640, 4599, 7868, 13209, 21843, 35581, 57222, 90882, 142769, 221910, 341649, 521196, 788460, 1183221, 1762462, 2606604
PARTITIONS INTO PARTS OF 3 KINDS. REF RS2 122.

1124 1, 3, 9, 22, 51, 111, 233, 474, 942, 1836, 3522, 6666, 12473, 23109, 42447, 77378, 140109, 252177, 451441, 804228, 1426380, 2519640, 4434420, 7777860
CONVOLVED FIBONACCI NUMBERS. REF RCI 101. FQ 8 163 70.

1125 1, 3, 9, 25, 65, 161, 385, 897, 2049, 4609, 10241, 22529, 49153, 106497, 229377, 491521, 1048577, 2228225, 4718593, 9961473, 20971521, 44040193, 92274689
CULLEN NUMBERS N.2N + 1. REF SI1 346.**

1126 1, 3, 9, 25, 69, 186, 503, 1353, 3651, 9865, 26748, 72729, 198447, 543159, 1491402, 4107152, 11342826, 31408719, 87180087, 242603070, 676524372
POWERS OF ROOTED TREE ENUMERATOR. REF R1 150.

1127 1, 3, 9, 25, 75, 231, 763, 2619, 9495, 35695, 140151, 568503, 2390479, 10349535, 46206735, 211799311, 997313823, 4809701439, 23758664095, 119952692895
PERMUTATIONS OF ORDER EXACTLY 2. REF CJM 7 159 55.

1128 1, 3, 9, 26, 75, 214, 612, 1747, 4995
PARTIALLY LABELED TREES. REF R1 138.

1129 1, 3, 9, 27, 81, 243, 729, 2187, 6561, 19683, 59049, 177147, 531441, 1594323, 4782969, 14348907, 43046721, 129140163, 387420489, 1162261467
POWERS OF THREE. REF BA1.

1130 1, 3, 9, 28, 90, 297, 1001, 3432, 11934, 41990, 149226, 534888, 1931540, 7020405, 25662825, 94287120, 347993910, 1289624490, 4796857230, 17902146600
LAPLACE TRANSFORM COEFFICIENTS. REF QAM 14 407 56.

1131 1, 3, 9, 29, 35, 42, 48, 113, 120, 126, 152, 185, 204, 224, 237, 243, 276, 302, 308, 321, 341, 386, 399, 419, 432, 477, 503, 510, 516, 542, 549, 588, 633, 659, 666, 705, 731
(N(N + 1) + 1)/13 IS PRIME. REF CU1 1 251.

1132 1, 3, 9, 29, 98, 343, 1230, 4489
PERMUTATIONS BY INVERSIONS. REF NET 96.

1133 1, 3, 9, 33, 139, 718, 4535
TOPOLOGIES OR UNLABELED TRANSITIVE DIGRAPHS. REF WR1.

1134 1, 3, 9, 35, 201, 1827
COEFFICIENTS OF BELLS FORMULA. REF NMT 10 65 62.

1135 1, 3, 9, 37, 153, 951, 5473, 42729, 353937, 3455083
SUMS OF LOGARITHMIC NUMBERS. REF MST 31 79 63.

1136 1, 3, 9, 42, 206, 1352
REGULAR SEMIGROUPS. REF PL1. MA4 2 2 67.

1137 1, 3, 9, 45, 225, 1575, 11025, 99225, 893025, 9823275, 108056025, 1404728325, 18261468225, 273922023375, 4108830350625, 69850115960625
PERMUTATIONS WITH ODD CYCLES. REF R1 87.

1138 1, 3, 9, 48, 504, 14188, 1351563
THRESHOLD FUNCTIONS. REF PGEC 19 821 70.

1139 1, 3, 9, 54, 450, 4725, 59535, 873180, 14594580
EXPANSION OF AN INTEGRAL. REF CO1 1 176.

1140 1, 3, 10, 13, 62, 75, 437, 512, 949, 6206, 13361, 73011, 597449, 1865358, 6193523, 26639450, 59472423, 383473988, 1593368375, 6756947488, 8350315863
CONVERGENTS TO CUBE ROOT OF 3. REF AMP 46 106 1866. LE1 67. HPR.

1141 1, 3, 10, 25, 56, 119, 246, 501, 1012, 2035, 4082, 8177, 16368, 32751, 65518, 131053, 262124, 524267, 1048554, 2097129, 4194280, 8388583, 16777190, 33554405
ASSOCIATED STIRLING NUMBERS. REF R1 76. DB1 296. CO1 2 58.

1142 1, 3, 10, 31, 97, 306, 961, 3020, 9489, 29809, 93648, 294204, 924269, 2903677, 9122171, 28658146, 90032221, 282844564, 888582403, 2791563950, 8769956796
POWERS OF PI. REF PE2 1(APPENDIX) 1. FMR 1 122.

1143 1, 3, 10, 33, 111, 379, 1312, 4596, 16266, 58082, 209010, 757259, 2760123, 10114131, 37239072, 137698584, 511140558, 1904038986, 7115422212, 26668376994
A SIMPLE RECURRENCE. REF IC 16 351 70.

1144 1, 3, 10, 35, 126, 462, 1716, 6435, 24310, 92378, 352716, 1352078, 5200300, 20058300, 77558760, 300540195, 1166803110, 4537567650, 17672631900
CENTRAL BINOMIAL COEFFICIENTS. REF RS1.

1145 1, 3, 10, 36, 137, 543, 2219, 9285, 39587, 171369, 751236, 3328218, 14878455, 67030785, 304036170, 1387247580, 6363044315, 29323149825, 135700543190
RESTRICTED HEXAGONAL POLYOMINOES. REF EMS 17 11 70. RE3.

1146 1, 3, 10, 38, 154, 654, 2871, 12925, 59345, 276835, 1308320, 6250832, 30142360, 146510216, 717061938, 3530808798, 17478955570, 86941210950, 434299921440
DISSECTIONS OF A POLYGON. REF EMN 32 6 40. BAMS 54 359 48.

1147 1, 3, 10, 38, 156, 692
SYMMETRIC PERMUTATIONS. REF LU1 1 222.

1148 1, 3, 10, 41, 196, 1057, 6322, 41393, 293608, 2237921, 18210094, 157329097, 1436630092, 13810863809, 139305550066, 1469959371233
TREES OF HEIGHT AT MOST 1. REF JCT 3 134 67, 5 102 68.

1149 1, 3, 10, 41, 206, 1237, 8660
RELATED TO EULER NUMBERS. REF MST 20 70 52.

1150 1, 3, 10, 42, 193, 966, 5215, 30170, 186234, 1222065, 8496274, 62395234, 482700052
MODIFIED BESSEL FUNCTIONS. REF AS1 429.

1151 1, 3, 10, 43, 225, 1393, 9976, 81201, 740785, 7489051, 83120346, 1004933203, 13147251985, 185066460993, 2789144166880, 44811373131073
A(N + 1) = NA(N) + A(N - 1). REF EUR 22 15 59.

1152 1, 3, 10, 44, 238, 1650
CONNECTED UNLABELED PARTIALLY ORDERED SETS. REF NAMS 17 646 70. WR1.

1153 1, 3, 10, 45, 272, 2548, 39632, 1104306, 56871880, 5463113568, 978181717680, 326167542296048, 202701136710498400, 235284321080559981952
SYMMETRIC RELATIONS. REF MI1 17 21 55. MAN 174 70 67. (DIVIDED BY 2.)

1154 1, 3, 10, 53, 265, 1700
SORTING NUMBERS. REF PSPM 19 173 71.

1155 1, 3, 10, 56, 468, 7123, 194066
NON-SEPARABLE GRAPHS. REF JCT 9 352 70.

1156 1, 3, 10, 70, 708, 15248, 543520
SELF-CONVERSE DIGRAPHS. REF MAT 13 157 66.

1157 1, 3, 11, 25, 137, 49, 363, 761, 7129, 7381, 83711, 86021, 1145993, 1171733, 1195757, 2436559, 42142223, 14274301, 275295799, 55835135, 18858053, 19093197
NUMERATORS OF HARMONIC NUMBERS. REF KN1 1 615.

1158 1, 3, 11, 29, 74, 167, 367, 755, 1515, 2931, 5551, 10263, 18677, 33409, 59024, 102984, 177016, 304458, 510909, 071180, 1458882, 2428548, 4021670, 6627515
TREES OF DIAMETER 6. REF IBMJ 4 476 60. KU1.

1159 1, 3, 11, 39, 131, 423, 1331
A PROBLEM IN PARITY. REF IJ1 11 162 69.

1160 1, 3, 11, 41, 153, 571, 2131, 7953, 29681, 110771, 413403, 1542841, 5757961, 21489003, 80198051, 299303201, 1117014753, 4168755811, 15558008491
A(N) = 4A(N - 1) - A(N - 2). REF EUL (1) 1 375 11. MMAG 40 78 67.

1161 1, 3, 11, 43, 683, 2731, 43691, 174763, 2796203, 715827883, 2932031007403, 768614336404564651, 201487636602438195784363
PRIMES OF FORM (2P + 1)/3. REF MMAG 27 157 54.**

1162 1, 3, 11, 44, 186, 814, 3652, 16689, 77359, 362671, 1716033, 8182213
FIXED HEXAGONAL POLYOMINOES. REF LU5.

1163 1, 3, 11, 45, 197, 903, 4279, 20793, 103049, 518859, 2646723, 13648869, 71039373, 372693519, 1968801519, 10463578353, 55909013009, 300159426963
DISSECTIONS OF A POLYGON, OR PARENTHESIZING A PRODUCT. REF EMN 32 6 40. BAMS 54 359 48. RCI 168. CO1 1 71.

1164 1, 3, 11, 49, 261, 1631, 11743, 95901, 876809, 8877691
BINOMIAL COEFFICIENT SUMS. REF CJM 22 26 70.

1165 1, 3, 11, 50, 274, 1764, 13068, 109584, 1026576, 10628640, 120543840, 1486442880, 19802759040, 283465647360, 4339163001600, 70734282393600
STIRLING NUMBERS OF FIRST KIND. REF AS1 833. DKB 226.

1166 1, 3, 11, 53, 309, 2119, 16687, 148329, 1468457, 16019531, 190899411, 2467007773, 34361893981, 513137616783, 8178130767479
A(N) = NA(N − 1) + (N − 1)A(N − 2). REF R1 188. DKB 263. MAG 52 381 68.

1167 1, 3, 11, 55, 330, 2345
SUBSPACES OF JORDAN ALGEBRAS. REF JL2 309.

1168 1, 3, 11, 56, 348, 2578, 22054, 213798, 2313638, 27627434, 360646314, 5107177312, 77954299144, 1275489929604, 22265845018412, 412989204564572
STOCHASTIC MATRICES OF INTEGERS. REF DMJ 35 659 68.

1169 1, 3, 11, 57, 361, 2763, 24611, 250737, 2873041, 36581523, 512343611, 7828053417, 129570724921, 2309644635483, 44110959165011, 898621108880097
GENERALIZED EULER NUMBERS. REF MTAC 21 693 67.

1170 1, 3, 11, 145, 197, 903, 4279, 20793, 103049
SCHRODERS SECOND PROBLEM. REF ZMP 15 366 1870.

1171 1, 3, 12, 29, 57, 99, 157, 234, 333, 456, 606, 786, 998, 1245
SERIES-REDUCED PLANTED TREES. REF RI1.

1172 1, 3, 12, 50, 27, 1323, 928, 1080, 48525, 3237113, 7587864, 23361540993, 770720657, 698808195, 179731134720, 542023437008852, 3212744374395
COTESIAN NUMBERS. REF QJM 46 63 14.

1173 1, 3, 12, 52, 241, 1173, 5929, 30880, 164796, 897380, 4970296, 27930828, 158935761, 914325657, 5310702819, 31110146416, 183634501753, 1091371140915
SIMPLE TRIANGULATIONS OF PLANE. REF CJM 15 268 63.

1174 1, 3, 12, 55, 273, 1428, 7752, 43263, 246675
DISSECTIONS OF A POLYGON. REF AMP 1 198 1841. EMN 32 5 40. CMA 2 25 70.

1175 1, 3, 12, 56, 288, 1584, 9152, 54912, 339456
ROOTED BICUBIC MAPS. REF CJM 15 269 63.

1176 1, 3, 12, 56, 321, 2175, 17008, 150504, 1485465, 16170035, 192384876, 2483177808, 34554278857, 515620794591, 8212685046336
PERMUTATIONS BY NUMBER OF PAIRS. REF DKB 264.

1177 1, 3, 12, 58, 325, 2143
COMMUTATIVE SEMIGROUPS. REF PL1. MA4 2 2 67.

1178 1, 3, 12, 60, 358, 2471, 19302, 167894, 1606137, 16733779, 188378402, 2276423485, 29367807524, 402577243425, 5840190914957, 89345001017415
COEFFICIENTS OF ITERATED EXPONENTIALS. REF SMA 11 353 45.

1179 1, 3, 12, 60, 360, 2520, 20160, 181440, 1814400, 19958400, 239500800, 3113510400, 43589145600, 653837184000, 10461394944000, 177843714048000
GENERALIZED STIRLING NUMBERS. REF PEF 77 26 62.

1180 1, 3, 12, 60, 420, 4620, 60060, 180180, 360360, 6126120, 116396280, 2677114440, 77636318760, 2406725881560, 89048857617720, 3651003162326520
A HIGHLY COMPOSITE SEQUENCE. REF BSMF 97 152 69.

1181 1, 3, 12, 70, 465, 3507, 30016, 286884, 3026655, 34944085, 438263364, 5933502822, 86248951243, 1339751921865, 22148051088480, 388246725873208
POLYGONS FORMED FROM N LINES. REF CO1 2 120.

1182 1, 3, 13, 27, 52791, 482427, 124996631
ASYMPTOTIC EXPANSION OF AN INTEGRAL. REF MTAC 19 114 65.

1183 1, 3, 13, 31, 43, 67, 71, 83, 89, 107, 151, 157, 163, 191, 197, 199, 227, 283, 293, 307, 311, 347, 359, 373, 401, 409, 431, 439, 443, 467, 479, 523, 557, 563, 569, 587, 599
10 IS A QUADRATIC RESIDUE MODULO P. REF KR1 1 61.

1184 1, 3, 13, 63, 321, 1683, 8989, 48639, 265729, 1462563, 8097453, 45046719, 251595969, 1409933619, 7923848253, 44642381823, 252055236609, 1425834724419
A SQUARE RECURRENCE. REF MES 54 75 24. SIAMR 12 277 70.

1185 1, 3, 13, 63, 326, 1761
ROOTED PLANAR MAPS. REF CJM 15 542 63.

1186 1, 3, 13, 65, 403, 2885, 23515, 214805
THE GAME OF MOUSETRAP. REF QJM 15 241 1878.

1187 1, 3, 13, 68, 399, 2530, 16965, 118668, 857956, 6369838
PLANAR TRIANGULATIONS. REF CJM 14 32 62.

1188 1, 3, 13, 70, 462, 3592, 32056, 322626, 3611890, 44491654, 597714474, 8693651092, 136059119332, 2279212812480, 40681707637888, 770631412413148
STOCHASTIC MATRICES OF INTEGERS. REF DMJ 35 659 68.

1189 1, 3, 13, 71, 465, 3539, 30637, 296967, 3184129, 37401155, 477471021, 6581134823, 97388068753, 1539794649171, 25902759280525, 461904032857319
A(N) = NA(N − 1) + (N − 3)A(N − 2). REF R1 188.

1190 1, 3, 13, 73, 501, 4051, 37633, 394353, 4596553, 58941091, 824073141, 12470162233, 202976401213, 3535017524403, 65573803186921, 1290434218669921
FORESTS OF GREATEST HEIGHT. REF RCI 194. PSPM 19 172 71.

1191 1, 3, 13, 75, 541, 4683, 47293, 545835, 7087261, 102247563, 1622632573, 28091567595, 526858348381, 10641342970443, 230283190977853
PREFERENTIAL ARRANGEMENTS. REF CAY 4 113. PLMS 22 341 1891. AMM 69 7 62. PSPM 19 172 71.

1192 1, 3, 13, 81, 721, 9153, 165313
COLORED GRAPHS. REF CJM 12 413 60 (DIVIDED BY 2). JCT 6 17 69.

1193 1, 3, 13, 83, 592, 4821, 43979, 444613, 4934720, 59661255, 780531033, 10987095719, 165586966816, 2660378564777, 45392022568023, 819716784789193
MENAGE NUMBERS. REF LU1 1 495.

1194 1, 3, 13, 87, 841, 11643
GRADED PARTIALLY ORDERED SETS. REF JCT 6 17 69.

1195 1, 3, 13, 87, 1053, 28576, 2141733, 508147108
INCIDENCE MATRICES. REF CPM 89 217 64.

1196 1, 3, 13, 146, 40422
SWITCHING NETWORKS. REF JFI 276 317 63.

1197 1, 3, 13, 183, 33673, 1133904603, 1285739649838492213, 1653126447166808570252515315100129583
A NONLINEAR RECURRENCE. REF DMJ 4 325 38.

1198 1, 3, 13, 308, 1476218
SWITCHING NETWORKS. REF JFI 276 317 63.

1199 1, 3, 13, 781, 137257, 28531167061, 25239592216021, 5170251636789604776 1, 10991220309223964384022 1, 94911218181126872883431967775 3
(PP - 1)/(P - 1) WHERE P IS PRIME. REF MTAC 16 421 62. PSPM 19 174 71.**

1200 1, 3, 14, 39, 91, 173, 307, 502, 779
PARTITIONS INTO NON-INTEGRAL POWERS. REF PCPS 47 215 51.

1201 1, 3, 14, 42, 128, 334, 850, 2010, 4625, 10201, 21990, 46108, 94912, 191562, 380933, 746338, 1444676, 2763931, 5235309, 9822686, 18275648, 33734658, 61826344
TREES OF DIAMETER 7. REF IBMJ 4 476 60. KU1.

1202 1, 3, 14, 78, 504, 3720, 30960, 287280, 2943360, 33022080, 402796800, 5308934400, 75203251200, 1139544806400, 18394619443200, 315149522688000
DIFFERENCES OF FACTORIAL NUMBERS. REF JRAM 198 61 57.

1203 1, 3, 14, 80, 518, 3647, 27274, 213480, 1731652
HAMILTONIAN POLYGONS. REF CJM 14 417 62.

1204 1, 3, 14, 240, 63488, 4227858432, 18302628885633695744, 338953138925153547590470800371487866880
SELF-COMPLEMENTARY BOOLEAN FUNCTIONS. REF PGEC 12 561 63.

1205 1, 3, 15, 21, 15, 33, 1365, 3, 255, 399, 165, 69, 1365, 3, 435, 7161
DENOMINATORS OF COSECANT NUMBERS. REF NO1 458. ANN 36 640 35. DA2 2 187.

1206 1, 3, 15, 27, 51, 147, 243, 267, 347, 471
17.2N + 1 IS PRIME. REF PAMS 9 674 58.**

1207 1, 3, 15, 35, 315, 693, 3003, 6435, 109395, 230945
COEFFICIENTS OF LEGENDRE POLYNOMIALS. REF PR1 156. AS1 798.

1208 1, 3, 15, 60, 260, 1092, 4641, 19635, 83215, 352440, 1493064, 6324552
FROM FIBONACCI IDENTITIES. REF FQ 6 82 68.

1209 1, 3, 15, 69, 309, 1365, 5973, 25941, 112065, 482067, 2066583
WALKS ON A TRIANGULAR LATTICE. REF AIP 9 354 60.

1210 1, 3, 15, 75, 363, 1767, 8463, 40695, 193983, 926943, 4404939, 20967075, 99421371
WALKS ON A CUBIC LATTICE. REF PPS 92 649 67.

1211 1, 3, 15, 75, 435, 3045, 24465
PERMUTATIONS BY NUMBER OF CYCLES. REF R1 85.

1212 1, 3, 15, 79, 474, 3207, 24087, 198923, 1791902, 17484377
COEFFICIENTS OF HANKEL FUNCTIONS. REF CL1 XXXV.

1213 1, 3, 15, 86, 534, 3481
FIXED POLYOMINOES MADE FROM CUBES. REF LU6.

1214 1, 3, 15, 93, 639, 4653, 35169
WALKS ON A HONEYCOMB. REF AIP 9 345 60.

1215 1, 3, 15, 104, 164, 194, 255, 495, 584, 975, 2204, 2625, 2834, 3255, 3705, 5186, 5187
RELATED TO EULERS TOTIENT FUNCTION. REF AMM 56 22 49.

1216 1, 3, 15, 105, 315, 6930, 18018, 90090, 218790, 2078505, 4849845, 22309287, 50702925, 1825305300, 4071834900, 18032411700, 39671305740, 347123925225
COEFFICIENTS OF LEGENDRE POLYNOMIALS. REF PR1 156. AS1 798.

1217 1, 3, 15, 105, 945, 10395, 135135, 2027025, 34459425, 654729075, 13749310575, 316234143225, 7905853580625, 213458046676875, 6190283353629375
DOUBLE FACTORIALS. REF AMM 55 425 48. MTAC 24 231 70.

1218 1, 3, 15, 105, 947, 10472, 137337, 2085605, 36017472, 697407850, 14969626900, 352877606716, 9064191508018, 252024567201300, 7542036496650006
COEFFICIENTS OF ITERATED EXPONENTIALS. REF SMA 11 353 45.

1219 1, 3, 16, 51, 126, 266
SEQUENCES BY NUMBER OF INCREASES. REF JCT 1 372 66.

1220 1, 3, 16, 67, 251, 888, 3023, 10038, 32722
PARTIALLY LABELED TREES. REF R1 138.

1221 1, 3, 16, 95, 672, 5397, 48704
DISCORDANT PERMUTATIONS. REF SMA 19 118 53.

1222 1, 3, 16, 96, 675, 5413, 48800, 488592, 5379333, 64595975, 840192288, 11767626752, 176574062535, 2825965531593, 48052401132800, 865108807357216
SUMS OF MENAGE NUMBERS. REF AH2 79. CJM 10 478 58. R1 198.

1223 1, 3, 16, 101, 756, 6607, 65794, 733833, 9046648
FORESTS OF ROOTED TREES. REF JCT 5 102 68.

1224 1, 3, 16, 111, 2548, 14385, 672360, 10351845, 270594968, 2631486186, 310710613080
COEFFICIENTS FOR STEP-BY-STEP INTEGRATION. REF JACM 11 231 64.

1225 1, 3, 16, 125, 1176, 12847, 160504, 2261289, 35464816
FORESTS OF ROOTED TREES. REF JCT 5 102 68.

1226 1, 3, 16, 125, 1296, 16087, 229384, 3687609, 66025360
FORESTS OF ROOTED TREES. REF JCT 5 102 68.

1227 1, 3, 16, 125, 1296, 16807, 262144, 4782969, 100000000, 2357947691, 61917364224, 1792160394037, 56693912375296, 1946195068359375
(N + 1)(N − 1). REF BA1. R1 128.**

1228 1, 3, 16, 139, 1750, 29388, 623909
BICOVERINGS. REF SMH 3 147 68.

1229 1, 3, 16, 218, 9608, 1540944, 882033440, 1793359192848, 13027956824399552, 341260431952972580352, 32522909385055886111197440
DIGRAPHS OR REFLEXIVE RELATIONS. REF MI1 17 20 55. MAN 174 70 67. HA5 225.

1230 1, 3, 16, 547, 538811, 620245817465
CONVERGENTS TO LEHMERS CONSTANT. REF DMJ 4 334 38.

1231 1, 3, 17, 99, 577, 3363, 19601, 114243, 665857, 3880899, 22619537, 131836323, 768398401, 4478554083, 26102926097, 152139002499, 886731088897
A(N) = 6A(N − 1) − A(N − 2). REF NCM 4 166 1878. QJM 45 14 14. ANN 36 644 35. AMM 75 683 68.

1232 1, 3, 17, 142, 1569, 21576, 355081, 6805296, 148869153, 3660215680, 99920609601, 2998836525312, 98139640241473, 3478081490967552
CONNECTED GRAPHS BY POINTS. REF AMS 26 515 55.

1233 1, 3, 17, 155, 2073, 38227, 929569, 28820619, 1109652905, 51943281731, 2905151042481, 191329672483963, 14655626154768697, 1291885088448017715
GENOCCHI NUMBERS. REF MTAC 1 386 45. FMR 1 73.

1234 1, 3, 17, 577, 665857, 886731088897, 1572584048032918633353217, 4946041176255201878775086487573351061418968498177
A NONLINEAR RECURRENCE. REF AMM 61 424 54.

1235 1, 3, 18, 61, 225, 716, 2272
ALKYLS. REF ZFK 93 437 36.

1236 1, 3, 18, 110, 795, 6489, 59332, 600732, 6674805, 80765135, 1057289046, 14890154058, 224497707343, 3607998868005
3-LINE LATIN RECTANGLES. REF R1 210 (DIVIDED BY 2). DKB 263.

1237 1, 3, 18, 190, 3285, 88851, 3640644, 220674924
PRECOMPLETE POST FUNCTIONS. REF SMD 10 619 69. RO3.

1238 1, 3, 18, 1200, 33601536
SWITCHING NETWORKS. REF JFI 276 317 63.

1239 1, 3, 18, 5778, 192900153618, 7177905237579946589743592924684178
EXTRACTING A SQUARE ROOT. REF AMM 44 645 37.

1240 1, 3, 19, 193, 2721, 49171, 1084483, 28245729, 848456353, 28875761731, 1098127402131, 46150226651233, 2124008553358849, 106246577894593683
CONVERGENTS TO E. REF BA4 17 1871. MTAC 2 69 46.

1241 1, 3, 19, 195, 3031, 67263
GRADED PARTIALLY ORDERED SETS. REF JCT 6 17 69.

1242 1, 3, 19, 211, 3651, 90921, 3081513, 136407699, 7642177651, 528579161353, 44237263696473, 4405990782649369, 515018848029036937
COEFFICIENTS OF A BESSEL FUNCTION. REF AMM 71 493 64.

1243 1, 3, 19, 219, 3991, 106623
GRADED PARTIALLY ORDERED SETS. REF JCT 6 17 69.

1244 1, 3, 19, 219, 4231, 130023, 6129859
LABELED PARTIALLY ORDERED SETS. REF CACM 10 296 67. PURB 19 240 68. CO1 1 74.

1245 1, 3, 19, 233, 4851, 158175, 7724333
CONNECTED LABELED TOPOLOGIES. REF WR1.

1246 1, 3, 20, 35, 126, 231, 3432, 6435, 24310, 46189, 352716, 676039, 2600150, 5014575, 155117520, 300540195, 1166803110
COEFFICIENTS OF LEGENDRE POLYNOMIALS. REF PR1 157. FMR 1 362.

1247 1, 3, 20, 119, 696, 4059, 23660, 137903, 803760, 4684659, 27304196, 159140519, 927538920, 5406093003, 31509019100, 183648021599, 1070379110496
PYTHAGOREAN TRIANGLES. REF MLG 2 322 10. FQ 6(3) 104 68.

1248 1, 3, 20, 130, 924, 7308, 64224, 623376, 6636960, 76998240, 967524480, 13096736640, 190060335360, 2944310342400, 48503818137600, 846795372595200
ASSOCIATED STIRLING NUMBERS. REF R1 75. CO1 2 98.

1249 1, 3, 20, 996, 9333312
POST FUNCTIONS. REF ZML 7 198 61.

1250 1, 3, 21, 282, 6210, 202410, 9135630, 545007960, 41514583320, 3930730108200, 452785322266200, 62347376347779600, 10112899541133589200
STOCHASTIC MATRICES OF INTEGERS. REF PSAM 15 101 63. ST2.

1251 1, 3, 21, 651, 457653, 210065930571, 44127887745696109598901, 1947270476915296449559659317606103024276803403
A NONLINEAR RECURRENCE. REF PRSE 59(2) 159 39. SA2.

1252 1, 3, 21, 6615, 64595475
STABLE FEEDBACK SHIFT REGISTERS. REF RO1 238.

1253 1, 3, 22, 192, 2046, 24853, 329406
PARTITION FUNCTION FOR CUBIC LATTICE. REF JMP 3 185 62.

1254 1, 3, 22, 207, 2412, 31754, 452640, 6840774
WALKS ON A CUBIC LATTICE. REF JMP 3 188 62.

1255 1, 3, 22, 333, 355, 103993, 104348, 208341, 312689, 833719, 1146408, 4272943, 5419351, 80143857, 165707065, 245850922, 411557987, 1068966896
CONVERGENTS TO PI. REF ELM 2 7 47.

1256 1, 3, 23, 165, 3802, 21385, 993605, 15198435, 394722916, 3814933122, 447827009070
COEFFICIENTS FOR STEP-BY-STEP INTEGRATION. REF JACM 11 231 64.

1257 1, 3, 23, 177, 1553, 14963, 157931
POLYGONS. REF IDM 26 118 19.

1258 1, 3, 24, 216, 1824, 15150
CARD MATCHING. REF R1 193.

1259 1, 3, 24, 1676, 22920064
SWITCHING NETWORKS. REF JFI 276 317 63.

1260 1, 3, 25, 155, 1005, 7488, 64164, 619986, 6646750, 78161249, 999473835, 13801761213, 204631472475, 3241541125110
PERMUTATIONS BY NUMBER OF PAIRS. REF DKB 264.

1261 1, 3, 25, 299, 4785, 95699, 2296777, 64309755, 2057912161, 74084837795, 2963393511801, 130389314519243, 6258687096923665, 325451729040030579
PERMUTATIONS WITH NO CYCLES OF LENGTH 4. REF R1 83.

1262 1, 3, 25, 765, 3121, 233275, 823537, 117440505, 387420481, 89999999991
PILE OF COCOANUTS PROBLEM. REF AMM 35 48 28.

1263 1, 3, 27, 143, 3315, 20349, 260015, 1710855, 92116035, 631165425, 8775943605, 61750730457
COEFFICIENTS OF LEGENDRE POLYNOMIALS. REF MTAC 3 17 48.

1264 1, 3, 29, 289, 1627, 27769, 18044381, 145511171, 1514611753, 142324922009
RELATED TO NUMERICAL INTEGRATION FORMULAS. REF MTAC 11 198 57.

1265 1, 3, 29, 322, 3571, 39603, 439204, 4870847, 54018521, 599074578, 6643838879, 73681302247, 817138163596, 9062201101803, 100501350283429
RELATED TO BERNOULLI NUMBERS. REF RCI 141.

1266 1, 3, 30, 70, 315, 693, 12012, 25740, 109395, 230945, 1939938, 4056234, 16900975, 35102025, 1163381400, 2404321560, 9917826435, 20419054425
COEFFICIENTS OF LEGENDRE POLYNOMIALS. REF PR1 156. AS1 798.

1267 1, 3, 30, 175, 4410, 29106, 396396, 2760615, 156434850
COEFFICIENTS OF LEGENDRE POLYNOMIALS. REF PR1 157. FMR 1 362.

1268 1, 3, 31, 8401, 100130704103
TERNARY TREES. REF CMB 11 90 68.

1269 1, 3, 32, 225, 1320, 7007, 34944, 167076, 775200, 3517470, 15690048
PARTITIONS OF A POLYGON BY NUMBER OF PARTS. REF CAY 13 95.

1270 1, 3, 33, 338, 3580, 39525, 452865, 5354832, 65022840, 807560625, 10224817515, 131631305614
C-NETS. REF JCT 4 275 68.

1271 1, 3, 33, 564, 8976, 155124, 2791300, 51395172
SPECIFIC HEAT FOR CUBIC LATTICE. REF PRV 129 102 63.

1272 1, 3, 33, 903, 46113, 3784503, 455538993, 75603118503, 16546026500673, 4616979073434903, 1599868423237443153, 674014138103352845703
MULTIPLES OF GLAISHERS I NUMBERS. REF PLMS 31 229 1899. FMR 1 76.

1273 1, 3, 37, 1, 13, 638
QUEENS PROBLEM. REF SL1 49.

1274 1, 3, 38, 135, 4315, 48125, 950684, 7217406, 682590930
COEFFICIENTS FOR STEP-BY-STEP INTEGRATION. REF JACM 11 231 64.

1275 1, 3, 40, 336, 2304, 14080, 79872, 430080, 2228224, 11206656, 55050240, 265289728, 1258291200
COEFFICIENTS OF CHEBYSHEV POLYNOMIALS. REF LA4 518.

1276 1, 3, 43, 95, 12139, 25333, 81227, 498233, 121563469, 246183839, 32808117961
COEFFICIENTS FOR NUMERICAL DIFFERENTIATION. REF PHM 33 13 42.

1277 1, 3, 45, 252, 28350, 1496880, 3405402000, 17513496000, 7815397590000, 5543722023840000, 235212205868640000, 206559082608278400000
COEFFICIENTS FOR REPEATED INTEGRATION. REF JM2 28 56 49.

1278 1, 3, 48, 765, 12192, 194307, 3096720, 49353213, 786554688, 12535521795, 199781794032, 3183973182717, 50743789129440, 808716652888323
A(N) = 16A(N − 1) − A(N − 2). REF NCM 4 167 1878. TH2 281.

1279 1, 3, 53, 680, 8064, 96370, 1200070, 15778800, 220047400, 3257228485, 51125192475, 849388162448
PERMUTATIONS BY NUMBER OF PAIRS. REF DKB 264.

1280 1, 3, 55, 8103, 8886111, 72004899337, 4311231547115195, 19073465724950996905525, 6235149080811616882909238709
EXP (N2). REF MNAS 14(5) 14 25. FW1. FMR 1 230.**

1281 1, 3, 57, 2763, 250737, 36581523, 7828053417, 2309644635483, 898621108880097, 445777636063460643, 274613643571568682777
GENERALIZED EULER NUMBERS. REF QJM 45 201 14. MTAC 21 689 67.

1282 1, 3, 59, 131, 251, 419, 659, 1019, 971, 1091, 2099, 1931, 1811, 3851, 3299, 2939, 3251, 4091, 4259, 8147, 5099, 9467, 6299, 6971, 8291, 8819
PRIMES BY CLASS NUMBER. REF MTAC 24 492 70.

1283 1, 3, 60, 630, 5040, 34650, 216216
COEFFICIENTS FOR EXTRAPOLATION. REF SE2 93.

1284 1, 3, 60, 1197, 23880, 476403, 9504180, 189607197, 3782639760, 75463188003, 1505481120300, 30034159217997, 599177703239640, 11953519905574803
A(N) = 20A(N − 1) − A(N − 2). REF NCM 4 167 1878. MTS 65(4, SUPPLEMENT) 8 56.

1285 1, 3, 70, 3783
FINITE AUTOMATA. REF CJM 17 112 65.

1286 1, 3, 73, 8599, 400091364
CONTINUED COTANGENT FOR PI. REF DMJ 4 339 38.

1287 1, 3, 196, 3406687200
INVERTIBLE BOOLEAN FUNCTIONS. REF JACM 10 27 63.

1288 1, 3, 567, 43659, 392931, 1724574159, 2498907956391, 1671769422825579, 88417613265912513891, 21857510418232875496803
COEFFICIENTS OF LEMNISCATE FUNCTION. REF MTAC 16 477 62.

1289 1, 3, 840, 54486432000
INVERTIBLE BOOLEAN FUNCTIONS. REF JACM 10 27 63.

SEQUENCES BEGINNING 1, 4

1290 1, 4, 0, 0, 8, 60, 144, 416, 1248, 4200, 13248, 42936, 138072
SUSCEPTIBILITY FOR CUBIC LATTICE. REF PHA 28 947 62.

1291 1, 4, 1, 4, 2, 1, 3, 5, 6, 2, 3, 7, 3, 0, 9, 5, 0, 4, 8, 8, 0, 1, 6, 8, 8, 7, 2, 4, 2, 0, 9, 6, 9, 8, 0, 7, 8, 5, 6, 9, 6, 7, 1, 8, 7, 5, 3, 7, 6, 9, 4, 8, 0, 7, 3, 1, 7, 6, 6, 7, 9, 7, 3, 7, 9, 9, 0, 7, 3
SQUARE ROOT OF 2. REF PNAS 37 65 51. MTAC 22 899 68.

1292 1, 4, 1, 6, 4, 8, 1, 13, 6, 12, 4, 14, 8, 24, 1, 18, 13, 20, 6, 32, 12, 24, 4, 31, 14, 40, 8, 30, 24, 32, 1, 48, 18, 48, 13, 38, 20, 56, 6, 42, 32, 44, 12, 78, 24, 48, 4, 57, 31, 72, 14
SUM OF ODD DIVISORS OF N. REF RCI 187.

1293 1, 4, 1, 12, 186, 4, 86, 4860
QUEENS PROBLEM. REF SL1 49.

1294 1, 4, 1, 16, 16, 120, 8, 728
QUEENS PROBLEM. REF SL1 49.

1295 1, 4, 2, 7, 5, 15, 6, 37, 13, 36, 32, 37, 34, 73, 58, 183, 150, 262, 186, 1009, 420, 707, 703, 760, 1180, 4639
POLYGONAL GRAPHS. REF SL1 21.

1296 1, 4, 2, 8, 5, 4, 10, 8, 9, 0, 14, 16, 10, 4, 0, 8, 14, 20, 2, 0, 11, 20, 32, 16, 0, 4, 14, 8, 9, 20, 26, 0, 2, 28, 0, 16, 16, 28, 22, 0, 14, 16, 0, 40, 0, 28, 26, 32, 17, 0, 32, 16, 22, 0, 10
COEFFICIENTS OF A MODULAR FORM. REF KNAW 59 207 56.

1297 1, 4, 2, 8, 13, 28, 26, 56, 69, 48, 134, 80, 182, 84, 312, 280, 204, 332, 142, 816, 91, 196, 780, 224, 526
RELATED TO REPRESENTATION AS SUMS OF SQUARES. REF QJM 38 56 07.

1298 1, 4, 3, 2, 3, 1, 2, 2, 1, 2, 3, 1, 3, 2, 3, 1, 2, 1, 2, 2, 2, 2, 2, 1, 3, 3, 2, 2, 3, 1, 2, 2, 3, 2, 2, 1, 3, 2, 3, 2, 3, 2, 3, 2, 1, 2, 3, 1, 3, 2, 2, 3, 3, 2, 3, 2, 2, 3, 4, 1, 2, 2, 2, 3, 3, 1, 3, 2, 2
FIBONACCI FREQUENCIES. REF HM1. MTAC 23 460 69. ACA 16 109 69.

1299 1, 4, 3, 4, 2, 9, 4, 4, 8, 1, 9, 0, 3, 2, 5, 1, 8, 2, 7, 6, 5, 1, 1, 2, 8, 9, 1, 8, 9, 1, 6, 6, 0, 5, 0, 8, 2, 2, 9, 4, 3, 9, 7, 0, 0, 5, 8, 0, 3, 6, 6, 6, 5, 6, 6, 1, 1, 4, 4, 5, 3, 7, 8, 3, 1, 6, 5, 8, 6
COMMON LOGARITHM OF E. REF PNAS 26 211 40.

1300 1, 4, 3, 4, 4, 8, 11, 4, 4, 12, 48, 12, 8, 16, 25, 16, 4, 20, 0, 32, 12, 24, 248, 4, 12, 4, 208, 28, 16, 32, 41, 48, 16, 32, 400, 36, 20, 48, 88, 40, 32, 44, 544, 16, 24, 48, 732, 8, 4
COEFFICIENTS OF A DIRICHLET SERIES. REF LEM 6 38 60.

1301 1, 4, 3, 11, 15, 13, 17, 24, 23, 73, 3000, 11000, 15000, 101, 104, 103, 111, 115, 113, 117, 124, 123, 173, 473, 373, 1104, 1103, 1111, 1115, 1113, 1117, 1124, 1123, 1173
SMALLEST NUMBER REQUIRING N LETTERS IN ENGLISH.

1302 1, 4, 3, 32, 75, 216, 3577, 5888, 15741, 106300, 13486539, 9903168, 42194238652, 710986864, 796661595, 127626606592, 450185515446285
COTESIAN NUMBERS. REF QJM 46 63 14.

1303 1, 4, 3, 192, 20, 11520, 315, 573440, 36288, 928972800, 1663200, 54499737600, 74131200, 1322526965760, 68108040000
VALUES OF AN INTEGRAL. REF PHM 36 295 45. MTAC 19 114 65.

1304 1, 4, 4, 2, 2, 4, 9, 5, 7, 0, 3, 0, 7, 4, 0, 8, 3, 8, 2, 3, 2, 1, 6, 3, 8, 3, 1, 0, 7, 8, 0, 1, 0, 9, 5, 8, 8, 3, 9, 1, 8, 6, 9, 2, 5, 3, 4, 9, 9, 3, 5, 0, 5, 7, 7, 5, 4, 6, 4, 1, 6, 1, 9, 4, 5, 4, 1, 6, 8
CUBE ROOT OF 3. REF SMA 18 175 52.

1305 1, 4, 4, 32, 16, 56, 80, 192, 98, 740, 704, 96, 224, 2440, 3520, 2624, 351, 780, 10632, 2688, 2960, 9496, 18176, 14208, 3934, 12552, 9856, 24608, 9760, 2720, 25344
RELATED TO REPRESENTATION AS SUMS OF SQUARES. REF QJM 38 320 07.

1306 1, 4, 5, 0, 4, 6, 0, 6, 8, 7, 0, 1, 8, 3, 5, 5, 1, 7, 7, 6, 6, 2, 3, 3, 8, 2, 3, 1, 1, 9, 7, 0, 9, 2, 2, 1, 1, 6, 6, 9, 3, 0, 1, 0, 1, 8, 7, 3, 6, 9, 3, 5, 2, 0, 8, 9, 0, 9, 8, 4, 8, 6, 9, 5, 4, 2, 3, 3, 6
NATURAL LOGARITHM OF EULERS CONSTANT. REF RS4 XVIII.

1307 1, 4, 5, 6, 4, 8, 13, 13, 6, 12, 20, 14, 8, 24, 29
GENERALIZED DIVISOR FUNCTION. REF PLMS 19 111 19.

1308 1, 4, 5, 6, 10, 15, 21, 31, 46, 67, 98, 144, 211, 309, 453, 664, 973, 1426, 2090, 3063, 4489, 6579, 9642, 14131, 20710, 30352, 44483, 65193, 95545, 140028, 205221
A(N) = A(N − 1) + A(N − 3). REF JA2 91. FQ 6(3) 68 68.

1309 1, 4, 5, 9, 14, 23, 37, 60, 97, 157, 254, 411, 665, 1076, 1741, 2817, 4558, 7375, 11933, 19308, 31241, 50549, 81790, 132339, 214129, 346468, 560597, 907065, 1467662
A(N) = A(N − 1) + A(N − 2). REF FQ 3 129 63.

1310 1, 4, 5, 10, 14, 41, 94, 154, 158
13.4N + 1 IS PRIME. REF PAMS 9 674 58.**

1311 1, 4, 5, 11, 16, 29, 45, 76, 121, 199, 320, 521, 841, 1364, 2205, 3571, 5776, 9349, 15125, 24476, 39601, 64079, 103680, 167761, 271441, 439204, 710645, 1149851
ASSOCIATED MERSENNE NUMBERS. REF EUR 11 22 49.

1312 1, 4, 5, 11, 20, 36, 65, 119, 218, 412, 770, 1466, 2784, 5322, 10226, 19691, 38048, 73665, 142927, 277822, 540851, 1054502, 2058507, 4023164
POPULATION OF 2U2 + 5V**2. REF MTAC 20 563 66.**

1313 1, 4, 5, 11, 22, 57, 51, 156
THE NO-THREE-IN-LINE PROBLEM. REF GU3. WE1 124.

1314 1, 4, 5, 15, 19, 45, 52, 118, 137, 281, 316, 625, 695, 1331
COEFFICIENTS OF MODULAR FUNCTIONS. REF PLMS 9 385 59.

1315 1, 4, 5, 16, 17, 20, 21, 64, 65, 68, 69, 80, 81, 84, 85, 256, 257, 260, 261, 272, 273, 276, 277, 320, 321, 324, 325, 336, 337, 340, 341, 1024, 1025, 1028, 1029, 1040, 1041
SUMS OF DISTINCT POWERS OF 4. REF MMAG 35 37 62. MTAC 18 537 64.

1316 1, 4, 5, 29, 34, 63, 286, 349, 635, 5429, 6064, 90325, 96389, 1054215, 2204819, 3259034, 15240955, 186150494, 387541943, 1348776323, 3085094589
CONVERGENTS TO CUBE ROOT OF 2. REF AMP 46 105 1866. LE1 67. HPR.

1317 1, 4, 6, 4, 3, 12, 16, 16, 6, 8, 18, 28, 26, 20, 2, 12, 23, 32, 36, 28, 6, 4, 22, 20, 39, 32, 32, 12, 2, 16, 12, 24, 40, 28, 34, 0, 6, 16, 0, 40, 6, 36, 26, 32, 5, 0, 20
COEFFICIENTS OF A MODULAR FORM. REF JLMS 39 435 64.

1318 1, 4, 6, 7, 7, 8, 9, 9, 10, 10, 10, 11, 11, 12, 12, 12, 13, 13, 13, 13, 14, 14, 14, 15, 15, 15, 15, 16, 16, 16, 16, 16, 17, 17, 17, 17, 18, 18, 18, 18, 18, 19, 19, 19, 19, 19, 19, 20
CHROMATIC NUMBERS. REF CJM 4 480 52.

1319 1, 4, 6, 7, 9, 10, 11, 12, 14, 15, 16, 17, 18, 19, 20, 22, 23, 24, 25, 26, 27, 28, 29, 30, 31, 32, 33, 35, 36, 37, 38, 39, 40, 41, 42, 43, 44, 45, 46, 47, 48, 49, 50, 51, 52, 53, 54
NON-FIBONACCI NUMBERS. REF FQ 3 183 65.

1320 1, 4, 6, 7, 9, 10, 15, 16, 22, 24, 25, 28, 31, 33, 36, 40, 42, 49
OF THE FORM X2 + 6Y**2. REF EUL (1) 1 425 11.**

1321 1, 4, 6, 7, 13, 14, 16, 20, 21, 23, 25, 27, 29, 32, 34, 42, 45, 49, 51, 53, 59, 60, 70, 75, 78, 81, 84, 85, 86, 87, 88, 90, 93, 95, 96, 104, 109, 114, 115, 116, 124, 125, 135, 137
ELLIPTIC CURVES. REF JRAM 212 23 63.

1322 1, 4, 6, 8, 9, 10, 12, 14, 15, 16, 18, 20, 21, 22, 24, 25, 26, 27, 28, 30, 32, 33, 34, 35, 36, 38, 39, 40, 42, 44, 45, 46, 48, 49, 50, 51, 52, 54, 55, 56, 57, 58, 60, 62, 63, 64, 65
COMPOSITE NUMBERS.

1323 1, 4, 6, 9, 10, 14, 15, 21, 22, 25, 26, 33, 34, 35, 38, 39, 46, 49, 51, 55, 57, 58, 62, 65, 69, 74, 77, 82, 85, 86, 87, 91, 93, 94, 95, 106, 111, 115, 118, 119, 121, 122, 123, 129
PRODUCT OF TWO PRIMES. REF EUR 17 8 54.

1324 1, 4, 6, 11, 14, 21, 24, 26, 29, 31, 39, 44, 46, 51, 54, 76, 79, 89, 94, 99, 101, 111, 119, 124, 129, 131, 136, 146, 149, 154, 156, 164, 176, 179, 181, 194, 201, 211, 214, 229
(4N2 + 1)/2 IS PRIME. REF EUL (1) 3 24 17.**

1325 1, 4, 6, 19, 49, 150, 442, 1424, 4522, 14924, 49536, 167367, 570285, 1965058
FLEXAGONS. REF AMM 64 153 57.

1326 1, 4, 6, 24, 66, 214, 676, 2209, 7296, 24460, 82926, 284068, 981882, 3421318, 12007554, 42416488, 150718770, 538421590, 1932856590
ROOTED C-NETS. REF CJM 15 265 63.

1327 1, 4, 7, 8, 9, 10, 11, 12, 12, 13, 13, 14, 15, 15, 16, 16, 16, 17, 17, 18, 18, 19, 19, 19, 20, 20, 20, 21, 21, 21, 22, 22, 22, 23, 23, 23, 24, 24, 24, 24, 25, 25, 25, 25, 26, 26, 26
CHROMATIC NUMBERS IF 4 COLOR CONJECTURE TRUE. REF PNAS 60 438 68.

1328 1, 4, 7, 8, 10, 26, 32, 70, 74, 122, 146, 308, 314, 386, 512, 554, 572, 626, 635, 728, 794, 842, 910, 914, 1015, 1082
RELATED TO EULERS TOTIENT FUNCTION. REF AMM 56 22 49.

1329 1, 4, 7, 8, 11, 17, 20, 20, 23, 29, 35, 38, 39, 45, 51, 51, 54, 63, 69, 72, 78, 84, 87, 87, 90, 99, 111, 115, 115, 127, 133, 133, 136, 142, 151, 157, 163, 169, 178, 178, 184, 199
POPULATION OF U2 + V**2 + W**2. REF PNISI 13 39 47.**

1330 1, 4, 7, 10, 13, 17, 22, 25, 30, 35, 40, 46, 53, 57, 61
ZARANKIEWICZS PROBLEM. REF TI1 132. CO1 2 138.

1331 1, 4, 7, 10, 13, 19, 28, 31, 34, 40, 43, 52, 70, 73, 76, 82, 85, 91, 97, 103, 112, 115, 124, 127, 136, 145, 148, 157, 166, 175, 187, 190, 199, 202, 223, 241, 244, 259, 265, 271
(N(N + 1) + 1)/3 IS PRIME. REF CU1 1 248.

1332 1, 4, 7, 13, 25, 49, 94, 181, 349, 673, 1297, 2500, 4819, 9289, 17905, 34513, 66526, 128233, 247177, 476449, 918385, 1770244, 3412255, 6577333, 12678217
TETRANACCI NUMBERS. REF FQ 2 260 64.

1333 1, 4, 7, 31, 871, 756031, 571580604871, 326704387862983487112031, 106735757048926752040856495274871386126283608871
A NONLINEAR RECURRENCE. REF AMM 70 403 63.

1334 1, 4, 8, 1, 2, 1, 1, 8, 2, 5, 0, 5, 9, 6, 0, 3, 4, 4, 7, 4, 9, 7, 7, 5, 8, 9, 1, 3, 4, 2, 4, 3, 6, 8, 4, 2, 3, 1, 3, 5, 1, 8, 4, 3, 3, 4, 3, 8, 5, 6, 6, 0, 5, 1, 9, 6, 6, 1, 0, 1, 8, 1, 6, 8, 8, 4, 0, 1, 6
NATURAL LOGARITHM OF THE GOLDEN RATIO. REF RS4 XVIII.

1335 1, 4, 8, 9, 16, 25, 27, 32, 36, 49, 64, 72, 81, 100, 108, 121, 125, 128, 144, 169, 196, 200, 216, 225, 243, 256, 288, 289, 324, 343, 361, 392, 400, 432, 441, 484, 500, 512, 529
POWERFUL NUMBERS. REF AMM 77 848 70.

1336 1, 4, 8, 9, 16, 25, 27, 32, 36, 49, 64, 81, 100, 121, 125, 128, 144, 169, 196, 216, 225, 243, 256, 289, 324, 343, 361, 400, 441, 484, 512, 529, 576, 625, 676, 729, 784, 841
PERFECT POWERS. REF FQ 8 268 70.

1337 1, 4, 8, 12, 13, 11, 28, 16, 23, 20, 1, 20, 12, 60, 11, 28, 59, 56, 52, 16, 28, 83, 84, 13, 112, 121, 61, 96, 144, 71, 160, 136, 140, 44, 76, 88, 181, 152, 179, 24, 47, 120, 220
QUADRATIC PARTITIONS OF PRIME-SQUARES. REF CU3 79. LE1 60.

1338 1, 4, 8, 12, 17, 21, 25, 30, 34, 38, 43, 47, 51
A BEATTY SEQUENCE. REF CMB 2 188 59.

1339 1, 4, 8, 12, 17, 22, 27, 32
DAVENPORT-SCHINZEL NUMBERS. REF PL2 1 250 70.

1340 1, 4, 8, 16, 25, 37, 53, 71, 94, 122
POSTAGE STAMP PROBLEM. REF CJ1 12 379 69.

1341 1, 4, 8, 17, 39, 89, 211, 507, 1238, 3057, 7638
UNIVALENT RADICALS. REF CAY 9 545.

1342 1, 4, 8, 18, 32, 58, 94, 151, 227, 338, 480, 676, 920, 1242, 1636
RESTRICTED PARTITIONS. REF CAY 2 279.

1343 1, 4, 8, 20, 92, 2744, 950998216
BOOLEAN FUNCTIONS. REF JACM 13 153 66.

1344 1, 4, 8, 22, 51, 136
BENZOLS. REF ZFK 93 422 36.

1345 1, 4, 8, 25, 53, 164, 348, 1077, 2285, 7072, 40051, 46437, 98521, 304920, 646920, 2002201, 4247881, 13147084, 27892928, 86327905
A TERNARY CONTINUED FRACTION. REF TOH 37 441 33.

1346 1, 4, 8, 38, 209, 1400, 10849, 95516
HIT POLYNOMIALS. REF RI3.

1347 1, 4, 8, 48, 10, 224, 80, 448, 231, 40, 248, 1408, 1466, 2240, 80, 1280, 4766, 924, 1944, 480, 9600, 6944, 2704, 8704, 15525, 5864, 3984, 14080, 25498, 2240, 10816
RELATED TO REPRESENTATION AS SUMS OF SQUARES. REF QJM 38 190 07.

1348 1, 4, 9, 11, 16, 29, 49, 76, 121, 199, 324, 521, 841, 1364, 2209, 3571, 5776, 9349, 15129, 24476, 39601, 64079, 103684, 167761, 271441, 439204, 710649, 1149851
A FIELDER SEQUENCE. REF FQ 6(3) 68 68.

1349 1, 4, 9, 16, 22, 36, 65, 112, 186, 309, 522, 885, 1492, 2509, 4225, 7124, 12010, 20236, 34094, 57453, 96823, 163163, 274946, 463316, 780755, 1315687, 2217112
A FIELDER SEQUENCE. REF FQ 6(3) 68 68.

1350 1, 4, 9, 16, 25, 36, 49, 64, 81, 100, 121, 144, 169, 196, 225, 256, 289, 324, 361, 400, 441, 484, 529, 576, 625, 676, 729, 784, 841, 900, 961, 1024, 1089, 1156, 1225, 1296
SQUARES. REF BA1.

1351 1, 4, 9, 16, 27, 36, 53, 70, 90, 113, 147, 173, 213, 260, 303, 355, 419, 477, 549, 634, 715, 806, 903, 1013, 1128, 1255, 1383, 1525, 1679, 1842, 2011, 2189, 2383, 2585
POSTAGE STAMP PROBLEM. REF CJ1 12 379 69. LU2.

1352 1, 4, 9, 16, 28, 43, 73, 130, 226, 386, 660, 1132, 1947, 3349, 5753, 9878, 16966, 29147, 50074, 86020, 147764, 253829, 436036, 749041, 1286728, 2210377, 3797047
A FIELDER SEQUENCE. REF FQ 6(3) 68 68.

1353 1, 4, 9, 19, 37, 73, 143, 279, 548, 1079, 2132, 4223, 8384, 16673, 33203, 66190, 132055, 263619, 526502, 1051899, 2102137, 4201783, 8399828, 16794048, 33579681
FROM WARINGS PROBLEM. REF HAR 1 668.

1354 1, 4, 9, 21, 40, 74, 125, 209, 330, 515, 778, 1160, 1690, 2439, 3457, 4857, 6735, 9264, 12607, 17040, 22826, 30391, 40165, 52788, 68938, 89589, 115778, 148957
BIPARTITE PARTITIONS. REF NI1 11.

1355 1, 4, 9, 22, 46, 102, 206, 427, 841, 1658, 3173, 6038, 11251, 20807, 37907, 68493, 122338, 216819, 380637, 663417, 1147033, 1969961, 3359677, 5694592, 9592063
SOLID PARTITIONS. REF MTAC 24 956 70.

1356 1, 4, 9, 28, 71, 202
CONNECTED GRAPHS WITH ONE CYCLE. REF R1 150.

1357 1, 4, 9, 61, 52, 63, 94, 46, 18, 1, 121, 441, 961, 691, 522, 652, 982, 423, 163, 4, 144, 484, 925, 675, 526, 676, 927, 487, 148, 9, 169, 4201, 9801, 6511, 5221, 6921, 9631
SQUARES WRITTEN BACKWARDS.

1358 1, 4, 9, 121, 484, 676, 10201, 12321, 14641, 40804, 44944, 69696, 94249, 698896, 1002001, 1234321, 4008004, 5221225, 6948496, 100020001, 102030201
PALINDROMIC SQUARES. REF JRM 3 94 70.

1359 1, 4, 10, 14, 20, 24, 30, 36, 40, 46, 50, 56, 60, 66, 72, 76, 82, 86, 92, 96, 102, 108, 112, 118, 122, 128, 132, 138, 150, 160, 169, 176, 186, 192, 196, 202, 206, 212, 218, 222
FIBONACCI NIM. REF FQ 3 62 65.

1360 1, 4, 10, 17, 18, 30, 34, 69, 109, 111, 189, 192, 193, 194, 195, 311, 763, 898, 900, 2215, 2810, 2811, 2812, 2813, 3417, 4260, 6000, 6002, 6003, 6004, 23331, 31569, 31601
RELATED TO GAPS BETWEEN PRIMES. REF MTAC 13 122 59.

1361 1, 4, 10, 17, 27, 40, 54, 71, 100, 121, 144, 170, 207, 237, 270, 314, 351, 400, 441, 484, 540, 587, 647, 710, 764, 831, 1000, 1061, 1134, 1210, 1277, 1357, 1440, 1524
SQUARES WRITTEN IN BASE 9. REF TH2 98.

1362 1, 4, 10, 20, 34, 56, 80, 120, 154, 220
COMPOSITIONS INTO RELATIVELY PRIME PARTS. REF FQ 2 250 64.

1363 1, 4, 10, 20, 35, 56, 84, 120, 165, 220, 286, 364, 455, 560, 680, 816, 969, 1140, 1330, 1540, 1771, 2024, 2300, 2600, 2925, 3276, 3654, 4060, 4495, 4960, 5456, 5984
TETRAHEDRAL NUMBERS OR BINOMIAL COEFFICIENTS C(N, 3). REF DI2 2 4. RS1. BE3 194. AS1 828.

1364 1, 4, 10, 20, 36, 64, 120, 240, 496, 952
EXPANSION OF BRACKET FUNCTION. REF FQ 2 254 64.

1365 1, 4, 10, 21, 40, 72, 125, 212
HIT POLYNOMIALS. REF RI3.

1366 1, 4, 10, 23, 40, 68, 108, 167, 241, 345, 482, 653, 869
FROM STORMERS PROBLEM. REF IJM 8 66 64.

1367 1, 4, 10, 23, 45, 83, 142, 237, 377, 588, 892, 1330, 1943, 2804, 3982, 5595, 7768, 10686, 14555, 19674, 26371, 35112, 46424, 61015, 79705, 103579, 133883, 172243
BIPARTITE PARTITIONS. REF NI1 19.

1368 1, 4, 10, 23, 48, 94, 166, 285, 464, 734, 1109, 1646
RESTRICTED PARTITIONS. REF CAY 2 280.

1369 1, 4, 10, 24, 49, 94, 169, 289, 468, 734, 1117, 1656
RESTRICTED PARTITIONS. REF CAY 2 280.

1370 1, 4, 10, 24, 70, 208, 700, 2344, 8230, 29144, 104968, 381304, 1398500, 5162224, 19175140, 71582944, 268439590, 1010580544, 3817763740, 14467258264
NECKLACES OF 4 COLORS. REF R1 162. IJM 5 658 61.

1371 1, 4, 10, 26, 59, 140, 307, 684, 1464, 3122, 6500, 13426, 27248, 54804, 108802, 214071, 416849, 805124, 1541637, 2930329, 5528733, 10362312, 19295226, 35713454
SOLID PARTITIONS. REF MTAC 24 956 70.

1372 1, 4, 10, 26, 59, 141, 310, 692, 1483, 3162, 6583, 13602, 27613, 55579, 110445, 217554, 424148, 820294, 1572647, 2992892, 5652954, 10605608, 19765082
RELATED TO SOLID PARTITIONS. REF PNISI 26 135 60. PCPS 63 1100 67.

1373 1, 4, 10, 27, 74, 202, 548, 1490, 4052, 11013, 29937, 81377, 221207, 601302, 1634509, 4443055, 12077476, 32829985, 89241150, 242582598, 659407867
SINH(N). REF AMP 3 33 1843. MNAS 14(5) 14 25. HA4. LF1 93.

1374 1, 4, 10, 30, 65, 173, 343, 778, 1518, 3088, 5609
RESTRICTED PARTITIONS. REF JCT 9 373 70.

1375 1, 4, 10, 30, 100, 354, 1300, 4890, 18700, 72354, 282340, 1108650, 4373500, 17312754, 68711380, 273234810, 1088123500, 4338079554, 17309140420
1N + 2**N + ··· + 4**N. REF AS1 813.**

1376 1, 4, 10, 56, 29, 332, 30, 1064, 302, 1940, 288, 1960, 1071, 1192, 1938, 736, 2000, 1488, 5014, 7288, 4170, 10644, 8482, 11184, 12647, 15544
RELATED TO REPRESENTATION AS SUMS OF SQUARES. REF QJM 38 56 07.

1377 1, 4, 11, 13, 23, 20, 24, 37, 61, 40, 71, 56, 97, 107, 73, 80, 143, 84, 131, 157, 191, 193, 112, 169, 132, 143, 140, 156, 229, 179, 176, 181, 241, 251, 359, 349, 347, 204, 313
QUADRATIC PARTITIONS OF PRIME-SQUARES. REF CU3 79. LE1 60.

1378 1, 4, 11, 20, 31, 44, 61, 100, 121, 144, 171, 220, 251, 304, 341, 400, 441, 504, 551, 620, 671, 744, 1021, 1100, 1161, 1244, 1331, 1420, 1511, 1604, 1701, 2000, 2101, 2204
SQUARES WRITTEN IN BASE 8. REF TH2 95.

1379 1, 4, 11, 23, 79, 148, 533, 977, 3553, 6484, 23627, 43079, 157039, 286276
A(2N) = A(2N − 1) + 3A(2N − 2), A(2N + 1) = 2A(2N) + 3A(2N − 1). REF MQET 1 12 16.

1380 1, 4, 11, 26, 52, 98, 171, 289, 467, 737, 1131, 1704, 2515, 3661, 5246, 7430, 10396, 14405, 19760, 26884, 36269, 48583, 64614, 85399, 112170, 146526, 190362
BIPARTITE PARTITIONS. REF NI1 11.

1381 1, 4, 11, 26, 56, 114, 223, 424, 789, 1444
ARRAYS OF DUMBBELLS. REF JMP 11 3098 70.

1382 1, 4, 11, 26, 57, 120, 247, 502, 1013, 2036, 4083, 8178, 16369, 32752, 65519, 131054, 262125, 524268, 1048555, 2097130, 4194281, 8388584, 16777191, 33554406
EULERIAN NUMBERS 2N − N − 1. REF R1 215. DB1 151.**

1383 1, 4, 11, 29, 54, 99, 163, 239, 344, 486, 648, 847, 1069, 1355, 1680, 2046, 2446, 2911, 3443, 4022, 4662, 5395, 6145, 6998, 7913, 8913, 10006, 11194, 12437, 13751
POPULATION OF U2 + V**2 + W**2. REF PNISI 13 37 47.**

1384 1, 4, 11, 29, 76, 199, 521, 1364, 3571, 9349, 24476, 64079, 167761, 439204, 1149851, 3010349, 7881196, 20633239, 54018521, 141422324, 370248451, 969323029
BISECTION OF LUCAS SEQUENCE. REF FQ 9 284 71.

1385 1, 4, 11, 31, 83, 227, 616, 1674, 4550, 12367, 33617, 91380, 248397, 675214, 1835421, 4989191, 13562027, 36865412, 100210581, 272400600
FROM PARTIAL SUMS OF HARMONIC SERIES. REF AMM 78 870 71.

1386 1, 4, 11, 35, 101, 290, 804, 2256, 6296, 17689, 49952, 142016, 406330, 1169356, 3390052
PARAFFINS. REF JACS 54 1544 32.

1387 1, 4, 11, 41, 162, 715, 3425, 17722, 98253, 580317, 3633280, 24011157, 166888165, 1216070380, 9264071767, 73600798037, 608476008122, 5224266196935
EXPANSION OF EXP(EXP(X) - 1 - X).

1388 1, 4, 11, 60, 362, 2987
CUBIC GRAPHS. REF RE4.

1389 1, 4, 11, 64, 5276
SWITCHING NETWORKS. REF JFI 276 317 63.

1390 1, 4, 11, 79, 7621
SWITCHING NETWORKS. REF JFI 276 317 63.

1391 1, 4, 12, 15, 21, 35, 40, 45, 60, 55, 80, 72, 99, 91, 112, 105, 140, 132, 165, 180, 168, 195, 221, 208, 209, 255, 260, 252, 231, 285, 312, 308, 288, 299, 272, 275, 340, 325
QUADRATIC PARTITIONS OF PRIME-SQUARES. REF CU3 77. LE1 60.

1392 1, 4, 12, 22, 34, 51, 100, 121, 144, 202, 232, 264, 331, 400, 441, 514, 562, 642, 1024, 1111, 1200, 1261, 1354, 1452, 1552, 1654, 2061, 2200, 2311, 2424, 2542, 2662
SQUARES WRITTEN IN BASE 7. REF TH2 93.

1393 1, 4, 12, 25, 44, 70, 104, 147, 200, 264, 340, 429, 532, 650, 784, 935, 1104, 1292, 1500, 1729, 1980, 2254, 2552, 2875, 3224, 3600, 4004, 4437, 4900, 5394, 5920, 6479
POWERS OF ROOTED TREE ENUMERATOR. REF R1 150.

1394 1, 4, 12, 28, 68, 164, 396, 940, 2244, 5324, 12668, 29940, 71012, 167468, 396204
WALKS ON A SQUARE LATTICE. REF JCP 34 1261 61.

1395 1, 4, 12, 30, 70, 159, 339, 706, 1436, 2853
PARTITIONS INTO NON-INTEGRAL POWERS. REF PCPS 47 215 51.

1396 1, 4, 12, 31, 71, 147, 285, 519, 902, 1502
RESTRICTED PARTITIONS. REF CAY 2 281.

1397 1, 4, 12, 32, 76, 168, 352, 704
COEFFICIENTS OF AN ELLIPTIC FUNCTION. REF CAY 9 128.

1398 1, 4, 12, 32, 80, 192, 448, 1024, 2304, 5120, 11264, 24576, 53248, 114688, 245760, 524288, 1114112, 2359296, 4980736
COEFFICIENTS OF CHEBYSHEV POLYNOMIALS. REF PRSE 62 190 46. AS1 796.

1399 1, 4, 12, 32, 88, 240, 652, 1744, 4616, 12208, 32328, 85408, 224608, 588832
WALKS ON A KAGOME LATTICE. REF PRV 114 53 59.

1400 1, 4, 12, 36, 96, 264, 648, 1584, 3576, 7872, 15360, 29184, 51120, 90384, 158448, 286296, 509808, 904296, 1556304
STRONGLY ASYMMETRIC SEQUENCES. REF MTAC 25 159 71.

1401 1, 4, 12, 36, 100, 276, 740, 1972, 5172, 13492, 34876, 89764, 229628, 585508, 1486308, 3763460, 9497380
SUSCEPTIBILITY FOR SQUARE LATTICE. REF PHA 28 924 62.

1402 1, 4, 12, 36, 100, 284, 780, 2172, 5916, 16268, 44100, 120292, 324932, 881500, 2374444, 6416596, 17245332, 46466676, 124658732, 335116620, 897697164
WALKS ON A SQUARE LATTICE. REF JCP 34 1537 61. MFS.

1403 1, 4, 12, 36, 108, 324, 948, 2796, 8196, 24060, 70188, 205284, 597996, 1744548, 5073900, 14774652, 42922308
WALKS ON A DIAMOND LATTICE. REF PHA 29 381 63.

1404 1, 4, 12, 44, 172, 772, 3308, 14924, 64956, 294252, 1301044
WALKS ON A CUBIC LATTICE. REF PCPS 58 99 62.

1405 1, 4, 12, 80, 3984, 37333248, 25626412338274304
BOOLEAN FUNCTIONS. REF HA2 147.

1406 1, 4, 12, 132, 3156, 136980, 10015092, 1199364852
COLORED GRAPHS. REF CJM 22 596 70.

1407 1, 4, 13, 36, 87, 190, 386, 734, 1324
INCIDENCE MATRICES. REF CPM 89 217 64.

1408 1, 4, 13, 36, 93, 225, 528, 1198, 2666, 5815, 12517, 26587, 55933, 116564, 241151, 495417, 1011950, 2055892, 4157514, 8371318, 16792066, 33564256, 66875221
TREES OF HEIGHT 4. REF IBMJ 4 475 60. KU1.

1409 1, 4, 13, 41, 131, 428, 1429, 4861, 16795, 58785, 208011, 742899, 2674439, 9694844, 35357669, 129644789, 477638699, 1767263189, 6564120419, 24466267019
PERMUTATIONS BY SUBSEQUENCES. REF MTAC 22 390 68.

1410 1, 4, 13, 42, 131, 402
PARAFFINS. REF ZFK 93 437 36.

1411 1, 4, 13, 50, 203, 1154, 6627, 49352, 403273, 3862376
SUMS OF LOGARITHMIC NUMBERS. REF MST 31 78 63.

1412 1, 4, 14, 40, 101, 236, 518, 1080, 2162, 4180, 7840, 14328, 25591, 44776, 76918, 129952, 216240, 354864, 574958
COEFFICIENTS OF AN ELLIPTIC FUNCTION. REF CAY 9 128.

1413 1, 4, 14, 40, 105, 256, 594, 1324, 2860, 6020, 12402, 25088
CONVOLVED FIBONACCI NUMBERS. REF RCI 101.

1414 1, 4, 14, 44, 133, 388, 1116, 3168, 8938, 25100, 70334, 196824, 550656, 1540832, 4314190, 12089368, 33911543, 95228760, 267727154, 753579420, 2123637318
POWERS OF ROOTED TREE ENUMERATOR. REF R1 150.

1415 1, 4, 14, 48, 165, 572, 2002, 7072, 25194, 90440, 326876, 1188640
PARTITIONS OF A POLYGON BY NUMBER OF PARTS. REF CAY 13 95.

1416 1, 4, 14, 49, 174, 628, 2298, 8504
PERMUTATIONS BY INVERSIONS. REF NET 96.

1417 1, 4, 14, 56, 331, 1324, 12284, 49136
RELATED TO EULER NUMBERS. REF JIMS 14 146 22.

1418 1, 4, 15, 54, 193, 690, 2476, 8928, 32358, 117866, 431381, 1585842, 5853849, 21690378, 80650536, 300845232, 1125555054, 4222603968, 15881652606
A SIMPLE RECURRENCE. REF IC 16 351 70.

1419 1, 4, 15, 55, 58, 74, 109, 110, 119, 140, 175, 245, 294, 418, 435, 452, 474, 492, 528, 535, 550, 562, 588, 644, 688, 702, 714, 740, 747, 753, 818, 868, 908, 918, 1098
SOLUTION OF A DIOPHANTINE EQUATION. REF MTAC 16 484 62. AT1 112.

1420 1, 4, 15, 56, 209, 780, 2911, 10864, 40545, 151316, 564719, 2107560, 7865521, 29354524, 109552575, 408855776, 1525870529, 5694626340, 21252634831
A(N) = 4A(N - 1) - A(N - 2). REF NCM 4 167 1878. AMM 24 82 17. MMAG 40 78 67. MTAC 24 180 70, 25 799 71.

1421 1, 4, 15, 56, 210, 792, 3003, 11440, 43758, 167960, 646646, 2496144
COEFFICIENTS OF CHEBYSHEV POLYNOMIALS. REF LA4 517. AS1 795. PL2 1 292 70.

1422 1, 4, 15, 76, 373, 2676, 17539, 152860, 1383561, 14658148
SUMS OF LOGARITHMIC NUMBERS. REF MST 31 79 63.

1423 1, 4, 15, 76, 455, 3186, 25487, 229384, 2293839
THE GAME OF MOUSETRAP. REF QJM 15 241 1878.

1424 1, 4, 15, 92, 653, 5897, 9323, 84626, 433832, 795028, 841971, 6939937, 51058209, 74944592, 307816406, 2862089986, 28034825342, 117067982148
DIGITS OF PI. REF MTAC 16 80 62.

1425 1, 4, 15, 276, 5534533
SWITCHING NETWORKS. REF JFI 276 317 63.

1426 1, 4, 16, 25, 37, 46, 58, 88, 109, 130, 142, 151, 184, 193, 205, 247, 268, 298, 310, 319, 331, 340, 382, 394, 403, 415, 424, 457, 478, 487, 541, 550, 604, 613, 688, 697, 709
(N(N + 1) + 1)/21 IS PRIME. REF CU1 1 252.

1427 1, 4, 16, 46, 106, 316, 1324, 5356, 18316, 63856, 272416, 1264264, 5409496, 22302736, 101343376, 507711376, 2495918224, 11798364736, 58074029056
EVEN PERMUTATIONS OF ORDER 2. REF CJM 7 168 55.

1428 1, 4, 16, 64, 256, 1024, 4096, 16384, 65536, 262144, 1048576, 4194304, 16777216, 67108864, 268435456, 1073741824, 4294967296, 17179869184
POWERS OF FOUR. REF BA1.

1429 1, 4, 16, 64, 736, 11584, 43072, 607232, 50435584, 1204185088
SUSCEPTIBILITY FOR DIAMOND LATTICE. REF PPS 86 13 65.

1430 1, 4, 16, 69, 348, 2016, 13357, 99376, 822040, 7477161, 74207208, 797771520, 9236662345, 114579019468, 1516103040832, 21314681315997
PERMUTATIONS BY LENGTH OF RUNS. REF DKB 261.

1431 1, 4, 16, 78, 457, 2938, 20118
TRIANGULATIONS OF THE DISK. REF PLMS 14 765 64.

1432 1, 4, 16, 80, 672, 4896, 49920, 460032, 5598720, 62584320, 885381120, 11644323840, 187811205120, 2841958748160, 51481298534400, 881192033648640
RESTRICTED PERMUTATIONS. REF MU1 3 468.

1433 1, 4, 16, 392, 1966074
SWITCHING NETWORKS. REF JFI 276 317 63.

1434 1, 4, 17, 72, 305, 1292, 5473, 23184, 98209, 416020, 1762289, 7465176, 31622993, 133957148, 567451585, 2403763488, 10182505537, 43133785636
A(N) = 4A(N - 1) + A(N - 2). REF TH2 282.

1435 1, 4, 18, 89, 466, 2537
ROOTED PLANAR MAPS. REF CJM 15 542 63.

1436 1, 4, 18, 96, 600, 4320, 35280, 322560, 3265920, 36288000, 439084800, 5748019200, 80951270400, 1220496076800, 19615115520000, 334764638208000
DIFFERENCES OF FACTORIAL NUMBERS. REF JRAM 198 61 57.

1437 1, 4, 18, 112, 820, 6912, 66178, 708256, 8372754, 108306280, 1521077404, 23041655136, 374385141832, 6493515450688, 119724090206940
STOCHASTIC MATRICES OF INTEGERS. REF DMJ 35 659 68.

1438 1, 4, 18, 126, 1160, 15973
SEMIGROUPS. REF PL1. MA4 2 2 67.

1439 1, 4, 18, 166, 7579, 7828352, 2414682040996
MONOTONE BOOLEAN FUNCTIONS, OR DEDEKINDS PROBLEM. REF HA2 188. BI1 63. CO1 2 116. WE1 181.

1440 1, 4, 19, 66, 219, 645, 1813, 4802, 12265, 30198, 72396, 169231, 387707, 871989, 1930868, 4215615, 9091410, 19389327, 40944999, 85691893, 177898521
TREES OF DIAMETER 8. REF IBMJ 4 476 60. KU1.

1441 1, 4, 19, 556, 2945786
SWITCHING NETWORKS. REF JFI 276 317 63.

1442 1, 4, 19, 632, 19245637
SWITCHING NETWORKS. REF JFI 276 317 63.

1443 1, 4, 19, 5779, 192900153619, 7177905237579946589743592924684179
A(N) = A(N - 1)3 - 3A(N - 1)**2 + 3. REF CR 83 1287 1876. DI2 1 397.**

1444 1, 4, 20, 56, 120, 220, 364, 560, 816, 1140, 1540, 2024, 2600, 3276, 4060, 4960, 5984, 7140, 8436, 9880, 11480, 13244, 15180, 17296, 19600, 22100, 24804, 27720
2N(N + 1)(2N + 1)/3. REF MTAC 4 23 50.

1445 1, 4, 20, 120, 840, 6720, 60480, 604800, 6652800, 79833600, 1037836800, 14529715200, 217945728000, 3487131648000, 59281238016000
GENERALIZED STIRLING NUMBERS. REF PEF 77 44 62.

1446 1, 4, 20, 124, 920, 7940, 78040
RELATED TO GAMMA FUNCTION. REF SE2 78.

1447 1, 4, 20, 148, 1348, 15104, 198144
FROM THE TRACE OF A MATRIX. REF MA3.

1448 1, 4, 20, 264, 80104
SWITCHING NETWORKS. REF JFI 276 317 63.

1449 1, 4, 21, 122, 849, 6719, 59873
HIT POLYNOMIALS. REF RI3.

1450 1, 4, 21, 134, 1001, 8544, 81901, 870274, 10146321, 128718044, 1764651461, 25992300894, 409295679481, 6860638482424, 121951698034461
A(N) = NA(N - 1) + (N - 4)A(N - 2). REF R1 188.

1451 1, 4, 21, 1531, 44782251
SWITCHING NETWORKS. REF JFI 276 317 63.

1452 1, 4, 21, 2914, 4379140552
SWITCHING NETWORKS. REF JFI 276 317 63.

1453 1, 4, 22, 107, 486, 2075, 8548, 33851, 130365, 489387, 1799700, 6499706, 23118465, 81134475, 281454170
CONNECTED GRAPHS BY POINTS AND LINES. REF ST1.

1454 1, 4, 22, 140, 969, 7084, 53820, 420732, 3362260
DISSECTIONS OF A POLYGON. REF AMP 1 198 1841.

1455 1, 4, 22, 154, 1304, 12915, 146115, 1855570, 26097835, 402215465, 6734414075, 121629173423, 2355470737637, 48664218965021, 1067895971109199
COEFFICIENTS OF ITERATED EXPONENTIALS. REF SMA 11 353 45.

1456 1, 4, 22, 166, 1726, 24814, 494902
RELATED TO PARTIALLY ORDERED SETS. REF JCT 6 17 69.

1457 1, 4, 22, 190, 3250, 136758, 17256831
INCIDENCE MATRICES. REF CPM 89 217 64.

1458 1, 4, 24, 120, 560, 2520, 11088, 48048
ALMOST CUBIC MAPS. REF PL2 1 292 70.

1459 1, 4, 24, 160, 1440, 18304, 330624
COLORED GRAPHS. REF CJM 12 412 60.

1460 1, 4, 24, 176, 1456, 13056, 124032, 1230592, 12629760, 133186560, 1436098560
ROOTED MAPS. REF CJM 14 416 62.

1461 1, 4, 24, 188, 1368, 10572
WALKS ON A DIAMOND LATTICE. REF PCPS 58 100 62.

1462 1, 4, 24, 188, 1705, 16980, 180670, 2020120, 23478426, 281481880, 3461873536, 43494961412, 556461655783
C-NETS. REF JCT 4 275 68.

1463 1, 4, 24, 192, 1920, 23040, 322560, 5160960, 92897280
SORTING NUMBERS. REF PSPM 19 172 71.

1464 1, 4, 26, 234, 2696, 37919, 630521, 12111114, 264051201, 6445170229, 174183891471, 5164718385337, 166737090160871, 5822980248613990
COEFFICIENTS OF ITERATED EXPONENTIALS. REF SMA 11 353 45.

1465 1, 4, 26, 236, 2752, 39208, 660032, 12818912, 282137824, 6939897856, 188666182784, 5617349020544, 181790703209728, 6353726042486272
SCHRODERS FOURTH PROBLEM. REF RCI 197. CO1 2 60.

1466 1, 4, 26, 236, 2760, 39572, 672592, 13227804, 295579520, 7398318500, 205075286784, 6236796259916, 206489747516416, 7393749269685300
TREES BY TOTAL HEIGHT. REF JA1 10 281 69.

1467 1, 4, 26, 260, 3368, 53744, 1022320, 22522960
BISHOPS PROBLEM. REF AH1 271.

1468 1, 4, 27, 248, 2830, 37782
A PROBLEM OF CONFIGURATIONS. REF CJM 4 25 52.

1469 1, 4, 27, 256, 3125, 46656, 823543, 16777216, 387420489, 10000000000, 285311670611, 8916100448256, 302875106592253, 11112006825558016
NN. REF BA1.**

1470 1, 4, 27, 14056, 104751025086
SWITCHING NETWORKS. REF JFI 276 317 63.

1471 1, 4, 28, 188, 1428, 10708
WALKS ON A DIAMOND LATTICE. REF PCPS 58 100 62.

1472 1, 4, 28, 196, 1324, 8980, 60028, 402412, 2675860, 17826340, 118145548, 784024780, 5193810940
WALKS ON A CUBIC LATTICE. REF PPS 92 649 67.

1473 1, 4, 28, 256, 2716, 31504, 387136
WALKS ON A DIAMOND LATTICE. REF AIP 9 345 60.

1474 1, 4, 28, 2272, 67170304
SWITCHING NETWORKS. REF JFI 276 321 AND 588 63.

1475 1, 4, 29, 206, 1708, 15702
HIT POLYNOMIALS. REF RI3.

1476 1, 4, 29, 355, 6942, 209527, 9535241
TOPOLOGIES OR LABELED TRANSITIVE DIGRAPHS. REF CACM 10 296 67. PURB 19 240 68. JA1 8 194 68.

1477 1, 4, 30, 220, 1855, 17304, 177996, 2002440, 24474285, 323060540, 4581585866, 69487385604, 1122488536715
PERMUTATIONS BY NUMBER OF PAIRS. REF DKB 263.

1478 1, 4, 30, 336, 5040, 95040, 2162160, 57657600
DISSECTIONS OF A BALL. REF CMA 2 26 70.

1479 1, 4, 31, 244, 1921, 15124, 119071, 937444, 7380481, 58106404, 457470751, 3601659604, 28355806081, 223244789044, 1757602506271, 13837575261124
A(N) = 8A(N − 1) − A(N − 2). REF NCM 4 167 1878.

1480 1, 4, 31, 362, 5676
MIXED HUSIMI TREES. REF PNAS 42 532 56.

1481 1, 4, 32, 200, 1120, 5880, 29568, 144144
ALMOST CUBIC MAPS. REF PL2 1 292 70.

1482 1, 4, 32, 292, 2672, 24780, 232512, 2201948
WALKS ON A CUBIC LATTICE. REF PCPS 58 100 62.

1483 1, 4, 32, 336, 4096, 54912, 786432, 11824384
ALMOST CUBIC MAPS. REF PL2 1 292 70.

1484 1, 4, 33, 456, 9460, 274800, 10643745, 530052880, 32995478376, 2510382661920, 229195817258100, 24730000147369440, 3113066087894608560
RELATED TO BESSEL FUNCTIONS. REF PAMS 14 2 63.

1485 1, 4, 33, 480, 11010, 367560, 16854390, 1016930880
FROM A DISTRIBUTION PROBLEM. REF DMJ 33 761 66.

1486 1, 4, 34, 113, 268, 524, 905, 1437, 2145, 3054, 4189, 5575, 7238, 9203, 11494, 14137, 17157, 20580, 24429, 28731, 33510, 38792, 44602, 50965, 57906, 65450, 73622
RELATED TO LATTICE POINTS IN SPHERES. REF PNISI 13 37 47.

1487 1, 4, 34, 496, 11056, 349504, 14873104, 819786496, 56814228736, 4835447317504, 495812444583424, 60283564499562496, 8575634961418940416
RELATED TO TANGENT NUMBERS. REF JFI 239 67 45. MTAC 1 385 45.

1488 1, 4, 34, 8900, 15320103918
SWITCHING NETWORKS. REF JFI 276 317 63.

1489 1, 4, 36, 308, 2764, 25404, 237164, 2237948
WALKS ON A CUBIC LATTICE. REF PCPS 58 100 62.

1490 1, 4, 36, 400, 4900, 63504, 853776
WALKS ON A SQUARE LATTICE. REF AIP 9 345 60.

1491 1, 4, 36, 480, 8400, 181440, 4656960, 138378240, 4670265600, 176432256000, 7374868300800, 337903056691200
COEFFICIENTS OF ORTHOGONAL POLYNOMIALS. REF MTAC 9 174 55.

1492 1, 4, 36, 576, 14400, 518400, 25401600, 1625702400, 131681894400, 13168189440000, 1593350922240000, 229442532802560000, 38775788043632640000
SQUARES OF FACTORIALS. REF RCI 217.

1493 1, 4, 36, 624, 18256, 814144, 51475776
COEFFICIENTS OF SINH X/ COS X. REF CMB 13 306 70.

1494 1, 4, 36, 3178, 298908192
SWITCHING NETWORKS. REF JFI 276 317 63.

1495 1, 4, 37, 559, 11776, 318511, 10522639, 410701432, 18492087079, 943507142461, 53798399207356, 3390242657205889, 233980541746413697
FROM BESSEL POLYNOMIALS. REF RCI 77. RI1.

1496 1, 4, 38, 728, 26704, 1866256, 251548592, 66296291072, 34496488594816, 35641657548953344, 73354596206766622208, 301272202649664088951808
CONNECTED LABELED GRAPHS. REF CJM 8 407 56. CA3.

1497 1, 4, 40, 468, 5828
SETS WITH A CONGRUENCE PROPERTY. REF MFC 15 315 65.

1498 1, 4, 40, 3264, 45826304
SWITCHING NETWORKS. REF JFI 276 317 63.

1499 1, 4, 40, 12096, 604800, 760320, 217945728000, 697426329600, 16937496576000, 30964207376793600, 187333454629601280000
COEFFICIENTS FOR CENTRAL DIFFERENCES. REF JM2 42 162 63.

1500 1, 4, 41, 614, 12281, 307024, 9210721, 322375234, 12895009361, 580275421244, 29013771062201, 1595757408421054, 95745444505263241
PERMUTATIONS WITH NO CYCLES OF LENGTH 5. REF R1 83.

1501 1, 4, 44, 408, 3688, 33212, 298932
COEFFICIENTS OF ELLIPTIC FUNCTIONS. REF TM1 4 92.

1502 1, 4, 46, 1064, 35792, 1673792, 103443808, 8154999232, 798030483328
RELATED TO LATIN RECTANGLES. REF BU2 33 125 41.

1503 1, 4, 46, 1322, 112519, 32267168, 34153652752
SELF-DUAL THRESHOLD FUNCTIONS. REF PGEC 17 806 68.

1504 1, 4, 48, 224, 448, 40, 1408, 2240, 1280, 924, 480, 6944, 8704, 5864, 14080, 2240, 33772, 19064, 11088, 54432, 4480, 38400, 43648, 75712, 124928, 62100, 70368
RELATED TO REPRESENTATION AS SUMS OF SQUARES. REF QJM 38 191 07.

1505 1, 4, 49, 273, 1023, 3003, 7462, 16422, 32946, 61446, 108031, 180895, 290745, 451269, 679644, 997084, 1429428, 2007768, 2769117, 3757117, 5022787, 6625311
CENTRAL FACTORIAL NUMBERS. REF RCI 217.

1506 1, 4, 51, 46218, 366543984720
SWITCHING NETWORKS. REF JFI 276 317 63.

1507 1, 4, 55, 2008, 153040, 20933840, 4662857360, 1579060246400, 772200774683520, 523853880779443200, 477360556805016931200
STOCHASTIC MATRICES OF INTEGERS. REF ST2.

1508 1, 4, 56, 9408, 16942080, 535281401856
REDUCED LATIN SQUARES. REF R1 210. RYS 53. FY1 22. RMM 193. JCT 3 98 67.

1509 1, 4, 60, 550, 4004, 25480, 148512, 813960, 4263600, 18573816
PARTITIONS OF A POLYGON BY NUMBER OF PARTS. REF CAY 13 95.

1510 1, 4, 64, 2304, 147456, 14745600, 2123366400, 416179814400, 106542032486400, 34519618525593600
CENTRAL FACTORIAL NUMBERS. REF OP1 7. FMR 1 110. RCI 217.

1511 1, 4, 74, 63440, 244728561176
SWITCHING NETWORKS. REF JFI 276 324 AND 588 63.

1512 1, 4, 80, 3904, 354560, 51733504, 11070525440, 3266330312704, 1270842139934720, 630424777638805504, 388362339077351014400
MULTIPLES OF EULER NUMBERS. REF QJM 44 110 13. FMR 1 75.

1513 1, 4, 80, 4752, 440192, 59245120
RESTRICTED PERMUTATIONS. REF R1 187.

1514 1, 4, 108, 27648, 86400000, 4031078400000, 3319766398771200000, 55696437941726556979200000, 21577941222941856209168026828800000
PRODUCTS OF POWERS. REF FMR 1 50.

1515 1, 4, 120, 3024, 151200, 79200, 1513512000, 1513512000, 51459408000, 74662922880, 18068427336960, 133196739984000, 1215553449093984000
COEFFICIENTS FOR NUMERICAL DIFFERENTIATION. REF JM2 22 120 43.

1516 1, 4, 120, 12096, 3024000, 1576143360, 1525620096000, 2522591034163200, 6686974460694528000, 27033456071346536448000
SPECIAL DETERMINANTS. REF BMG 6 105 65.

1517 1, 4, 128, 16384, 4456448
GENERALIZED TANGENT NUMBERS. REF MTAC 21 690 67.

1518 1, 4, 129, 43968
COMMUTATIVE GROUPOIDS. REF JL2 246.

1519 1, 4, 136, 44224, 179228736, 9383939974144
RELATIONAL SYSTEMS. REF OB1.

1520 1, 4, 140, 4056, 129360, 4381848
SPECIFIC HEAT FOR CUBIC LATTICE. REF PRV 129 102 63.

1521 1, 4, 272, 55744, 23750912, 17328937984, 19313964388352, 30527905292468224, 64955605537174126592, 179013508069217017790464
GENERALIZED TANGENT NUMBERS. REF MTAC 21 690-67.

1522 1, 4, 302, 2569966041123963092
INVERTIBLE BOOLEAN FUNCTIONS. REF PGEC 13 350 64.

1523 1, 4, 32896, 3002399885885440, 14178431955039103827204744901417762816
RELATIONAL SYSTEMS. REF OB1.

SEQUENCES BEGINNING 1, 5

1524 1, 5, 1, 0, 5, 2, 8, 18, 19, 7, 16, 13, 6, 34, 27, 56, 12, 69, 11, 73, 20, 70, 70, 72, 57, 1, 30, 95, 71, 119, 56, 67, 94, 86, 151, 108, 21, 106, 48, 72, 159, 35, 147, 118, 173, 180
WILSON REMAINDERS. REF JLMS 28 253 53. AFM 4 481 61.

1525 1, 5, 1, 7, 8, 5, 19, 11, 23, 35, 27, 64, 61, 85, 137, 133, 229, 275, 344, 529, 599, 875, 1151, 1431, 2071, 2560, 3481, 4697, 5953, 8245, 10649, 14111, 19048, 24605
A SLOWLY INCREASING SEQUENCE. REF JLMS 8 166 33.

1526 1, 5, 1, 41, 31, 461, 895, 6481, 22591, 107029, 604031, 1964665, 17669471, 37341149, 567425279, 627491489, 19919950975, 2669742629, 759627879679
RELATED TO LATIN RECTANGLES. REF JMSJ 1(4) 240 50. R1 209.

1527 1, 5, 3, 251, 95, 19087, 5257, 1070017, 25713, 26842253, 4777223, 703604254357, 106364763817, 1166309819657, 25221445, 8092989203533249
FROM BERNOULLI POLYNOMIALS. REF JM2 22 49 43.

1528 1, 5, 4, 8, 7, 6, 11, 8, 9, 14, 18, 13, 11, 17, 16, 12, 13, 14, 28, 19, 14, 18, 16, 27, 22, 31, 16, 17, 26, 19, 24, 24, 23, 22, 28, 37, 41, 27, 32, 21, 26, 22, 23, 31, 22, 44, 48, 23
QUADRATIC PARTITIONS OF PRIMES. REF CU2 1. LE1 55.

1529 1, 5, 5, 10, 15, 6, 5, 25, 15, 20, 9, 45, 5, 25, 20, 10, 15, 20, 50, 35, 30, 55, 50, 15, 80, 1, 50, 35, 45, 15, 5, 50, 25, 55, 85, 51, 50, 10, 40, 65, 10, 10, 115, 50, 115, 100, 85, 80
COEFFICIENTS OF A MODULAR FORM. REF KNAW 59 207 56.

1530 1, 5, 6, 7, 9, 53, 60, 66, 83, 136, 185, 312, 3064, 3718, 8096, 9826, 12384, 16602, 16760, 182424, 323392
CLASS NUMBERS OF QUADRATIC FIELDS. REF MTAC 24 445 70.

1531 1, 5, 7, 4, 11, 8, 1, 5, 7, 17, 19, 13, 2, 20, 23, 19, 14, 25, 7, 23, 11, 13, 28, 22, 17, 29, 26, 32, 16, 35, 1, 5, 37, 35, 13, 29, 34, 31, 19, 2, 28, 10, 23, 25, 32, 43, 29, 1, 31, 11
QUADRATIC PARTITIONS OF PRIMES. REF CU2 1. LE1 55.

1532 1, 5, 7, 7, 2, 1, 5, 6, 6, 4, 9, 0, 1, 5, 3, 2, 8, 6, 0, 6, 0, 6, 5, 1, 2, 0, 9, 0, 0, 8, 2, 4, 0, 2, 4, 3, 1, 0, 4, 2, 1, 5, 9, 3, 3, 5, 9, 3, 9, 9, 2, 3, 5, 9, 8, 8, 0, 5, 7, 6, 7, 2, 3, 4, 8, 8, 4, 8, 6
EULERS CONSTANT. REF MTAC 17 175 63.

1533 1, 5, 7, 8, 9, 10, 11, 12, 12, 13, 13, 14, 15, 15, 16, 16, 16, 17, 17, 18, 18, 19, 19, 19, 20, 20, 20, 21, 21, 21, 22, 22, 22, 23, 23, 23, 24, 24, 24, 24, 25, 25, 25, 25, 26, 26, 26
CHROMATIC NUMBERS IF 4 COLOR CONJECTURE FALSE. REF PNAS 60 438 68.

1534 1, 5, 7, 9, 11, 12, 13, 16, 17, 17, 19, 19, 22, 21, 23, 24, 26, 27, 29, 27, 28, 29, 32, 31, 31, 33, 32, 34, 33, 37, 37, 37, 39, 41, 39, 41, 43, 41, 41, 42, 43, 44, 46, 43, 44, 47, 49
QUADRATIC PARTITIONS OF PRIMES. REF CU2 1. LE1 55.

1535 1, 5, 7, 9, 53, 73, 83, 157, 185, 1927, 2295, 2273, 5313, 7173, 9529, 18545, 22635, 66011, 121725, 344909
CLASS NUMBERS OF QUADRATIC FIELDS. REF MTAC 24 445 70.

1536 1, 5, 7, 13, 11, 23, 15, 29, 25, 35
RELATED TO PLANAR PARTITIONS. REF MES 54 115 24.

1537 1, 5, 7, 19, 31, 53, 67
(10P + 1)/11 IS PRIME. REF SE1.**

1538 1, 5, 7, 21, 33, 429, 715, 2431, 4199, 29393, 52003, 185725, 334305
FROM DOUBLE FACTORIALS. REF RG1 415.

1539 1, 5, 8, 11, 15, 18, 22, 25, 29, 32, 35, 39, 42
WYTHOFF GAME. REF CMB 2 188 59.

1540 1, 5, 8, 31, 55, 203, 368, 1345, 2449, 8933, 16280, 59359, 108199
A(2N) = A(2N - 1) + 3A(2N - 2), A(2N + 1) = 2A(2N) + 3A(2N - 1). REF MQET 1 12 16.

1541 1, 5, 9, 17, 21, 29, 45, 177
7.2N - 1 IS PRIME. REF MTAC 22 421 68.**

1542 1, 5, 9, 17, 33, 65, 129, 253, 497, 977, 1921, 3777, 7425, 14597, 28697, 56417, 110913, 218049, 428673, 842749, 1656801, 3257185, 6403457, 12588865, 24749057
PENTANACCI NUMBERS. REF FQ 2 260 64.

1543 1, 5, 9, 21, 37, 69, 69, 89, 137, 177, 421, 481, 657, 749, 885, 1085, 1305, 1353, 1489, 1861, 2617, 2693, 3125, 5249, 5761, 7129, 8109, 9465, 9465, 10717, 12401, 12401
LATTICE POINTS IN CIRCLES. REF MTAC 20 306 66.

1544 1, 5, 9, 49, 2209, 4870849, 23725150497409, 562882766124611619513723649, 316837008400094222150776738483768236006420971486980609
A NONLINEAR RECURRENCE. REF AMM 70 403 63.

1545 1, 5, 9, 251, 475, 19087, 36799, 1070017, 2082753, 134211265
NUMERATORS OF GENERALIZED BERNOULLI NUMBERS. REF MT1 136.

1546 1, 5, 10, 10, 0, 19, 35, 40, 25, 10, 45, 75, 80, 60, 15, 45, 85, 115, 115, 90, 21, 35, 95, 130, 135, 135, 70, 35, 65, 105, 146, 120, 150, 90, 65, 25, 90, 115, 150, 125, 130, 45
COEFFICIENTS OF A MODULAR FORM. REF JLMS 39 435 64.

1547 1, 5, 10, 14, 18, 22, 27, 31, 35, 40, 44, 48, 53
WYTHOFF GAME. REF CMB 2 188 59.

1548 1, 5, 10, 15, 20, 26, 31, 36, 41, 47, 52, 57, 62
A BEATTY SEQUENCE. REF CMB 2 189 59.

1549 1, 5, 10, 17, 16, 32, 22, 41, 37, 50
RELATED TO PLANAR PARTITIONS. REF MES 54 115 24.

1550 1, 5, 10, 21, 21, 38, 29, 53, 46, 65
RELATED TO PLANAR PARTITIONS. REF MES 54 115 24.

1551 1, 5, 10, 21, 26, 50, 50, 85, 91, 130, 122, 210, 170, 250, 260, 341, 290, 455, 362, 546, 500, 610, 530, 850, 651, 850, 820, 1050, 842, 1300, 962, 1365, 1220, 1450, 1300
SUM OF SQUARES OF DIVISORS OF N. REF AS1 827.

1552 1, 5, 10, 21, 26, 53, 50, 85, 91, 130
RELATED TO PLANAR PARTITIONS. REF MES 54 115 24.

1553 1, 5, 10, 30, 74, 199, 515, 1355, 3540, 9276, 24276, 63565, 166405, 435665, 1140574, 2986074, 7817630, 20466835, 53582855, 140281751
FROM A DEFINITE INTEGRAL. REF EMS 10 184 57.

1554 1, 5, 10, 40, 150, 624, 2580, 11160, 48750, 217000
IRREDUCIBLE POLYNOMIALS, OR NECKLACES. REF AMM 77 744 70.

1555 1, 5, 11, 13, 19, 23, 29, 37, 47, 53, 59, 61, 67, 71, 83, 97, 101, 107, 131, 139, 149, 163, 167, 173, 179, 181, 191, 193, 197, 211, 227, 239, 263, 269, 293, 307, 311, 313, 317
SOLUTION OF A CONGRUENCE. REF KR1 1 63.

1556 1, 5, 11, 15, 16, 17, 18, 23, 25, 27, 32, 35, 36, 39, 45, 46, 47, 48, 49, 50, 51, 52, 53, 54, 55, 57, 61, 65, 68, 73, 75, 77, 79, 82, 85, 89, 91, 95, 96, 101, 105, 106, 110, 111
SQUARES CONTAIN A TWO. REF EUR 18 17 55.

1557 1, 5, 11, 17, 23, 29, 30, 36, 42, 48, 54, 60, 61, 67, 73, 79, 85, 91, 92, 98, 104, 110, 116, 122, 123, 129, 135, 141, 147, 153, 154, 155
FACTORIALS ENDING IN ZEROS. REF MMAG 27 55 53.

1558 1, 5, 11, 19, 29, 41, 71, 89, 109, 131, 181, 239, 271, 379, 419, 461, 599, 701, 811, 929, 991, 1259, 1481, 1559, 1721, 1979, 2069, 2161, 2351, 2549, 2861, 2969, 3079
PRIMES OF FORM N(N – 1) – 1. REF PO1 249, LE1 46.

1559 1, 5, 11, 27, 45, 71, 109, 163
POSTAGE STAMP PROBLEM. REF CJ1 12 379 69.

1560 1, 5, 11, 29, 97, 149, 211, 127, 1847, 541, 907, 1151, 1693, 2503, 2999, 4327, 5623, 1361, 9587, 30631, 19373, 16183, 15727, 81509, 28277, 31957, 19661, 35671
INCREASING GAPS BETWEEN PRIMES. REF MTAC 21 485 67.

1561 1, 5, 11, 82, 257, 130638, 130895, 785113, 4056460, 4841573, 8898033, 13739606, 36377245, 50116851, 86494096, 2125975155, 2212469251, 4338444406
CONVERGENTS TO CUBE ROOT OF 6. REF AMP 46 107 1866. LE1 67. HPR.

1562 1, 5, 12, 22, 35, 51, 70, 92, 117, 145, 176, 210, 247, 287, 330, 376, 425, 477, 532, 590, 651, 715, 782, 852, 925, 1001, 1080, 1162, 1247, 1335, 1426, 1520, 1617, 1717
PENTAGONAL NUMBERS N(3N – 1)/2. REF DI2 2 1. BE3 189, FQ 8 84 70.

1563 1, 5, 12, 23, 39, 62, 91, 127
PARTITIONS INTO NON-INTEGRAL POWERS. REF PCPS 47 214 51.

1564 1, 5, 12, 28, 54, 100, 170, 284, 450, 702, 1062, 1583, 2308, 3329, 4720, 6628, 9190, 12634, 17189, 23219, 31092, 41371, 54651, 71782, 93695, 121684, 157169
BIPARTITE PARTITIONS. REF NI1 26.

1565 1, 5, 12, 29, 57, 109, 189, 323, 522, 831, 1279, 1941, 2876, 4215, 6066, 8644, 12151, 16933, 23336, 31921, 43264, 58250, 77825, 103362, 136371, 178975, 233532
BIPARTITE PARTITIONS. REF PCPS 49 72 53. NI1 1.

1566 1, 5, 13, 17, 29, 37, 41, 53, 61, 73, 89, 97, 101, 109, 113, 137, 149, 157, 173, 181, 193, 197, 229, 233, 241, 257, 269, 277, 281, 293, 313, 317, 337, 349, 353, 373, 389, 397
PRIMES OF THE FORM 4N + 1. REF AS1 870.

1567 1, 5, 13, 25, 41, 61, 85, 113, 145, 181, 221, 265, 313, 365, 421, 481, 545, 613, 685, 761, 841, 925, 1013, 1105, 1201, 1301, 1405, 1513, 1625, 1741, 1861, 1985, 2113, 2245
N2 + (N + 1)**2. REF MMAG 35 162 62. SIAMR 12 277 70.**

1568 1, 5, 13, 25, 45, 72, 115, 166, 235, 327, 428, 548, 709, 874, 1095
POSTAGE STAMP PROBLEM. REF CJ1 12 379 69.

1569 1, 5, 13, 27, 48, 78, 118, 170, 235, 315, 411
TRIANGLES CONTAINED IN A CERTAIN FIGURE. REF MAG 46 55 62.

1570 1, 5, 13, 29, 49, 81, 113, 149, 197, 253, 317, 377, 441, 529, 613, 709, 797, 901, 1009, 1129, 1257, 1373, 1517, 1653, 1793, 1961, 2121, 2289, 2453, 2629, 2821, 3001
LATTICE POINTS IN CIRCLES. REF PNISI 13 37 47. MTAC 16 287 62.

1571 1, 5, 13, 30, 59, 109, 187, 312, 497, 775, 1176, 1753, 2561, 3694, 5245, 7366, 10223, 14056, 19137, 25853, 34637, 46092, 60910, 80009, 104462, 135674, 175274
BIPARTITE PARTITIONS. REF NI1 32.

1572 1, 5, 13, 33, 73, 151, 289, 526, 910, 1514
RESTRICTED PARTITIONS. REF CAY 2 281.

1573 1, 5, 14, 27, 41, 44, 65, 76, 90
C(2N, N)/(N + 1)2 IS AN INTEGER. REF JIMS 18 97 29.**

1574 1, 5, 14, 30, 55, 91, 140, 204, 285, 385, 506, 650, 819, 1015, 1240, 1496, 1785, 2109, 2470, 2870, 3311, 3795, 4324, 4900, 5525, 6201, 6930, 7714, 8555, 9455, 10416
SQUARE PYRAMIDAL NUMBERS. REF DI2 2 2. BE3 194. AS1 813.

1575 1, 5, 14, 1026, 4324, 311387, 6425694, 579783114, 4028104212, 7315072725560
RELATED TO ZEROS OF BESSEL FUNCTION. REF MTAC 1 406 45.

1576 1, 5, 15, 35, 70, 125, 200, 255, 275
EXPANSION OF BRACKET FUNCTION. REF FQ 2 254 64.

1577 1, 5, 15, 35, 70, 125, 210, 325, 495
COMPOSITIONS INTO RELATIVELY PRIME PARTS. REF FQ 2 250 64.

1578 1, 5, 15, 35, 70, 126, 210, 330, 495, 715, 1001, 1365, 1820, 2380, 3060, 3876, 4845, 5985, 7315, 8855, 10626, 12650, 14950, 17550, 20475, 23751, 27405, 31465
FIGURATE NUMBERS OR BINOMIAL COEFFICIENTS C(N, 4). REF DI2 2 7. RS1. BE3 196. AS1 828.

1579 1, 5, 15, 40, 98, 237, 534, 1185, 2554, 5391
PARTITIONS INTO NON-INTEGRAL POWERS. REF PCPS 47 215 51.

1580 1, 5, 15, 45, 120, 326, 835, 2145, 5345, 13220, 32068, 76965, 181975, 425490, 982615, 2245444, 5077090, 11371250
4-DIMENSIONAL PARTITIONS. REF PCPS 63 1099 67.

1581 1, 5, 15, 45, 120, 331, 855, 2214, 5545, 13741
RELATED TO 4-DIMENSIONAL PARTITIONS. REF PCPS 63 1100 67.

1582 1, 5, 15, 45, 165, 629, 2635, 11165, 48915, 217045, 976887, 4438925, 20346485, 93900245, 435970995, 2034505661, 9536767665, 44878791365, 211927736135
NECKLACES OF 5 COLORS. REF R1 162. IJM 5 658 61.

1583 1, 5, 15, 55, 140, 448, 1022, 2710, 6048, 14114, 28831
RESTRICTED PARTITIONS. REF JCT 9 373 70.

1584 1, 5, 15, 55, 225, 979, 4425, 20515, 96825, 462979, 2235465, 10874275, 53201625, 261453379, 1289414505, 6376750435, 31605701625, 156925970179
1N + 2**N + ··· + 5**N. REF AS1 813.**

1585 1, 5, 16, 42, 99, 219, 466, 968, 1981
RADON PARTITIONS. REF MFM 73 18 69.

1586 1, 5, 16, 86, 448, 3580
CUBIC GRAPHS. REF RE4.

1587 1, 5, 17, 49, 129, 321, 769, 1793, 4097, 9217, 20481, 45057, 98305, 212993, 458753, 983041, 2097153, 4456449, 9437185, 19922945, 41943041, 88080385
GENUS OF THE N-CUBE. REF HSG 16.

1588 1, 5, 17, 83, 593, 2893, 36101, 172195
A SIMPLE RECURRENCE. REF DMJ 26 580 59.

1589 1, 5, 18, 42, 75, 117, 168, 228, 297, 375, 462, 558, 663, 777, 900, 1032, 1173, 1323, 1482, 1650, 1827, 2013, 2208, 2412, 2625, 2847, 3078, 3318, 3567
DISCORDANT PERMUTATIONS. REF SMA 20 23 54.

1590 1, 5, 18, 45, 100, 185, 323, 522, 804
PARTITIONS INTO NON-INTEGRAL POWERS. REF PCPS 47 214 51.

1591 1, 5, 18, 56, 160, 432, 1120, 2816, 6912, 16640, 39424, 92160, 212992, 487424, 1105920, 2490368, 5570560
COEFFICIENTS OF CHEBYSHEV POLYNOMIALS. REF PRSE 62 190 46. AS1 795.

1592 1, 5, 18, 58, 179, 543, 1636, 4916, 14757, 44281, 132854, 398574, 1195735, 3587219
PERMUTATIONS BY LENGTH OF RUNS. REF DKB 260.

1593 1, 5, 18, 82, 643, 15182, 7848984
PRECOMPLETE POST FUNCTIONS. REF SMD 10 619 69. RO3.

1594 1, 5, 19, 61, 180, 498, 1323, 3405, 8557, 21103, 51248, 122898, 291579, 685562, 1599209, 3705122, 8532309, 19543867, 44552066, 101124867, 228640542
TREES OF HEIGHT 5. REF IBMJ 4 475 60. KU1.

1595 1, 5, 19, 65, 210, 654, 1985, 5911, 17345, 50305
FROM CONVOLVED FIBONACCI NUMBERS. REF RI1.

1596 1, 5, 19, 65, 211, 665, 2059, 6305, 19171, 58025, 175099, 527345, 1586131, 4766585, 14316139, 42981185, 129009091, 387158345, 1161737179, 3485735825
3N − 2**N. REF EUR 24 20 61. CR 268 579 69.**

1597 1, 5, 19, 67, 236, 797, 2678, 8833, 28908, 93569, 300748, 959374, 3042808, 9597679, 30134509
CONNECTED GRAPHS BY POINTS AND LINES. REF ST1.

1598 1, 5, 19, 71, 265, 989, 3691, 13775, 51409, 191861, 716035, 2672279, 9973081, 37220045, 138907099, 518408351, 1934726305, 7220496869, 26947261171
A(N) = 4A(N - 1) - A(N - 2). REF EUL (1) 1 375 11. MMAG 40 78 67.

1599 1, 5, 20, 65, 185, 481, 1165, 2665, 5820, 12220, 24802, 48880, 93865, 176125, 323685, 583798, 1035060, 1806600, 3108085
COEFFICIENTS OF AN ELLIPTIC FUNCTION. REF CAY 9 128.

1600 1, 5, 20, 65, 190, 511, 1295, 3130, 7288, 16438, 36128
CONVOLVED FIBONACCI NUMBERS. REF RCI 101.

1601 1, 5, 20, 70, 230, 721, 2200, 6575, 19385, 56575, 163952, 472645, 1357550, 3888820, 11119325, 31753269, 90603650, 258401245, 736796675, 2100818555
POWERS OF ROOTED TREE ENUMERATOR. REF R1 150.

1602 1, 5, 20, 75, 275, 1001, 3640, 13260, 48450, 177650, 653752, 2414425, 8947575, 33266625, 124062000, 463991880, 1739969550, 6541168950, 24647883000
LAPLACE TRANSFORM COEFFICIENTS. REF QAM 14 407 56.

1603 1, 5, 20, 84, 354, 1540, 6704, 29610, 131745, 591049, 2669346
RESTRICTED HEXAGONAL POLYOMINOES. REF EMS 17 11 70.

1604 1, 5, 20, 96, 469, 3135, 20684, 173544, 1557105, 16215253
SUMS OF LOGARITHMIC NUMBERS. REF MST 31 78 63.

1605 1, 5, 20, 206, 54155
SWITCHING NETWORKS. REF JFI 276 317 63.

1606 1, 5, 20, 300, 9980, 616260, 65814020, 11878194300
COLORED GRAPHS. REF CJM 22 596 70.

1607 1, 5, 21, 84, 330, 1287, 5005, 19448, 75582, 293930, 1144066, 4457400
PARTITIONS OF A POLYGON BY NUMBER OF PARTS. REF CAY 13 95.

1608 1, 5, 21, 85, 341, 1365, 5461
CENTRAL FACTORIAL NUMBERS. REF TH1 35. FMR 1 112. RCI 217.

1609 1, 5, 21, 119, 735, 4830, 33253
TRIANGULATIONS OF THE DISK. REF PLMS 14 759 64.

1610 1, 5, 22, 71, 186, 427, 888, 1704
PARTITIONS INTO NON-INTEGRAL POWERS. REF PCPS 47 214 51.

1611 1, 5, 22, 93, 386, 1586, 6476, 26333, 106762, 431910
ROOTED PLANAR MAPS. REF BAMS 74 74 68.

1612 1, 5, 22, 1001, 2882, 15251, 720027, 7081807, 7451547, 26811862, 54177145
PENTAGONAL PALINDROMES. REF AMM 48 211 41.

1613 1, 5, 23, 17, 719, 5039, 1753, 2999, 125131, 7853, 479001599, 3593203, 87178291199, 1510259, 6880233439, 256443711677, 108514808571661, 78143369
LARGEST FACTOR OF FACTORIAL (N) - 1. REF SMA 14 25 48.

1614 1, 5, 23, 119, 719, 5039, 40319, 362879, 3628799, 39916799, 479001599, 6227020799, 87178291199, 1307674367999, 20922789887999, 355687428095999
FACTORIAL (N) - 1. REF AS1 833.

1615 1, 5, 23, 1681, 257543, 67637281, 27138236663, 15442193173681
GLAISHERS T NUMBERS. REF FMR 1 76. JCPM.

1616 1, 5, 24, 84, 251, 653, 1543
PARTITIONS INTO NON-INTEGRAL POWERS. REF PCPS 47 214 51.

1617 1, 5, 24, 128, 835, 6423, 56410, 554306, 6016077, 71426225, 920484892, 12793635300, 190730117959, 3035659077083
PERMUTATIONS BY NUMBER OF PAIRS. REF DKB 263.

1618 1, 5, 24, 133, 846, 5661, 39556
TRIANGULATIONS OF THE DISK. REF PLMS 14 759 64.

1619 1, 5, 25, 29, 41, 49, 61, 65, 85, 89, 101, 125, 145, 149, 245, 265, 365, 385, 485, 505, 601, 605, 625, 649, 701, 725, 745, 749, 845, 865, 965, 985, 1105, 1205, 1249, 1345
FIBONACCI COINCIDENCES. REF FQ 4 156 66.

1620 1, 5, 25, 125, 625, 3125, 15625, 78125, 390625, 1953125, 9765625, 48828125, 244140625, 1220703125, 6103515625, 30517578125, 152587890625
POWERS OF FIVE. REF BA1.

1621 1, 5, 25, 129, 681, 3653, 19825, 108545, 598417, 3317445, 18474633, 103274625, 579168825, 3256957317
BINOMIAL COEFFICIENT SUMS. REF AMM 43 29 36.

1622 1, 5, 25, 149, 1081, 9365, 94585
EXPANSION OF AN INTEGRAL. REF SKA 11 95 28.

1623 1, 5, 25, 149, 1081, 9366, 94586, 1091670, 14174522, 204495125
COMBINATION LOCKS. REF MMAG 37 132 64.

1624 1, 5, 26, 97, 265, 362, 1351, 13775, 70226, 262087, 716035, 978122
RELATED TO BERNOULLI NUMBERS. REF ANN 36 645 35.

1625 1, 5, 26, 154, 1044, 8028, 69264, 663696, 6999840, 80627040, 1007441280, 13575738240, 196287356160, 3031488633600, 49811492505600
GENERALIZED STIRLING NUMBERS. REF PEF 77 7 62.

1626 1, 5, 27, 502, 2375, 95435, 1287965, 29960476, 262426878, 28184365650
COEFFICIENTS FOR STEP-BY-STEP INTEGRATION. REF JACM 11 231 64.

1627 1, 5, 27, 1204, 85617952
SWITCHING NETWORKS. REF JFI 276 317 63.

1628 1, 5, 29, 19, 2309, 30029, 8369, 929, 46027, 81894851, 876817, 38669, 304250263527209, 92608862041, 3219318233447599
LARGEST FACTORS OF A SEQUENCE. REF SMA 14 26 48.

1629 1, 5, 29, 118, 418, 1383, 4407, 13736, 42236, 128761, 390385, 1179354, 3554454
PERMUTATIONS BY LENGTH OF RUNS. REF DKB 260.

1630 1, 5, 29, 169, 985, 5741, 33461, 195025, 1136689, 6625109, 38613965, 225058681, 1311738121, 7645370045, 44560482149, 259717522849, 1513744654945
PYTHAGOREAN TRIANGLES. REF AMM 4 25 1897. MLG 2 322 10. FQ 6(3) 104 68.

1631 1, 5, 29, 233, 2329, 27949, 391285, 6260561, 112690097, 2253801941, 49583642701, 1190007424825, 30940193045449, 866325405272573
PERMUTATIONS WITH NO CYCLES OF LENGTH 2. REF LU1 1 223. R1 83.

1632 1, 5, 29, 23669, 1508789, 5025869, 7841261, 9636461, 18127229, 31839341, 37989701, 240511301, 23739440141, 44913466781, 60664576541, 123464393861
SEQUENCE OF PRESCRIBED QUADRATIC CHARACTER. REF MTAC 24 446 70.

1633 1, 5, 30, 115, 425, 1396, 4440
ALKYLS. REF ZFK 93 437 36.

1634 1, 5, 30, 210, 1680, 15120, 151200, 1663200, 19958400, 259459200, 3632428800, 54486432000, 871782912000, 14820309504000, 266765571072000
GENERALIZED STIRLING NUMBERS. REF PEF 77 61 62.

1635 1, 5, 31, 197, 1435, 11765, 107755
THE GAME OF MOUSETRAP. REF QJM 15 241 1878.

1636 1, 5, 31, 211, 1031, 2801, 4651, 5261, 6841, 8431, 14251, 17891, 20101, 21121, 22621, 22861, 26321, 30941, 33751, 36061, 41141, 46021, 48871, 51001, 58411, 61051
QUINTAN PRIMES. REF CU1 2 200.

1637 1, 5, 31, 227, 1909, 18089, 190435, 2203319, 27772873, 378673901, 5551390471, 87057596075, 1453986832381, 25762467303377, 482626240281739
A(N) = NA(N − 1) + (N − 5)A(N − 2). REF R1 188.

1638 1, 5, 31, 257, 2671, 33305, 484471, 8054177, 150635551, 3130287705
FROM FIBONACCI SUMS. REF FQ 5 48 67.

1639 1, 5, 32, 288, 3413, 50069, 873612, 17650828, 405071317, 10405071317, 295716741928, 9211817190184, 312086923782437, 11424093749340453
SUM OF NN. REF AMM 53 471 46.**

1640 1, 5, 33, 236, 1918, 17440
HIT POLYNOMIALS. REF RI3.

1641 1, 5, 33, 287, 3309, 50975, 1058493
RELATED TO PARTIALLY ORDERED SETS. REF JCT 6 17 69.

1642 1, 5, 34, 258, 2136, 19320, 190800, 2051280
TERMS IN CERTAIN DETERMINANTS. REF PLMS 10 122 1879.

1643 1, 5, 35, 140, 420, 1050, 2310, 4620, 8580
RELATED TO BINOMIAL MOMENTS. REF JO2 449.

1644 1, 5, 35, 189, 924, 4290, 19305
COEFFICIENTS FOR EXTRAPOLATION. REF SE2 93.

1645 1, 5, 35, 225, 67375, 66693, 955040625, 1861234375
FROM A HYPERGEOMETRIC FUNCTION. REF JACM 3 14 56.

1646 1, 5, 35, 285, 2530, 23751, 231880, 2330445, 23950355
DISSECTIONS OF A POLYGON. REF AMP 1 198 1841.

1647 1, 5, 35, 294, 2772, 28314, 306735, 3476330, 40831076, 493684828, 6114096716
HAMILTONIAN ROOTED MAPS. REF CJM 14 416 62.

1648 1, 5, 35, 315, 3455, 44590, 660665, 11035095, 204904830, 4183174520, 93055783320, 2238954627848, 57903797748386, 1601122732128779
COEFFICIENTS OF ITERATED EXPONENTIALS. REF SMA 11 353 45.

1649 1, 5, 35, 1260, 4620, 30030, 90090, 1021020, 2771340, 14549535, 37182145, 1487285800, 3650610600, 17644617900, 42075627300, 396713057400
COEFFICIENTS OF LEGENDRE POLYNOMIALS. REF PR1 156. AS1 798.

1650 1, 5, 35, 2266, 30564722
SWITCHING NETWORKS. REF JFI 276 317 63.

1651 1, 5, 36, 329, 3655, 47844, 721315, 12310199, 234615096, 4939227215, 113836841041, 2850860253240, 77087063678521, 2238375706930349
A(N) = (2N + 1)A(N - 1) + A(N - 2). REF CJM 8 308 56.

1652 1, 5, 36, 3406, 14694817, 727050997716715
CONTINUED COTANGENT FOR SQUARE ROOT OF 2. REF DMJ 4 339 38.

1653 1, 5, 40, 260, 1820, 12376, 85085, 582505, 3994320, 27372840, 187628376
FROM FIBONACCI IDENTITIES. REF FQ 6 82 68.

1654 1, 5, 40, 440, 6170, 105315, 2120610, 49242470, 1296133195, 38152216495, 1242274374380, 44345089721923, 1722416374173854, 72330102999829054
COEFFICIENTS OF ITERATED EXPONENTIALS. REF SMA 11 353 45.

1655 1, 5, 41, 73, 193, 1181, 6481, 16493, 21523361, 530713, 42521761, 570461, 769, 4795973261, 647753, 47763361, 128653413121, 109688713, 24127552321
LARGEST FACTOR OF 9N + 1. REF KR1 2 89.**

1656 1, 5, 41, 545, 11681, 402305, 22207361
COLORED GRAPHS. REF CJM 12 413 60 (DIVIDED BY 3).

1657 1, 5, 45, 385, 3710, 38934, 444990, 5506710, 73422855, 1049946755, 16035550531, 260577696015
PERMUTATIONS BY NUMBER OF PAIRS. REF DKB 263.

1658 1, 5, 46, 19930, 69945183326
SWITCHING NETWORKS. REF JFI 276 317 63.

1659 1, 5, 49, 485, 4801, 47525, 470449, 4656965, 46099201, 456335045, 4517251249, 44716177445, 442644523201, 4381729054565, 43374646022449
A(N) = 10A(N - 1) - A(N - 2). REF EUL (1) 1 374 11. TH2 281.

1660 1, 5, 49, 809, 20317, 722813, 34607305, 2145998417, 167317266613, 16020403322021, 1848020950359841, 252778977216700025, 40453941942593304589
MULTIPLES OF GLAISHERS I NUMBERS. REF PLMS 31 224 1899. FMR 1 76.

1661 1, 5, 49, 820, 21076, 773136, 38402064, 2483133696, 202759531776, 20407635072000, 2482492033152000, 359072203696128000, 60912644957448192000
CENTRAL FACTORIAL NUMBERS. REF RCI 217.

1662 1, 5, 52, 1522, 145984, 48464496, 56141454464, 229148550030864, 3333310786076963968, 174695272746749919580928
UNRESTRICTED RELATIONS. REF PAMS 4 494 53. MI1 17 19 55. MAN 174 66 67. (DIVIDED BY 2.)

1663 1, 5, 53, 173, 293, 437, 9173, 24653, 74093, 170957, 214037, 2004917, 44401013, 94948157, 154554077, 163520117, 261153653, 1728061733, 9447241877
SEQUENCE OF PRESCRIBED QUADRATIC CHARACTER. REF MTAC 24 449 70.

1664 1, 5, 57, 352, 1280, 3522, 7970, 15872, 29184, 49410, 79042
GENERALIZED CLASS NUMBERS. REF MTAC 21 689 67.

1665 1, 5, 58, 1274, 41728, 1912112, 116346400, 9059742176, 877746364288
RELATED TO LATIN RECTANGLES. REF BU2 33 125 41.

1666 1, 5, 61, 479, 3111, 18270, 101166, 540242, 2819266, 14494859, 73802835, 373398489, 1881341265
PERMUTATIONS BY LENGTH OF RUNS. REF DKB 260.

1667 1, 5, 61, 1385, 50521, 2702765, 199360981, 19391512145, 2404879675441, 370371188237525, 69348874393137901, 15514534163557086905
EULER NUMBERS. REF AS1 810. MTAC 21 675 67.

1668 1, 5, 70, 560, 3360, 16800, 73920, 295680, 1098240, 3843840, 12812800, 41000960, 126730240, 380190720, 1111326720, 3175219200, 8890613760
PRODUCT OF BINOMIAL COEFFICIENTS. REF MFM 74 62 70.

1669 1, 5, 93, 1030, 8885, 65954, 442610, 2762412, 16322085
ROOTED PLANAR MAPS. REF BAMS 74 74 68.

1670 1, 5, 111, 5232, 49910, 3527745, 76435695, 2673350008, 33507517680, 4954123399050
COEFFICIENTS FOR STEP-BY-STEP INTEGRATION. REF JACM 11 231 64.

1671 1, 5, 120, 1840, 27552, 421248, 6613504, 106441472, 1750927872
ALMOST CUBIC MAPS. REF PL2 1 292 70.

1672 1, 5, 205, 22265, 4544185, 1491632525, 718181418565, 476768795646785, 417370516232719345, 465849831125196593045, 645702241048404020542525
MULTIPLES OF EULER NUMBERS. REF MES 28 51 1898. FMR 1 75. HPR.

1673 1, 5, 229, 401, 577, 1129, 1297, 7057, 8761, 14401, 32401, 41617, 57601, 90001
INCREASING CLASS NUMBERS. REF MTAC 23 214 69.

1674 1, 5, 251, 19087, 1070017, 134211265, 703604254357
FROM HIGHER ORDER BERNOULLI NUMBERS. REF NO1 461.

1675 1, 5, 253, 39299, 13265939
COEFFICIENTS OF LEMNISCATE FUNCTION. REF HUR 2 372.

1676 1, 5, 259, 3229, 117469, 7156487, 2430898831, 60967921, 141433003757, 25587296781661
COEFFICIENTS FOR NUMERICAL DIFFERENTIATION. REF OP1 23. PHM 33 14 42.

1677 1, 5, 691, 7, 3617, 43867, 174611, 854513, 236364091, 8553103, 23749461029, 8615841276005, 7709321041217, 2577687858367, 26315271553053477373
NUMERATORS OF BERNOULLI NUMBERS. REF DA2 2 230. AS1 810.

SEQUENCES BEGINNING 1, 6

1678 1, 6, 1, 2, 6, 16, 18, 6, 22, 3, 28, 15, 2, 3, 6, 5, 21, 46, 42, 16, 13, 18, 58, 60, 6, 33, 22, 35, 8, 6, 13, 9, 41, 28, 44, 6, 15, 96, 2, 4, 34, 53, 108, 3, 112, 6, 48, 22, 5, 42, 21, 130
PERIODS OF RECIPROCALS OF INTEGERS. REF PCPS 3 204 1878. LE1 12.

1679 1, 6, 1, 8, 0, 3, 3, 9, 8, 8, 7, 4, 9, 8, 9, 4, 8, 4, 8, 2, 0, 4, 5, 8, 6, 8, 3, 4, 3, 6, 5, 6, 3, 8, 1, 1, 7, 7, 2, 0, 3, 0, 9, 1, 7, 9, 8, 0, 5, 7, 6, 2, 8, 6, 2, 1, 3, 5, 4, 4, 8, 6, 2, 2, 7, 0, 5, 2, 6
PHI, THE GOLDEN RATIO. REF FQ 4 161 66.

1680 1, 6, 2, 6, 16, 18, 22, 28, 15, 3, 5, 21, 46, 13, 58, 60, 33, 35, 8, 13, 41, 44, 96, 4, 34, 53, 108, 112, 42, 130, 8, 46, 148, 75, 78, 81, 166, 43, 178, 180, 95, 192, 98, 99, 30, 222
PERIODS OF RECIPROCALS OF PRIMES. REF RS6 22 203 1874. LE1 15.

1681 1, 6, 3, 82, 84, 444, 769, 1110, 2643, 860, 2901, 1176, 6277, 1170, 21315, 2308, 14244, 29442, 15540, 58194, 13338, 31886, 4080, 176682, 70715, 51240
RELATED TO REPRESENTATION AS SUMS OF SQUARES. REF QJM 38 312 07.

1682 1, 6, 6, 0, 0, 8, 42, 114, 66, 24, 123, 134
PARTITION FUNCTION FOR CUBIC LATTICE. REF AIP 9 279 60.

1683 1, 6, 6, 4, 6, 12, 28, 72, 198, 572, 1716, 5304, 16796, 54264, 178296, 594320, 2005830, 6843420
BINOMIAL COEFFICIENTS. REF TH1 164. FMR 1 55.

1684 1, 6, 7, 20, 27, 47, 74, 269, 6799, 7068, 35071, 112281, 371914, 2715679, 141587222, 144302901, 430193024, 1434881973, 3299956970, 50934236523
CONVERGENTS TO FIFTH ROOT OF 2. REF AMP 46 115 1866. LE1 67. HPR.

1685 1, 6, 8, 10, 12, 14, 15, 18, 20, 21, 22, 26, 27, 28
A TWO-WAY CLASSIFICATION OF INTEGERS. REF CMB 2 89 59.

1686 1, 6, 8, 40, 176, 1421, 10352, 93114, 912920, 9929997, 117970704, 1521176826, 21150414880, 315400444070, 5020920314016, 84979755347122
DISCORDANT PERMUTATIONS. REF SMA 20 23 54.

1687 1, 6, 8, 180, 32, 10080, 3456, 453600, 115200, 47900160, 71680, 217945728000, 36578304000, 2241727488000, 45984153600, 2000741783040000
COEFFICIENTS FOR REPEATED INTEGRATION. REF PHM 38 336 47.

1688 1, 6, 8, 262, 2448, 17997702, 44082372248, 582976662938638O698502, 256989942683351711945337288361248
A SIMPLE RECURRENCE. REF MMAG 37 167 64.

1689 1, 6, 9, 3, 1, 4, 7, 1, 8, 0, 5, 5, 9, 9, 4, 5, 3, 0, 9, 4, 1, 7, 2, 3, 2, 1, 2, 1, 4, 5, 8, 1, 7, 6, 5, 6, 8, 0, 7, 5, 5, 0, 0, 1, 3, 4, 3, 6, 0, 2, 5, 5, 2, 5, 4, 1, 2, 0, 6, 8, 0, 0, 0, 9, 4, 9, 3, 3, 9
NATURAL LOGARITHM OF 2. REF MTAC 17 177 63.

1690 1, 6, 9, 10, 12, 15, 18, 20, 21, 24, 27, 28, 30, 33, 35, 36, 39, 40, 42, 44, 45, 48, 50, 51, 52, 54, 56, 57, 60, 63, 66, 69, 70, 72, 75, 78, 80, 81, 84, 87, 88, 90, 93, 96, 98, 99, 100
EVEN ABUNDANT NUMBERS. REF QJM 44 274 13 (DIVIDED BY 2).

1691 1, 6, 9, 10, 30, 0, 11, 42, 0, 70, 18, 54, 49, 90, 0, 22, 60, 0, 110, 0, 81, 180, 78, 0, 130, 198, 0, 182, 30, 90, 121, 84, 0, 0, 210, 0, 252, 102, 270, 170, 0, 0, 69, 330, 0, 38
COEFFICIENTS OF A MODULAR FORM. REF KNAW 59 207 56.

1692 1, 6, 9, 12, 18, 27, 42, 66, 105, 168, 270, 435, 702, 1134, 1833, 2964, 4794, 7755, 12546, 20298, 32841, 53136, 85974, 139107, 225078, 364182, 589257, 953436, 1542690
RESTRICTED PERMUTATIONS. REF CMB 4 32 61.

1693 1, 6, 9, 13, 19, 37, 58, 97, 143, 227, 328, 492, 688, 992, 1364, 1903, 2551, 3473, 4586, 6097, 7911, 10333, 13226, 16988, 21454, 27172, 33938, 42437, 52423, 64833
A GENERALIZED PARTITION FUNCTION. REF PNISI 17 237 51.

1694 1, 6, 10, 22, 34, 48, 60, 78, 84, 90, 114, 144, 120, 168, 180, 234, 246, 288, 240, 210, 324, 300, 360, 474, 330, 528, 576, 390, 462, 480, 420, 570, 510, 672, 792, 756, 876
INVERSE GOLDBACH NUMBERS. REF WO1.

1695 1, 6, 11, 17, 22, 27, 32, 37, 43, 48, 53, 58, 64
WYTHOFF GAME. REF CMB 2 189 59.

1696 1, 6, 11, 20, 36, 65, 119, 218, 400, 735, 1351, 2484, 4568, 8401, 15451, 28418, 52268, 96135, 176819, 325220, 598172, 1100209, 2023599, 3721978, 6845784
RESTRICTED PERMUTATIONS. REF CMB 4 32 61 (DIVIDED BY 4).

1697 1, 6, 11, 21, 41, 81, 161, 321, 636, 1261, 2501, 4961, 9841, 19521, 38721, 76806, 152351, 302201, 599441, 1189041, 2358561, 4678401, 9279996, 18407641
HEXANACCI NUMBERS. REF FQ 2 302 64.

1698 1, 6, 11, 36, 85, 235, 600, 1590, 4140, 10866, 28416, 74431, 194821, 510096, 1335395, 3496170, 9153025, 23963005, 62735880
FROM A DEFINITE INTEGRAL. REF EMS 10 184 57.

1699 1, 6, 11, 71, 4691, 21982031, 483209576974811, 233491495280173380882643611671
A NONLINEAR RECURRENCE. REF AMM 70 403 63.

1700 1, 6, 12, 24, 60, 72, 168, 192, 324, 360, 660, 576, 1092, 1008, 1440, 1536, 2448, 1944, 3420, 2880, 4032, 3960, 6072, 4608, 7500, 6552, 8748, 8064, 12180, 8640, 14880
INDICES OF MODULAR GROUPS. REF GU6 15.

1701 1, 6, 12, 90, 360, 2040, 10080, 54810, 290640
WALKS ON A TRIANGULAR LATTICE. REF AIP 9 345 60.

1702 1, 6, 12, 156, 1680, 21264, 592032, 5712096, 390388992
PARTITION FUNCTION FOR DIAMOND LATTICE. REF PPS 86 10 65.

1703 1, 6, 15, 19, 24, 42, 73, 127, 208, 337, 528, 827, 1263, 1902, 2819, 4133, 5986, 8578, 12146, 17057, 23711, 32708, 44726, 60713, 81800, 109468, 145526, 192288
A GENERALIZED PARTITION FUNCTION. REF PNISI 17 236 51.

1704 1, 6, 15, 20, 9, 24, 65, 90, 75, 6, 90, 180, 220, 180, 66, 110, 264, 360, 365, 264, 66, 178, 375, 510, 496, 414, 180, 60, 330, 570, 622, 582, 390, 220, 96, 300, 621, 630, 705
COEFFICIENTS OF A MODULAR FORM. REF JLMS 39 435 64.

1705 1, 6, 15, 28, 45, 66, 91, 120, 153, 190, 231, 276, 325, 378, 435, 496, 561, 630, 703, 780, 861, 946, 1035, 1128, 1225, 1326, 1431, 1540, 1653, 1770, 1891, 2016, 2145, 2278
HEXAGONAL NUMBERS N(2N − 1). REF DI2 2 2. BE3 189.

1706 1, 6, 15, 36, 72, 127, 212
POSTAGE STAMP PROBLEM. REF CJ1 12 379 69.

1707 1, 6, 17, 37, 71, 127, 217, 346, 513, 798
POSTAGE STAMP PROBLEM. REF CJ1 12 379 69.

1708 1, 6, 17, 38, 70, 116, 185, 258, 384, 490, 686, 826, 1124, 1292, 1705, 1896, 2491, 2670, 3416, 3680, 4602, 4796, 6110, 6178, 7700, 7980, 9684, 9730, 12156, 11920, 14601
RELATED TO THE DIVISOR FUNCTION. REF SMA 19 39 53.

1709 1, 6, 18, 40, 75, 126, 196, 288, 405, 550, 726, 936, 1183, 1470, 1800, 2176, 2601, 3078, 3610, 4200, 4851, 5566, 6348, 7200, 8125, 9126, 10206, 11368, 12615, 13950
PENTAGONAL PYRAMIDAL NUMBERS. REF DI2 2 2. BE3 194.

1710 1, 6, 18, 40, 81, 201, 414, 916, 1899, 3973, 8059, 16402, 32561, 64520, 125986, 244448, 469195, 895077, 1692143, 3179406, 5929721, 10993373, 20250589, 37096872
SOLID PARTITIONS. REF PNISI 26 135 60.

1711 1, 6, 18, 54, 162, 474, 1398, 4074, 11898, 34554, 100302, 290334, 839466
WALKS ON A TRIANGULAR LATTICE. REF JCP 34 1261 61.

1712 1, 6, 18, 66, 208, 646, 1962, 5962, 18014, 54578, 165650, 504220, 1539330, 4712742, 14475936
PARAFFINS. REF JACS 54 1105 32.

1713 1, 6, 20, 42, 70, 900, 22, 352
QUEENS PROBLEM. REF SL1 49.

1714 1, 6, 20, 50, 105, 196, 336, 540, 825, 1210, 1716, 2366, 3185, 4200, 5440, 6936, 8721, 10830, 13300, 16170, 19481, 23276, 27600, 32500, 38025, 44226, 51156, 58870
4-DIMENSIONAL FIGURATE NUMBERS. REF BE3 195.

1715 1, 6, 20, 134, 915, 7324, 65784, 657180
FROM MENAGE POLYNOMIALS. REF R1 198.

1716 1, 6, 20, 135, 924, 7420, 66744, 667485, 7342280, 88107426, 1145396460, 16035550531, 240533257860, 3848532125880, 65425046139824
RENCONTRES NUMBERS. REF R1 65.

1717 1, 6, 20, 180, 1106, 9292, 82980, 831545, 9139482, 109595496, 1423490744, 19911182207, 298408841160, 4770598226296, 81037124739588
DISCORDANT PERMUTATIONS. REF SMA 20 23 54.

1718 1, 6, 21, 55, 120, 231, 406, 666, 1035, 1540, 2211, 3081, 4186, 5565, 7260, 9316, 11781, 14706, 18145, 22155, 26796, 32131, 38226, 45150, 52975, 61776, 71631, 82621
DOUBLY TRIANGULAR NUMBERS. REF TCPS 9 477 1856.

1719 1, 6, 21, 56, 126, 252, 462, 792, 1287, 2002, 3003, 4368, 6188, 8568, 11628, 15504, 20349, 26334, 33649, 42504, 53130, 65780, 80730, 98280, 118755, 142506
FIGURATE NUMBERS OR BINOMIAL COEFFICIENTS C(N, 5). REF DI2 2 7. RS1. BE3 196. AS1 828.

1720 1, 6, 21, 71, 216, 657, 1907, 5507, 15522, 43352, 119140, 323946, 869476, 2308071, 6056581
5-DIMENSIONAL PARTITIONS. REF PCPS 63 1099 67.

1721 1, 6, 21, 71, 216, 672, 1982, 5817, 16582, 46633
RELATED TO 5-DIMENSIONAL PARTITIONS. REF PCPS 63 1100 67.

1722 1, 6, 21, 91, 266, 994, 2562, 7764, 19482, 51212, 116028
RESTRICTED PARTITIONS. REF JCT 9 373 70.

1723 1, 6, 21, 91, 441, 2275, 12201, 67171, 376761, 2142595, 12313161, 71340451, 415998681, 2438235715, 14350108521, 84740914531, 501790686201
1N + 2**N + ··· + 6**N. REF AS1 813.**

1724 1, 6, 22, 64, 162, 374, 809, 1668, 3316, 6408, 12108, 22468, 41081, 74202, 132666, 235160, 413790, 723530, 1258225, 2177640, 3753096, 6444336, 11028792
FROM ROOK POLYNOMIALS. REF SMA 20 18 54.

1725 1, 6, 22, 64, 163, 382, 848, 1816
RADON PARTITIONS. REF MFM 73 18 69.

1726 1, 6, 22, 130, 822, 6202, 52552, 499194, 5238370, 60222844, 752587764, 10157945044, 147267180508, 2282355168060, 37655004171808, 658906772228668
MATRICES WITH 2 ROWS. REF PLMS 17 29 17. EMN 34 3 44.

1727 1, 6, 22, 159, 1044, 9121, 78132, 748719, 7161484, 70800861, 699869892
FINAL DIGITS OF SQUARES. REF AMM 67 1002 60.

1728 1, 6, 24, 45, 480, 10080, 24192, 907200, 1036800, 239500800, 106444800, 9906624000, 475517952000, 15692092416000, 4828336128000, 8002967132160000
COEFFICIENTS FOR REPEATED INTEGRATION. REF PHM 38 336 47.

1729 1, 6, 24, 80, 240, 672, 1792, 4608, 11520, 28160, 67584, 159744, 372736, 860160, 1966080, 4456448, 10027008, 22413312, 49807360, 110100480, 242221056
COEFFICIENTS OF CHEBYSHEV POLYNOMIALS. REF PRSE 62 190 46. AS1 796. MFM 74 62 70.

1730 1, 6, 24, 90, 318, 1098, 3696, 12270, 40224, 130650, 421176, 1348998, 4299018
SUSCEPTIBILITY FOR TRIANGULAR LATTICE. REF PRV 124 411 61.

1731 1, 6, 24, 90, 336, 1254, 4680, 17466, 65184, 243270, 907896, 3388314, 12645360, 47193126, 176127144, 657315450, 2453134656, 9155223174, 34167758040
A(N) = 4A(N − 1) − A(N − 2). REF MTAC 24 180 70.

1732 1, 6, 25, 60, 203, 3710, 21347
RELATED TO WEBER FUNCTIONS. REF KNAW 66 751 63.

1733 1, 6, 25, 90, 300, 954, 2929, 8840, 26185, 76490
FROM CONVOLVED FIBONACCI NUMBERS. REF RI1.

1734 1, 6, 25, 90, 301, 966, 3025, 9330, 28501, 86526, 261625, 788970, 2375101, 7141686, 21457825, 64439010, 193448101, 580606446, 1742343625, 5228079450
STIRLING NUMBERS OF SECOND KIND. REF AS1 835. DKB 223.

1735 1, 6, 26, 71, 155, 295, 511, 826, 1266, 1860, 2640, 3641, 4901, 6461, 8365, 10660, 13396, 16626, 20406, 24795, 29855, 35651, 42251, 49726, 58150, 67600, 78156
GENERALIZED STIRLING NUMBERS. REF PEF 77 7 62.

1736 1, 6, 26, 94, 308, 941, 2744, 7722, 21166, 56809, 149971, 390517, 1005491, 2564164, 6485901, 16289602, 40659669, 100934017, 249343899, 613286048
TREES OF HEIGHT 6. REF IBMJ 4 475 60. KU1.

1737 1, 6, 27, 98, 309, 882, 2330, 5784, 13644, 30826, 67107, 141444, 289746, 578646, 1129527, 2159774, 4052721, 7474806, 15063859
COEFFICIENTS OF AN ELLIPTIC FUNCTION. REF CAY 9 128.

1738 1, 6, 27, 98, 315, 924, 2534, 6588, 16410, 39436, 91974
CONVOLVED FIBONACCI NUMBERS. REF RCI 101.

1739 1, 6, 27, 104, 369, 1236, 3989, 12522, 38535, 116808, 350064, 1039896, 3068145, 9004182, 26314773, 76652582, 222705603, 645731148, 1869303857, 5404655358
POWERS OF ROOTED TREE ENUMERATOR. REF R1 150.

1740 1, 6, 27, 122, 516, 2148, 8792, 35622, 143079, 570830, 2264649
SUSCEPTIBILITY FOR HONEYCOMB. REF PHA 28 934 62.

1741 1, 6, 28, 120, 495, 2002, 8008, 31824, 125970, 497420, 1961256
COEFFICIENTS OF CHEBYSHEV POLYNOMIALS. REF LA4 517.

1742 1, 6, 28, 125, 527, 2168, 8781, 35155, 139531, 550068
SPHEROIDAL HARMONICS. REF MES 54 75 24.

1743 1, 6, 28, 140, 270, 496, 672, 1638, 2970, 6200, 8128, 8190, 18600, 18620, 27846, 30240, 32760, 55860, 105664, 117800, 167400, 173600, 237510, 242060, 332640
NUMBERS WITH INTEGRAL HARMONIC MEAN. REF AMM 61 95 54.

1744 1, 6, 28, 496, 8128, 33550336, 8589869056, 137438691328, 2305843008139952128, 2658455991569831744654692615953842176
EVEN PERFECT NUMBERS. REF SMA 19 128 53. REC 4 56 61. BE3 19. NAMS 18 608 71.

1745 1, 6, 29, 150, 841, 5166, 34649, 252750, 1995181, 16962726, 154624469
QUASI-ALTERNATING PERMUTATIONS. REF NET 113.

1746 1, 6, 30, 42, 30, 66, 2730, 6, 510, 798, 330, 138, 2730, 6, 870, 14322, 510, 6, 1919190, 6, 13530, 1806, 690, 282, 46410, 66, 1590, 798, 870, 354, 56786730, 6, 510
DENOMINATORS OF BERNOULLI NUMBERS. REF DA2 2 230. AS1 810.

1747 1, 6, 30, 84, 90, 132, 5460, 360, 1530, 7980, 13860, 8280, 81900, 1512, 3480, 114576
DENOMINATORS OF BERNOULLI NUMBERS. REF DA2 2 208.

1748 1, 6, 30, 126, 510, 2046, 8190, 32766, 131070, 524286, 2097150, 8388606, 33554430, 134217726, 536870910, 2147483646, 8589934590, 34359738366
RELATED TO EULER NUMBERS. REF QJM 47 110 16. FMR 1 112. DA2 2 283.

1749 1, 6, 30, 126, 534, 2214, 9246, 38142, 157974, 649086, 2674926
WALKS ON A CUBIC LATTICE. REF JCP 34 1261 61.

1750 1, 6, 30, 138, 606, 2586, 10818, 44574, 181542, 732678, 2935218, 11687202, 46296210
SUSCEPTIBILITY FOR TRIANGULAR LATTICE. REF SSP 3 268 70.

1751 1, 6, 30, 138, 618, 2730, 11946, 51882, 224130, 964134, 4133166, 17668938, 75355206, 320734686, 1362791250, 5781765582, 24497321682, 103673881482
WALKS ON A TRIANGULAR LATTICE. REF JCP 46 3481 67.

1752 1, 6, 30, 140, 630, 2772, 12012, 51480, 218790, 923780, 3879876, 16224936, 67603900, 280816200, 1163381400
PRODUCT OF BINOMIAL COEFFICIENTS. REF OP1 21. SE2 92. JO2 449. JM2 22 120 43. LA4 514.

1753 1, 6, 30, 150, 726, 3510, 16710, 79494, 375174, 1769686, 8306862, 38975286
SUSCEPTIBILITY FOR CUBIC LATTICE. REF PHA 28 942 62.

1754 1, 6, 30, 150, 726, 3534, 16926, 81390, 387966, 1853886, 8809878, 41934150, 198842742, 943974510, 4468911678, 21175146054, 100121875974
WALKS ON A CUBIC LATTICE. REF JCP 39 411 63. MFS.

1755 1, 6, 30, 174, 1158, 8742, 74046, 696750, 7219974
BINOMIAL COEFFICIENT SUMS. REF CJM 22 26 70.

1756 1, 6, 30, 180, 840, 5460, 30996, 209160, 1290960, 9753480, 69618120, 571627056, 4443697440, 40027718640, 346953934320, 3369416698080
PERMUTATIONS OF ORDER EXACTLY 4. REF CJM 7 159 55.

1757 1, 6, 32, 109, 288, 654, 1337
PARTITIONS INTO NON-INTEGRAL POWERS. REF PCPS 47 215 51.

1758 1, 6, 32, 175, 1012, 6230, 40819
GENERALIZED STIRLING NUMBERS OF SECOND KIND. REF FQ 5 366 67.

1759 1, 6, 35, 180, 921, 4626, 23215, 116160
AN INHOMOGENEOUS RECURRENCE. REF AMM 3 244 1896.

1760 1, 6, 35, 204, 1189, 6930, 40391, 235416, 1372105, 7997214, 46611179, 271669860, 1583407981, 9228778026, 53789260175, 313506783024, 1827251437969
A(N) = 6A(N − 1) − A(N − 2). REF DI2 2 10. MAG 47 237 63. BE3 193. FQ 9 95 71.

1761 1, 6, 35, 221, 1554, 12108, 104203, 986452, 10186669, 114173261, 1381629682, 17963567972
PERMUTATIONS BY LENGTH OF RUNS. REF DKB 262 (DIVIDED BY 2).

1762 1, 6, 35, 225, 1624, 13132, 118124, 1172700, 12753576, 150917976, 1931559552, 26596717056, 392156797824, 6165817614720, 102992244837120
STIRLING NUMBERS OF FIRST KIND. REF AS1 833. DKB 226.

1763 1, 6, 36, 150, 540, 1806, 5796, 18150, 55980, 171006, 519156, 1569750, 4733820, 14250606, 42850116, 128746950, 386634060, 1160688606, 3483638676
DIFFERENCES OF ZERO. REF VO1 31. DA2 2 212. R1 33.

1764 1, 6, 36, 200, 1170, 7392, 50568, 372528, 2936070
LABELED TREES OF HEIGHT 2. REF IBMJ 4 478 60.

1765 1, 6, 36, 216, 1296, 7776, 46656, 279936, 1679616, 10077696, 60466176, 362797056, 2176782336, 13060694016, 78364164096, 470184984576, 2821109907456
POWERS OF SIX. REF BA1.

1766 1, 6, 36, 240, 1800, 15120, 141120, 1451520, 16329600, 199584000, 2634508800, 37362124800, 566658892800, 9153720576000, 156920924160000
LAH NUMBERS. REF R1 44. CO1 1 166.

1767 1, 6, 37, 195, 979, 4663, 21474, 96496, 425365
POLYOMINOES WITH HOLES. REF PA1. JRM 2 182 69.

1768 1, 6, 40, 112, 1152, 2816, 13312, 30270, 557056, 1245184, 5505024, 12058624, 104857600, 226492416, 973078528, 2080374784
FROM DOUBLE FACTORIALS. REF RG1 414.

1769 1, 6, 40, 155, 456, 1128
SEQUENCES BY NUMBER OF INCREASES. REF JCT 1 372 66.

1770 1, 6, 40, 360, 4576, 82656
COLORED GRAPHS. REF CJM 12 412 60 (DIVIDED BY 4).

1771 1, 6, 41, 293, 2309, 19975, 189524, 1960041, 21993884, 266361634, 3465832370, 48245601976, 715756932697, 11277786883706, 188135296650845
PERMUTATIONS BY LENGTH OF RUNS. REF DKB 261.

1772 1, 6, 42, 336, 3024, 30240, 332640, 3991680, 51891840, 726485760, 10897286400, 174356582400, 2964061900800, 53353114214400, 1013709170073600
GENERALIZED STIRLING NUMBERS. REF PEF 107 5 63.

1773 1, 6, 44, 145, 336, 644, 1096, 1719, 2540, 3586, 4884, 6461, 8344, 10560, 13136, 16099, 19476, 23294, 27580, 32361, 37664, 43516, 49944, 56975, 64636, 72954, 81956
DISCORDANT PERMUTATIONS. REF SMA 20 23 54.

1774 1, 6, 44, 430, 5322, 79184, 1381144
TOTAL DIAMETER OF LABELED TREES. REF IBMJ 4 478 60.

1775 1, 6, 45, 420, 4725, 62370, 945945, 16216200, 310134825
VALUES OF BESSEL POLYNOMIALS. REF RCI 77. RI1.

1776 1, 6, 46, 450, 5650, 91866, 1957066
RELATED TO PARTIALLY ORDERED SETS. REF JCT 6 17 69.

1777 1, 6, 48, 390, 3216, 26844, 229584, 2006736, 17809008
SPECIFIC HEAT FOR CUBIC LATTICE. REF PRV 129 102 63.

1778 1, 6, 48, 528, 7920, 149856, 3169248, 77046528, 2231209728, 71938507776, 2446325534208
SUSCEPTIBILITY FOR CUBIC LATTICE. REF PRV 164 801 67.

1779 1, 6, 50, 225, 735, 1960, 4536, 9450, 18150, 32670, 55770, 91091, 143325, 218400, 323680, 468180, 662796, 920550, 1256850, 1689765, 2240315, 2932776
STIRLING NUMBERS OF FIRST KIND. REF AS1 833. DKB 226.

1780 1, 6, 51, 506, 5481, 62832, 749398, 9203634, 115607310
DISSECTIONS OF A POLYGON. REF AMP 1 198 1841.

1781 1, 6, 51, 561, 7556, 120196, 2201856, 45592666, 1051951026, 26740775306, 742069051906, 22310563733864, 722108667742546, 25024187820786357
COEFFICIENTS OF ITERATED EXPONENTIALS. REF SMA 11 353 45.

1782 1, 6, 57, 741, 12244, 245755, 5809875, 158198200, 4877852505, 168055077875, 6400217406500, 267058149580823, 12118701719205803, 594291742526530761
COEFFICIENTS OF ITERATED EXPONENTIALS. REF SMA 11 353 45.

1783 1, 6, 60, 90, 87360, 146361946186458562560000
UNITARY PERFECT NUMBERS. REF NAMS 18 630 71.

1784 1, 6, 60, 840, 15120, 332640, 8648640, 259459200, 8821612800, 335221286400, 14079294028800, 647647525324800, 32382376266240000
DISSECTIONS OF A BALL. REF MTAC 3 168 48, 9 174 55. CMA 2 25 70.

1785 1, 6, 60, 1368, 15552, 201240, 2016432, 21582624
FOLDING A MAP. REF CJ1 14 77 71.

1786 1, 6, 63, 616, 6678, 77868, 978978, 13216104, 190899423, 2939850914, 48106651593
PERMUTATIONS BY NUMBER OF PAIRS. REF DKB 263.

1787 1, 6, 66, 702, 7350, 76266, 786858, 8086074, 82848522, 846886962, 8640964782
WALKS ON A CUBIC LATTICE. REF PPS 92 649 67.

1788 1, 6, 72, 1320, 32760, 1028160, 39070080
DISSECTIONS OF A BALL. REF CMA 2 25 70.

1789 1, 6, 80, 30240, 1814400, 2661120, 871782912000, 3138418483200, 84687482880000, 170303140572364800, 1124000727777607680000
COEFFICIENTS FOR CENTRAL DIFFERENCES. REF JM2 42 162 63.

1790 1, 6, 90, 945, 9450, 93555, 638512875, 18243225, 325641566250, 38979295480125, 1531329465290625, 13447856940643125, 201919571963756521875
DENOMINATORS OF SUMS OF INVERSE POWERS S(2N). REF FMR 1 84.

1791 1, 6, 90, 1860, 44730, 1172556, 32496156
WALKS ON A CUBIC LATTICE. REF AIP 9 345 60.

1792 1, 6, 90, 2040, 67950, 3110940, 187530840, 14398171200, 1371785398200, 158815387962000, 21959547410077200, 3574340599104475200
STOCHASTIC MATRICES OF INTEGERS. REF ST2. DMJ 33 763 66.

1793 1, 6, 90, 2520, 113400, 7484400, 681080400, 81729648000, 12504636144000, 2375880867360000, 548828480360160000, 151476660579404160000
RELATED TO EULER NUMBERS. REF QJM 47 110 16. FMR 1 112. DA2 2 283. PSAM 15 101 63.

1794 1, 6, 96, 1200, 14400, 176400, 2257920, 30481920, 435456000, 6586272000, 105380352000
COEFFICIENTS OF LAGUERRE POLYNOMIALS. REF AS1 799.

1795 1, 6, 120, 1980, 32970, 584430, 11204676, 233098740, 5254404210, 127921380840, 3350718545460, 94062457204716, 2819367702529560
FROM BESSEL POLYNOMIALS. REF RCI 77. RI1.

1796 1, 6, 120, 5250, 395010, 45197460, 7299452160, 1580682203100, 441926274289500, 154940341854097800, 66565404923242024800
EXPANSION OF A SKEW DETERMINANT. REF EMN 34 4 44.

1797 1, 6, 130, 2380, 44100, 866250, 18288270, 416215800, 10199989800, 268438920750, 7562120816250, 227266937597700, 7262844156067500
ASSOCIATED STIRLING NUMBERS. REF TOH 37 259 33. JO2 152. CO1 2 98.

1798 1, 6, 168, 20160, 9999360, 20158709760
NONSINGULAR BINARY MATRICES. REF JSIAM 20 377 71.

1799 1, 6, 210, 223092870, 3217644767340672907899084554130
A HIGHLY COMPOSITE SEQUENCE. REF AMM 74 874 67.

1800 1, 6, 240, 1020, 78120, 279930, 40353600, 134217720, 31381059600
PILE OF COCOANUTS PROBLEM. REF AMM 35 48 28.

1801 1, 6, 350, 43260, 14591171
SINGULAR (0, 1)-MATRICES. REF JCT 3 198 67.

1802 1, 6, 360, 10080, 259200, 239500800, 145297152000, 15692092416000, 16005934264320000, 8515157028618240000, 3372002183332823040000
COEFFICIENTS FOR REPEATED INTEGRATION. REF JM2 28 56 49.

1803 1, 6, 425, 65625, 27894671
SINGULAR (0, 1)-MATRICES. REF JCT 3 198 67.

1804 1, 6, 438, 3962646
POST FUNCTIONS. REF JCT 4 298 68.

1805 1, 6, 522, 152166, 93241002, 97949265606, 157201459863882, 357802951084619046, 1096291279711115037162, 4350684698032741048452486
GENERALIZED TANGENT NUMBERS. REF MTAC 21 690 67.

1806 1, 6, 720, 1512000, 53343360000, 31052236723200000, 295415578275110092800000, 45669605890716810734764032000000
AN ILL-CONDITIONED DETERMINANT. REF MTAC 9 155 55. HPR.

1807 1, 6, 924, 81738720000, 256963707943061374889193111552000
INVERTIBLE BOOLEAN FUNCTIONS. REF PGEC 13 530 64.

1808 1, 6, 1230, 134355076
POST FUNCTIONS. REF JCT 4 296 68.

1809 1, 6, 2862, 537157696
POST FUNCTIONS. REF JCT 4 297 68.

SEQUENCES BEGINNING 1, 7

1810 1, 7, 1, 31, 1, 127, 17, 73, 31, 2047, 1, 8191, 5461, 4681, 257, 131071, 73, 524287
FROM GENERALIZED BERNOULLI NUMBERS. REF JM2 23 211 44.

1811 1, 7, 2, 1, 1, 3, 18, 5, 1, 1, 6, 30, 8, 1, 1, 9, 42, 11, 1, 1, 12, 54, 14, 1, 1, 15, 66, 17, 1, 1, 18, 78, 20, 1, 1, 21, 90, 23, 1, 1, 24, 102, 26, 1, 1, 27, 114, 29, 1, 1, 30, 126, 32, 1, 1
CONTINUED FRACTION EXPANSION OF E2. REF PE1 138.**

1812 1, 7, 3, 2, 0, 5, 0, 8, 0, 7, 5, 6, 8, 8, 7, 7, 2, 9, 3, 5, 2, 7, 4, 4, 6, 3, 4, 1, 5, 0, 5, 8, 7, 2, 3, 6, 6, 9, 4, 2, 8, 0, 5, 2, 5, 3, 8, 1, 0, 3, 8, 0, 6, 2, 8, 0, 5, 5, 8, 0, 6, 9, 7, 9, 4, 5, 1, 9, 3
SQUARE ROOT OF 3. REF PNAS 37 444 51. MTAC 22 234 68.

1813 1, 7, 5, 145, 5, 6095, 5815, 433025, 956375, 46676375, 172917875, 7108596625, 38579649875, 1454225641375, 10713341611375, 384836032842625
EXPANSION OF EXP(ARCTAN X). REF DMJ 26 573 59. HPR.

1814 1, 7, 7, 2, 4, 5, 3, 8, 5, 0, 9, 0, 5, 5, 1, 6, 0, 2, 7, 2, 9, 8, 1, 6, 7, 4, 8, 3, 3, 4, 1, 1, 4, 5, 1, 8, 2, 7, 9, 7, 5, 4, 9, 4, 5, 6, 1, 2, 2, 3, 8, 7, 1, 2, 8, 2, 1, 3, 8, 0, 7, 7, 8, 9, 8, 5, 2, 9, 1
SQUARE ROOT OF PI. REF RS4 XVIII.

1815 1, 7, 8, 23, 31, 54, 85, 309, 7810, 8119, 40286, 128977, 427217, 3119496, 162641009, 165760505, 494162019, 1648246562, 3790655143, 58508073707
CONVERGENTS TO FIFTH ROOT OF 2. REF AMP 46 115 1866. LE1 67. HPR.

1816 1, 7, 11, 19, 23, 31, 43, 47, 59, 67, 71, 83, 103, 107, 127, 131, 139, 151, 163, 167, 179, 191, 199, 211, 227, 239, 251, 263, 271, 283, 307, 311, 331, 347, 367, 379, 383, 419
QUADRATIC FORMS BY CLASS NUMBER. REF IAS 2 178 35.

1817 1, 7, 11, 26, 45, 83, 125, 140, 182, 197, 201, 216, 239, 258, 311, 330, 353, 444, 467, 482, 486, 524, 539, 558, 600, 752, 771, 843, 881, 885, 923, 980, 999, 1071, 1113
(N(N + 1) + 1)/19 IS PRIME. REF CU1 1 252.

1818 1, 7, 11, 27, 77, 107, 111, 127, 177, 777, 1127, 1177, 1777, 7777, 11777, 27777, 77777, 107777, 111777, 127777, 177777, 777777, 1127777, 1177777, 1777777, 7777777
SMALLEST NUMBER REQUIRING N SYLLABLES IN ENGLISH.

1819 1, 7, 13, 19, 31, 37, 43, 61, 67, 73, 79, 97, 103, 109, 127, 139, 151, 157, 163, 181, 193, 199, 211, 223, 229, 241, 271, 277, 283, 307, 313, 331, 337, 349, 367, 373, 379, 397
PRIMES OF THE FORM 3N + 1. REF RE2 1.

1820 1, 7, 13, 97, 8833, 77968897, 6079148431583233, 36956045653220845240164417232897
A NONLINEAR RECURRENCE. REF AMM 70 403 63.

1821 1, 7, 14, 7, 49, 21, 35, 41, 49, 133, 98, 21, 126, 112, 176, 105, 126, 140, 35, 147, 259, 98, 420, 224, 238, 455, 273, 14, 322, 406, 35, 7, 637, 196, 245, 181, 574, 462, 147
COEFFICIENTS OF A MODULAR FORM. REF KNAW 59 207 56.

1822 1, 7, 16, 49, 212, 1158, 7584, 57720, 499680, 4843440
BINOMIAL COEFFICIENT SUMS. REF CJM 22 26 70.

1823 1, 7, 17, 19, 23, 29, 47, 59, 61, 97, 109, 113, 131, 149, 167, 179, 181, 193, 223, 229, 233, 257, 263, 269, 313, 337, 367, 379, 383, 389, 419, 433, 461, 487, 491, 499, 503
PRIMES WITH 10 AS A PRIMITIVE ROOT. REF KR1 1 61.

1824 1, 7, 17, 23, 41, 47, 71, 79, 97, 103, 137, 167, 191, 193, 199, 239, 263, 271, 311, 313, 359, 367, 383, 401, 409, 449, 463, 479, 487, 503, 521, 569, 599, 607, 647, 719, 743
2 IS A QUADRATIC RESIDUE MODULO P. REF KR1 1 59.

1825 1, 7, 17, 31, 43, 79, 89, 113, 127, 137, 199, 223, 233, 257, 281, 283, 331, 353, 401, 449, 463, 487, 521, 569, 571, 593, 607, 617, 631, 641, 691, 739, 751, 809, 811, 823, 857
PRIMES WITH 3 AS SMALLEST PRIMITIVE ROOT. REF KR1 1 57. AS1 864.

1826 1, 7, 18, 34, 55, 81, 112, 148, 189, 235, 286, 342, 403, 469, 540, 616, 697, 783, 874, 970, 1071, 1177, 1288, 1404, 1525, 1651, 1782, 1918, 2059, 2205, 2356, 2512, 2673
HEPTAGONAL NUMBERS N(5N − 3)/2. REF DI2 2 2. BE3 189.

1827 1, 7, 18, 44, 88, 169, 296, 507, 824, 1314, 2029, 3083, 4578, 6714, 9676, 13795, 19408, 27053, 37302, 51029, 69180, 93139, 124447, 165259, 218021, 286068, 373207
BIPARTITE PARTITIONS. REF NI1 11.

1828 1, 7, 19, 37, 61, 127, 271, 331, 397, 547, 631, 919, 1657, 1801, 1951, 2269, 2437, 2791, 3169, 3571, 4219, 4447, 5167, 5419, 6211, 7057, 7351, 8269, 9241, 10267, 11719
CUBAN PRIMES. REF MES 41 144 12. CU1 1 259.

1829 1, 7, 19, 41, 751, 989, 2857, 16067, 2171465, 1364651, 6137698213, 90241897, 105930069, 15043611773, 55294720874657, 203732352169, 69028763155644023
COTESIAN NUMBERS. REF QJM 46 63 14.

1830 1, 7, 19, 47, 97, 189, 339, 589, 975, 1576, 2472, 3804, 5727, 8498, 12400, 17874, 25433, 35818, 49908, 68939, 94378, 128234, 172917, 231630, 308240, 407804, 536412
BIPARTITE PARTITIONS. REF PCPS 49 72 53. NI1 1.

1831 1, 7, 19, 53, 115, 217, 389
POSTAGE STAMP PROBLEM. REF CJ1 12 379 69.

1832 1, 7, 19, 53, 149, 421, 1193, 3387, 9627, 27383, 77923
KEYS. REF MAG 53 11 69.

1833 1, 7, 19, 57, 81, 251, 437, 691, 739, 1743, 3695, 6619, 8217, 9771, 14771, 15155, 16831, 18805, 26745, 30551, 41755, 46297, 54339, 72359, 86407, 96969, 131059
LATTICE POINTS IN SPHERES. REF MTAC 20 306 66.

1834 1, 7, 19, 73, 241, 847, 2899, 10033, 34561, 119287, 411379, 1419193, 4895281, 16886527, 58249459, 200931553, 693110401, 2390878567, 8247309139
A(N) = 2A(N − 1) + 5A(N − 2). REF MQET 1 11 16.

1835 1, 7, 21, 35, 28, 21, 105, 181, 189, 77, 140, 385, 546, 511, 252, 203, 693, 1029, 1092, 798, 203, 581, 1281, 1708, 1687, 1232, 413, 602, 1485, 2233, 2366, 2009, 1099
COEFFICIENTS OF A MODULAR FORM. REF JLMS 39 435 64.

1836 1, 7, 21, 53, 109, 212, 389
POSTAGE STAMP PROBLEM. REF CJ1 12 379 69.

1837 1, 7, 21, 112, 456, 2603, 13203
BAXTER PERMUTATIONS. REF MA4 2 25 67.

1838 1, 7, 21, 112, 588, 3360, 19544, 117648, 720300, 4483696
IRREDUCIBLE POLYNOMIALS, OR NECKLACES. REF AMM 77 744 70.

1839 1, 7, 22, 50, 95, 161, 252, 372, 525, 715, 946, 1222, 1547, 1925, 2360, 2856, 3417, 4047, 4750, 5530, 6391, 7337, 8372, 9500, 10725, 12051, 13482, 15022, 16675, 18445
HEXAGONAL PYRAMIDAL NUMBERS. REF DI2 2 2. BE3 194.

1840 1, 7, 22, 153, 15209
SWITCHING NETWORKS. REF JFI 276 317 63.

1841 1, 7, 23, 47, 71, 199, 167, 191, 239, 383, 311, 431, 647, 479, 983, 887, 719, 839, 1031, 1487, 1439, 1151, 1847, 1319, 3023, 1511
PRIMES BY CLASS NUMBER. REF MTAC 24 492 70.

1842 1, 7, 23, 61, 127, 199, 337, 479, 677, 937, 1193, 1511, 1871, 2267, 2707, 3251, 3769, 4349, 5009, 5711, 6451, 7321, 8231, 9173, 10151, 11197, 12343, 13487, 14779
A SPECIAL SEQUENCE OF PRIMES. REF ACA 6 372 61.

1843 1, 7, 23, 71, 311, 479, 1559, 5711, 10559, 18191, 31391, 366791, 4080359, 12537719, 30706079
PRIMES WITH LARGE LEAST NONRESIDUES. REF RS5 XV.

1844 1, 7, 25, 63, 129, 231, 377, 575, 833, 1159, 1561, 2047, 2625, 3303, 4089, 4991, 6017, 7175, 8473, 9919, 11521, 13287, 15225, 17343, 19649, 22151, 24857, 27775
A SQUARE RECURRENCE. REF SIAMR 12 277 70.

1845 1, 7, 25, 65, 140, 266, 462, 750, 1155, 1705, 2431, 3367, 4550, 6020, 7820, 9996, 12597, 15675, 19285, 23485, 28336, 33902, 40250, 47450, 55575, 64701, 74907, 86275
STIRLING NUMBERS OF SECOND KIND. REF AS1 835. DKB 223.

1846 1, 7, 25, 66, 143, 273, 476, 775, 1197, 1771, 2530, 3510, 4750, 6293, 8184, 10472, 13209, 16450, 20254, 24682, 29799, 35673, 42375, 49980, 58565, 68211, 79002
FERMAT COEFFICIENTS. REF MMAG 27 141 54.

1847 1, 7, 28, 84, 210, 462, 924, 1716, 3003, 5005, 8008, 12376, 18564, 27132, 38760, 54264, 74613, 100947, 134596, 177100, 230230, 296010, 376740, 475020, 593775
FIGURATE NUMBERS OR BINOMIAL COEFFICIENTS C(N, 6). REF DI2 2 7. RS1. BE3 196. AS1 828.

1848 1, 7, 28, 105, 357, 1197, 3857, 12300, 38430, 118874, 362670, 1095430, 3271751, 9673993
6-DIMENSIONAL PARTITIONS. REF PCPS 63 1099 67.

1849 1, 7, 28, 105, 357, 1232, 4067, 13301, 42357, 132845
RELATED TO 6-DIMENSIONAL PARTITIONS. REF PCPS 63 1100 67.

1850 1, 7, 28, 140, 784, 4676, 29008, 184820, 1200304, 7907396, 52666768, 353815700, 2393325424, 16279522916, 111239118928, 762963987380, 5249352196144
1N + 2**N + ··· + 7**N. REF AS1 813.**

1851 1, 7, 29, 93, 256, 638, 1586
RADON PARTITIONS. REF MFM 73 18 69.

1852 1, 7, 29, 94, 263, 667, 1577, 3538, 7622
ARRAYS OF DUMBBELLS. REF JMP 11 3098 70.

1853 1, 7, 31, 37, 109, 121, 127, 133, 151, 157, 403, 421, 511, 529, 631, 637, 661, 679, 1579, 1621, 1633, 1969, 1981, 2017, 2041, 2047, 2053, 2071, 2077, 2143, 2149, 2167
GOOD NUMBERS. REF MTAC 18 541 64.

1854 1, 7, 31, 43, 67, 73, 79, 103, 127, 163, 181, 223, 229, 271, 277, 307, 313, 337, 349, 409, 421, 439, 457, 463, 499
RELATED TO KUMMERS CONJECTURE. REF HA3 482.

1855 1, 7, 31, 127, 73, 23, 8191, 151, 131071, 524287, 337, 47, 601, 262657, 233, 2147483647, 599479, 71, 223, 79, 13367, 431, 631, 2351, 4432676798593, 103
SMALLEST PRIMITIVE FACTOR OF 2(2N + 1) – 1. REF KR1 2 84.**

1856 1, 7, 31, 127, 73, 89, 8191, 151, 131071, 524287, 337, 178481, 1801, 262657, 2089, 2147483647, 599479, 122921, 616318177, 121369, 164511353, 2099863, 23311
LARGEST FACTOR OF 2(2N + 1) – 1. REF KR1 2 84.**

1857 1, 7, 31, 127, 487, 1423, 1303, 2143, 2647, 4447, 5527, 5647, 6703, 5503, 11383, 8863, 13687, 13183, 12007, 22807, 18127, 21487, 22303, 29863, 25303, 27127
PRIMES BY CLASS NUMBER. REF MTAC 24 492 70.

1858 1, 7, 31, 127, 2555, 1414477, 57337, 118518239, 5749691557, 91546277357, 1792042792463, 1982765468311237, 286994504449393, 3187598676787461083
NUMERATORS OF COSECANT NUMBERS. REF NO1 458. ANN 36 640 35. DA2 2 187.

1859 1, 7, 32, 120, 400, 1232, 3584, 9984, 26880, 70400, 180224, 452608, 1118208, 2723840, 6553600
COEFFICIENTS OF CHEBYSHEV POLYNOMIALS. REF PRSE 62 190 46. AS1 795.

1860 1, 7, 33, 123, 257, 515, 925, 1419, 2109, 3071, 4169, 5575, 7153, 9171, 11513, 14147, 17077, 20479, 24405, 28671, 33401, 38911, 44473, 50883, 57777, 65267, 73525
LATTICE POINTS IN SPHERES. REF PNISI 13 37 47. MTAC 16 287 62.

1861 1, 7, 33, 715, 4199, 52003, 334305, 17678835, 119409675, 1641030105, 11435320455, 322476036831
COEFFICIENTS OF LEGENDRE POLYNOMIALS. REF MTAC 3 17 48.

1862 1, 7, 34, 136, 487, 1615, 5079, 15349, 45009, 128899, 362266, 1002681, 2740448, 7411408, 19865445, 52840977, 139624510, 366803313, 958696860, 2494322662
TREES OF HEIGHT 7. REF IBMJ 4 475 60. KU1.

1863 1, 7, 34, 1056, 5884954
SWITCHING NETWORKS. REF JFI 276 317 63.

1864 1, 7, 35, 140, 483, 1498, 4277, 11425, 28889, 69734, 161735, 362271, 786877, 1662927, 3428770, 6913760, 13660346, 26492361, 50504755
COEFFICIENTS OF AN ELLIPTIC FUNCTION. REF CAY 9 128.

1865 1, 7, 35, 140, 490, 1554, 4578, 12720, 33708, 85864, 211546
CONVOLVED FIBONACCI NUMBERS. REF RCI 101.

1866 1, 7, 35, 154, 637, 2548, 9996, 38760, 149226, 572033, 2187185, 8351070, 31865925, 121580760, 463991880, 1771605360, 6768687870, 25880277150
LAPLACE TRANSFORM COEFFICIENTS. REF QAM 14 407 56.

1867 1, 7, 37, 197, 1172, 8018, 62814, 556014, 5488059, 59740609, 710771275
CYCLES IN COMPLETE GRAPH. REF PIEE 115 763 68.

1868 1, 7, 37, 199, 40321, 5512813, 136601407, 32373535937, 4039314145093, 377880467185583, 123905113265594071
COEFFICIENTS FOR REPEATED INTEGRATION. REF JM2 28 56 49.

1869 1, 7, 41, 239, 1393, 8119, 47321, 275807, 1607521, 9369319, 54608393, 318281039, 1855077841, 10812186007, 63018038201, 367296043199, 2140758220993
A(N) = 6A(N - 1) - A(N - 2). REF AMM 4 25 1897. IDM 10 236 03. ANN 36 644 35.

1870 1, 7, 41, 479, 59, 266681, 63397, 514639, 178939, 10410343, 18500393, 40799043101, 1411432849
COEFFICIENTS FOR NUMERICAL DIFFERENTIATION. REF OP1 21. JM2 22 120 43.

1871 1, 7, 45, 323, 2621, 23811, 239653, 2648395, 31889517, 415641779, 5830753109, 87601592187, 1403439027805
PERMUTATIONS BY NUMBER OF PAIRS. REF DKB 263.

1872 1, 7, 46, 4336, 134281216
SWITCHING NETWORKS. REF JFI 276 317 63.

1873 1, 7, 47, 342, 2754, 24552, 241128, 2592720, 30334320, 383970240, 5231113920, 76349105280, 1188825724800, 19675048780800, 344937224217600
GENERALIZED STIRLING NUMBERS. REF PEF 77 26 62.

1874 1, 7, 49, 343, 2401, 16807, 117649, 823543, 5764801, 40353607, 282475249, 1977326743, 13841287201, 96889010407, 678223072849, 4747561509943
POWERS OF SEVEN. REF BA1.

1875 1, 7, 55, 529, 6192, 86580, 1425517, 27298231, 601580874, 15116315767, 429614643061, 13711655205088, 488332318973593, 19296579341940068
NEAREST INTEGER TO BERNOULLI NUMBERS. REF DA2 2 236. AS1 810.

1876 1, 7, 56, 504, 5040, 55440, 665280, 8648640, 121080960, 1816214400, 29059430400, 494010316800, 8892185702400, 168951528345600, 3379030566912000
GENERALIZED STIRLING NUMBERS. REF PEF 107 19 63.

1877 1, 7, 61, 661, 8953, 152917, 3334921
RELATED TO PARTIALLY ORDERED SETS. REF JCT 6 17 69.

1878 1, 7, 70, 819, 10472, 141778, 1997688, 28989675, 430321633
DISSECTIONS OF A POLYGON. REF AMP 1 198 1841.

1879 1, 7, 70, 910, 14532, 274778, 5995892, 148154860, 4085619622, 124304629050, 4133867297490, 149114120602860, 5796433459664946, 241482353893283349
COEFFICIENTS OF ITERATED EXPONENTIALS. REF SMA 11 353 45.

1880 1, 7, 71, 1001, 18089, 398959, 10391023, 312129649, 10622799089, 403978495031, 16977719590391, 781379079653017, 39085931702241241
CONVERGENTS TO E. REF BA4 17 1871. MTAC 2 69 46.

1881 1, 7, 74, 882, 11144
SYMMETRIC PERMUTATIONS. REF LU1 1 222.

1882 1, 7, 77, 1155, 21973, 506989, 13761937, 429853851, 15192078027, 599551077881, 26140497946017, 1248134313062231, 64783855286002573
COEFFICIENTS OF ITERATED EXPONENTIALS. REF SMA 11 353 45.

1883 1, 7, 85, 1660, 48076, 1942416, 104587344, 7245893376, 628308907776, 66687811660800, 8506654697548800
DIFFERENCES OF RECIPROCALS OF UNITY. REF DKB 228.

1884 1, 7, 85, 1777, 63601, 3882817
COLORED GRAPHS. REF CJM 12 413 60 (DIVIDED BY 4).

1885 1, 7, 104, 1455, 20272, 282359, 3932760, 54776287, 762935264, 10626317415, 148005508552, 2061450802319, 28712305723920, 399910829332567
SOLUTION OF A PELLIAN. REF AMM 53 465 46.

1886 1, 7, 106, 113, 33102, 33215, 66317, 99532, 265381, 364913, 1360120, 1725033, 25510582, 52746197, 78256779, 131002976, 340262731, 811528438, 1963319607
CONVERGENTS TO PI. REF ELM 2 7 47.

1887 1, 7, 107, 199, 6031, 5741, 1129981, 435569, 35661419, 1523489833, 45183033541, 12597680311, 19055094997949, 9331210633373, 104148936040729
COEFFICIENTS FOR REPEATED INTEGRATION. REF PHM 38 336 47.

1888 1, 7, 127, 247, 463, 487, 1423, 33247, 56743, 74743, 118903, 348727, 773767, 2430943, 242675623, 393292183, 1656835783, 2713676023, 4352137927, 8133814327
SEQUENCE OF PRESCRIBED QUADRATIC CHARACTER. REF MTAC 24 444 70.

1889 1, 7, 127, 4369, 243649, 20036983, 2280356863, 343141433761, 65967241200001, 15773461423793767, 4591227123230945407
FROM INVERSE ERROR FUNCTION. REF PJM 13 470 63.

1890 1, 7, 128, 975, 4608, 16340, 48384, 124303, 281600, 583746, 1146240, 2125108, 3691008, 6151880, 10055424, 15914895, 24136704, 35748899, 52583040
RELATED TO REPRESENTATION AS SUMS OF SQUARES. REF QJM 38 349 07.

1891 1, 7, 305, 33367, 6815585, 2237423527, 1077270776465, 715153093789687, 626055764653322945, 698774745485355051847, 968553361387420436695025
MULTIPLES OF EULER NUMBERS. REF MES 28 51 1898. FMR 1 75. HPR.

1892 1, 7, 322, 33385282, 37210469265847998489922
EXTRACTING A SQUARE ROOT. REF AMM 44 645 37.

1893 1, 7, 791, 3748629, 151648960887729
AN EXPANSION FOR PI. REF AMM 54 138 47.

1894 1, 7, 1734, 89512864
GROUPOIDS. REF JL2 246.

SEQUENCES BEGINNING 1, 8

1895 1, 8, 1, 26, 8, 48, 1, 73, 26, 120, 8, 170, 48, 208, 1, 290, 73, 360, 26, 384, 120, 528, 8, 651, 170, 656, 48, 842, 208, 960, 1, 960, 290, 1248, 73, 1370, 360, 1360, 26, 1682
RELATED TO THE DIVISORS OF N. REF QJM 20 164 1884.

1896 1, 8, 10, 80, 231, 248, 1466, 80, 4766, 1944, 9600, 2704, 15525, 3984, 25498, 10816, 29760, 800, 1994, 11728, 29362, 5560, 2310, 1952, 21649, 38128, 192854, 2480
RELATED TO REPRESENTATION AS SUMS OF SQUARES. REF QJM 38 190 07.

1897 1, 8, 11, 69, 88, 96, 101, 111, 181, 609, 619, 689, 808, 818, 888, 906, 916, 986, 1001, 1111, 1691, 1881, 1961, 6009, 6119, 6699, 6889, 6969, 8008, 8118, 8698, 8888
NUMBERS WHICH ARE THE SAME UPSIDE DOWN. REF MMAG 34 184 61.

1898 1, 8, 12, 64, 210, 96, 1016, 512, 2043, 1680, 1092, 768, 1382, 8128, 2520, 4096, 14706, 16344, 39940, 13440, 12192, 8736, 68712, 6144, 34025, 11056
RELATED TO REPRESENTATION AS SUMS OF SQUARES. REF QJM 38 198 07.

1899 1, 8, 15, 212, 865, 31560, 397285, 8760472, 73512810, 7619823960
COEFFICIENTS FOR STEP-BY-STEP INTEGRATION. REF JACM 11 231 64.

1900 1, 8, 20, 0, 70, 64, 56, 0, 125, 160, 308, 0, 110, 0, 520, 0, 57, 560, 0, 0, 182, 512, 880, 0, 1190, 448, 884, 0, 0, 0, 1400, 0, 1330, 1000, 1820, 0, 646, 1280, 0, 0, 1331, 2464
COEFFICIENTS OF A MODULAR FORM. REF KNAW 59 207 56.

1901 1, 8, 21, 40, 65, 96, 133, 176, 225, 280, 341, 408, 481, 560, 645, 736, 833, 936, 1045, 1160, 1281, 1408, 1541, 1680, 1825, 1976, 2133, 2296, 2465, 2640, 2821, 3008
OCTAGONAL NUMBERS N(3N − 2). REF DI2 2 1. BE3 189.

1902 1, 8, 23, 57, 119, 231, 415, 719, 1189, 1915, 2997, 4595, 6898, 10198, 14833, 21303, 30211, 42393, 58869, 81028, 110551, 149683, 201160, 268539, 356167, 469630
BIPARTITE PARTITIONS. REF NI1 11.

1903 1, 8, 24, 112, 560, 2976, 16464, 94016, 549648
WALKS ON A SQUARE LATTICE. REF JCP 34 1532 61.

1904 1, 8, 26, 60, 115, 196, 308, 456, 645, 880, 1166, 1508, 1911, 2380, 2920, 3536, 4233, 5016, 5890, 6860, 7931, 9108, 10396, 11800, 13325, 14976, 16758, 18676, 20735
HEPTAGONAL PYRAMIDAL NUMBERS. REF DI2 2 2. BE3 194.

1905 1, 8, 27, 64, 125, 216, 343, 512, 729, 1000, 1331, 1728, 2197, 2744, 3375, 4096, 4913, 5832, 6859, 8000, 9261, 10648, 12167, 13824, 15625, 17576, 19683, 21952, 24389
CUBES. REF BA1.

1906 1, 8, 28, 56, 62, 0, 148, 328, 419, 280, 140, 728, 1232, 1336, 848, 224, 1582, 2688, 3072, 2408, 742, 1568, 3836, 5264, 5306, 3744, 924, 2576, 5686, 7792, 8092, 6272
COEFFICIENTS OF A MODULAR FORM. REF JLMS 39 435 64.

1907 1, 8, 30, 80, 175, 336, 588, 960, 1485, 2200, 3146, 4368, 5915, 7840, 10200, 13056, 16473, 20520, 25270, 30800, 37191, 44528, 52900, 62400, 73125, 85176, 98658
4-DIMENSIONAL FIGURATE NUMBERS. REF BE3 195.

1908 1, 8, 30, 192, 1344, 10800, 97434, 976000
FROM MENAGE POLYNOMIALS. REF R1 197.

1909 1, 8, 33, 168, 962, 5928, 38907, 268056
PARTITION FUNCTION FOR CUBIC LATTICE. REF PHM 2 745 57.

1910 1, 8, 35, 211, 1459, 11584, 103605, 1030805
FROM MENAGE NUMBERS. REF MES 32 63 02. R1 198.

1911 1, 8, 36, 120, 330, 792, 1716, 3432, 6435, 11440, 19448, 31824, 50388, 77520, 116280, 170544, 245157, 346104, 480700, 657800, 888030, 1184040, 1560780, 2035800
FIGURATE NUMBERS OR BINOMIAL COEFFICIENTS C(N, 7). REF DI2 2 7. RS1. BE3 196. AS1 828.

1912 1, 8, 36, 148, 554, 2024, 7134, 24796, 84625, 285784, 953430, 3151332, 10314257
7-DIMENSIONAL PARTITIONS. REF PCPS 63 1099 67.

1913 1, 8, 36, 148, 554, 2094, 7624, 27428, 96231, 332159
RELATED TO 7-DIMENSIONAL PARTITIONS. REF PCPS 63 1100 67.

1914 1, 8, 36, 204, 1296, 8772, 61776, 446964, 3297456, 24684612, 186884496, 1427557524, 10983260016, 84998999652, 660994932816, 5161010498484
1N + 2**N + ··· + 8**N. REF AS1 813.**

1915 1, 8, 36, 229, 1625, 13208, 120288, 1214673, 13496897, 162744944, 2128047988, 29943053061
PERMUTATIONS BY LENGTH OF RUNS. REF PLMS 31 341 30. SPS 37-40-4 209 66.

1916 1, 8, 40, 160, 560, 1792, 5376, 15360, 42240, 112640, 292864, 745472, 1863680, 4587520, 11141120
COEFFICIENTS OF CHEBYSHEV POLYNOMIALS. REF PRSE 62 190 46. AS1 796.

1917 1, 8, 40, 176, 748, 3248, 14280, 63768, 285296, 1285688
WALKS ON A CUBIC LATTICE. REF PCPS 58 99 62.

1918 1, 8, 43, 188, 728, 2593, 8706, 27961, 86802, 262348, 776126, 2256418, 6466614, 18311915, 51334232, 142673720, 393611872, 1078955836, 2941029334
TREES OF HEIGHT 8. REF IBMJ 4 475 60. KU1.

1919 1, 8, 44, 152, 372, 824, 1544, 2712, 4448
THE NO-THREE-IN-LINE PROBLEM. REF GU3. WE1 124.

1920 1, 8, 44, 309, 2428, 21234, 205056, 2170680, 25022880, 312273360, 4196666880, 60451816320, 929459059200, 15196285843200, 263309095526400
DIFFERENCES OF FACTORIAL NUMBERS. REF JRAM 198 61 57.

1921 1, 8, 45, 220, 1001, 4368, 18564, 77520, 319770, 1307504
COEFFICIENTS OF CHEBYSHEV POLYNOMIALS. REF LA4 517.

1922 1, 8, 45, 416, 1685, 31032, 1603182, 13856896, 132843888, 6551143600
COEFFICIENTS FOR NUMERICAL INTEGRATION. REF MTAC 6 217 52.

1923 1, 8, 48, 256, 1280, 6144, 28672, 131072, 589824, 2621440, 11534336, 50331648
COEFFICIENTS OF CHEBYSHEV POLYNOMIALS. REF LA4 516.

1924 1, 8, 49, 288, 1681, 9800, 57121, 332928, 1940449, 11309768, 65918161, 384199200, 2239277041, 13051463048, 76069501249, 443365544448, 2584123765441
TRIANGULAR NUMBERS WHICH ARE SQUARES. REF DI2 2 10. MAG 47 237 63. BE3 193. FQ 9 95 71.

1925 1, 8, 50, 2908, 115125476
SWITCHING NETWORKS. REF JFI 276 317 63.

1926 1, 8, 52, 288, 1424, 6648, 29700, 128800, 545600
SERIES-PARALLEL NUMBERS. REF R1 142.

1927 1, 8, 54, 384, 3000, 25920, 246960, 2580480
TERMS IN CERTAIN DETERMINANTS. REF PLMS 10 122 1879.

1928 1, 8, 56, 392, 2648, 17864, 118760, 789032, 5201048, 34268104
SUSCEPTIBILITY FOR CUBIC LATTICE. REF PHA 28 942 62.

1929 1, 8, 56, 392, 2648, 17960, 120056, 804824, 5351720, 35652680, 236291096, 1568049560, 10368669992, 68626647608, 453032542040, 2992783648424
WALKS ON A CUBIC LATTICE. REF PRV 114 53 59. MFS.

1930 1, 8, 56, 464, 3520, 27768
WALKS ON A DIAMOND LATTICE. REF PCPS 58 100 62.

1931 1, 8, 57, 1292, 7135, 325560, 4894715, 125078632, 1190664342, 137798986920
COEFFICIENTS FOR STEP-BY-STEP INTEGRATION. REF JACM 11 231 64.

1932 1, 8, 58, 444, 3708, 33976, 341064
FROM SUMS OF PRODUCTS OF POWERS. REF SE2 83. NMT 7 16 59.

1933 1, 8, 60, 416, 2791, 18296, 118016, 752008
SUSCEPTIBILITY FOR SQUARE LATTICE. REF JCP 38 811 63.

1934 1, 8, 60, 444, 3599, 32484, 325322, 3582600, 43029621, 559774736, 7841128936, 117668021988, 1883347579515
PERMUTATIONS BY NUMBER OF PAIRS. REF DKB 263.

1935 1, 8, 61, 5020, 128541455
AN EXPANSION FOR PI. REF AMM 54 138 47.

1936 1, 8, 63, 496, 3905, 30744, 242047, 1905632, 15003009, 118118440, 929944511, 7321437648, 57641556673, 453811015736, 3572846569215, 28128961537984
A(N) = 8A(N - 1) - A(N - 2). REF NCM 4 167 1878.

1937 1, 8, 64, 512, 4096, 32768, 262144, 2097152, 16777216, 134217728, 1073741824, 8589934592, 68719476736, 549755813888, 4398046511104, 35184372088832
POWERS OF EIGHT. REF BA1.

1938 1, 8, 67, 602, 5811, 60875, 690729, 8457285, 111323149, 1569068565, 23592426102, 377105857043, 6387313185590, 114303481217895, 2155348564851616
PERMUTATIONS BY LENGTH OF RUNS. REF DKB 261.

1939 1, 8, 72, 2160, 15504, 220248, 1564920, 89324640, 640807200, 9246847896, 67087213336, 1957095947664
COEFFICIENTS OF LEGENDRE POLYNOMIALS. REF MTAC 3 17 48.

1940 1, 8, 78, 944, 13800, 237432, 4708144, 105822432, 2660215680, 73983185000, 2255828154624, 74841555118992, 2684366717713408, 103512489775594200
TREES BY TOTAL HEIGHT. REF JA1 10 281 69.

1941 1, 8, 80, 1088, 19232, 424400
BICOVERINGS. REF SMH 3 145 68.

1942 1, 8, 80, 4374, 9800, 123200, 336140, 11859210, 11859210, 177182720, 1611308699, 3463199999, 63927525375
FROM STORMERS PROBLEM. REF IJM 8 66 64.

1943 1, 8, 96, 1664, 36800, 1008768, 32626560, 1221399040, 51734584320, 2459086364672, 129082499311616
SUSCEPTIBILITY FOR CUBIC LATTICE. REF PRV 164 801 67.

1944 1, 8, 102, 948, 7900, 62928, 491832
COLORED SERIES-PARALLEL NETWORKS. REF R1 159.

1945 1, 8, 104, 1092, 12376, 136136, 1514513, 16776144, 186145312, 2063912136
FROM FIBONACCI IDENTITIES. REF FQ 6 82 68.

1946 1, 8, 105, 1456, 20273, 282360, 3932761, 54776288, 762935265, 10626317416, 148005508553, 2061450802320, 28712305723921, 399910829332568
A(N) = 14A(N - 1) - A(N - 2) - 6. REF AMM 53 465 46.

1947 1, 8, 106, 49008, 91901007752
SWITCHING NETWORKS. REF JFI 276 317 63.

1948 1, 8, 120, 16880, 1791651440
SWITCHING NETWORKS. REF JFI 276 317 63.

1949 1, 8, 127, 2024, 32257, 514088, 8193151, 130576328, 2081028097, 33165873224, 528572943487, 8424001222568, 134255446617601, 2139663144659048
A(N) = 16A(N - 1) - A(N - 2). REF NCM 4 167 1878. TH2 281.

1950 1, 8, 136, 3968, 176896, 11184128, 951878656, 104932671488, 14544442556416, 2475749026562048, 507711943253426176, 123460740095103991808
TANGENT NUMBERS. REF MTAC 21 672 67 (DIVIDED BY 2).

1951 1, 8, 176, 265728, 2199038984192
SWITCHING NETWORKS. REF JFI 276 317 63.

1952 1, 8, 216, 8000, 343000, 16003008, 788889024
WALKS ON A CUBIC LATTICE. REF AIP 9 345 60.

1953 1, 8, 288, 366080, 1468180471808
SWITCHING NETWORKS. REF JFI 276 317 63.

1954 1, 8, 343, 1331, 1030301, 1367631, 1003003001, 10662526601, 1000300030001, 1030607060301, 1334996994331
PALINDROMIC CUBES. REF JRM 3 97 70.

1955 1, 8, 352, 38528, 7869952, 2583554048, 1243925143552, 825787662368768, 722906928498737152, 806875574817679474688, 1118389087843083461066752
GENERALIZED EULER NUMBERS. REF MTAC 21 689 67.

1956 1, 8, 1368, 300608, 186086600
FOLDING A MAP. REF CJ1 14 77 71.

1957 1, 8, 1920, 193536, 154828800, 1167851520, 892705701888000, 1428329123020800, 768472460034048000, 4058540589291090739200
COEFFICIENTS FOR CENTRAL DIFFERENCES. REF JM2 42 162 63.

1958 1, 8, 2080, 22386176, 11728394650624, 314824619911446167552
RELATIONAL SYSTEMS. REF OB1.

SEQUENCES BEGINNING 1, 9

1959 1, 9, 7, 3, 2, 5, 3, 3, 7, 6, 5, 2, 0, 1, 3, 5, 8, 6, 3, 4, 6, 7, 3, 5, 4, 8, 7, 6, 8, 0, 9, 5, 9, 0, 9, 1, 1, 7, 3, 9, 2, 9, 2, 7, 4, 9, 4, 5, 3, 7, 5, 4, 2, 0, 4, 8, 0, 5, 6, 4, 8, 9, 4, 7, 4, 2, 9, 6, 2
A RANDOM SEQUENCE. REF RA1.

1960 1, 9, 8, 6, 1, 2, 2, 8, 8, 6, 6, 8, 1, 0, 9, 6, 9, 1, 3, 9, 5, 2, 4, 5, 2, 3, 6, 9, 2, 2, 5, 2, 5, 7, 0, 4, 6, 4, 7, 4, 9, 0, 5, 5, 7, 8, 2, 2, 7, 4, 9, 4, 5, 1, 7, 3, 4, 6, 9, 4, 3, 3, 3, 6, 3, 7, 4, 9, 4
NATURAL LOGARITHM OF 3. REF RS4 2.

1961 1, 9, 8, 6, 9, 6, 0, 4, 4, 0, 1, 0, 8, 9, 3, 5, 8, 6, 1, 8, 8, 3, 4, 4, 9, 0, 9, 9, 9, 8, 7, 6, 1, 5, 1, 1, 3, 5, 3, 1, 3, 6, 9, 9, 4, 0, 7, 2, 4, 0, 7, 9, 0, 6, 2, 6, 4, 1, 3, 3, 4, 9, 3, 7, 6, 2, 2, 0, 0
PI SQUARED. REF RS4 XVIII.

1962 1, 9, 14, 21, 27, 34, 43, 50, 61
ZARANKIEWICZS PROBLEM. REF TI1 146. CO1 2 138.

1963 1, 9, 18, 162, 2520, 33192, 1019088, 7804944, 723961728, 2596523904
SPECIFIC HEAT FOR CUBIC LATTICE. REF PRV 164 801 67.

1964 1, 9, 28, 73, 126, 252, 344, 585, 757, 1134, 1332, 2044, 2198, 3096, 3528, 4681, 4914, 6813, 6860, 9198, 9632, 11988, 12168, 16380, 15751, 19782, 20440, 25112, 24390
SUM OF CUBES OF DIVISORS OF N. REF AS1 827.

1965 1, 9, 30, 69, 133, 230, 369, 560, 814, 1143, 1560, 2079, 2715, 3484, 4403, 5490, 6764, 8245, 9954, 11913, 14145, 16674, 19525, 22724, 26298, 30275, 34684, 39555
POWERS OF ROOTED TREE ENUMERATOR. REF R1 150.

1966 1, 9, 30, 70, 135, 231, 364, 540, 765, 1045, 1386, 1794, 2275, 2835, 3480, 4216, 5049, 5985, 7030, 8190, 9471, 10879, 12420, 14100, 15925, 17901, 20034, 22330, 24795
OCTAGONAL PYRAMIDAL NUMBERS. REF DI2 2 2. BE3 194.

1967 1, 9, 30, 180, 980, 8326, 70272, 695690, 7518720, 89193276, 1148241458, 15947668065, 237613988040, 3780133322620, 63945806121448
DISCORDANT PERMUTATIONS. REF SMA 20 23 54.

1968 1, 9, 34, 95, 210, 406, 740, 1161, 1920, 2695, 4116, 5369, 7868, 9690, 13640, 16116, 22419, 25365, 34160, 38640, 50622, 55154, 73320, 77225, 100100, 107730
RELATED TO THE DIVISOR FUNCTION. REF SMA 19 39 53.

1969 1, 9, 34, 104, 283, 957, 3033, 9519
IMPERFECT SQUARED RECTANGLES. REF GA1 207. BO4.

1970 1, 9, 35, 95, 210, 406, 714, 1170, 1815, 2695, 3861, 5369, 7280, 9660, 12580, 16116, 20349, 25365, 31255, 38115, 46046, 55154, 65550, 77350, 90675, 105651
4-DIMENSIONAL FIGURATE NUMBERS. REF BE3 195.

1971 1, 9, 36, 84, 117, 54, 177, 540, 837, 755, 54, 1197, 2535, 3204, 2520, 246, 3150, 6426, 8106, 7011, 2844, 3549, 10359, 15120, 15804, 11403, 2574, 8610, 18972, 25425
COEFFICIENTS OF A MODULAR FORM. REF JLMS 39 435 64.

1972 1, 9, 36, 100, 225, 441, 784, 1296, 2025, 3025, 4356, 6084, 8281, 11025, 14400, 18496, 23409, 29241, 36100, 44100, 53361, 64009, 76176, 90000, 105625, 123201
SUMS OF CUBES. REF AS1 813.

1973 1, 9, 39, 1141, 12721, 804309
COEFFICIENTS OF SINH X/ COS X. REF CMB 13 309 70.

1974 1, 9, 41, 129, 321, 681, 1289, 2241, 3649, 5641, 8361, 11969, 16641, 22569, 29961, 39041, 50049, 63241, 78889, 97281, 118721, 143529, 172041, 204609, 241601
A SQUARE RECURRENCE. REF SIAMR 12 277 70.

1975 1, 9, 42, 132, 334, 728, 1428, 2584, 4389, 7084, 10963, 16380, 23751, 33563, 46376, 62832, 83657, 109668, 141778, 181001, 228459, 285384, 353127, 433160
FERMAT COEFFICIENTS. REF MMAG 27 141 54.

1976 1, 9, 45, 165, 495, 1287, 3003, 6435, 12870, 24310, 43758, 75582, 125970, 203490, 319770, 490314, 735471, 1081575, 1562275, 2220075, 3108105, 4292145
FIGURATE NUMBERS OR BINOMIAL COEFFICIENTS C(N, 8). REF DI2 2 7. RS1. BE3 196. AS1 828.

1977 1, 9, 45, 285, 2025, 15333, 120825, 978405, 8080425, 67731333, 574304985, 4914341925, 42364319625, 367428536133, 3202860761145, 28037802953445
1N + 2**N + ··· + 9**N. REF AS1 813.**

1978 1, 9, 46, 177, 571, 1632, 4270, 10446, 24244, 53942, 115954, 242240, 494087, 987503, 1939634, 3753007, 7167461, 13532608, 25293964, 46856332, 86110792
FROM ROOK POLYNOMIALS. REF SMA 20 18 54.

1979 1, 9, 50, 1225, 7938, 106722, 736164, 41409225, 295488050
COEFFICIENTS OF LEGENDRE POLYNOMIALS. REF PR1 157. FMR 1 362.

1980 1, 9, 53, 362, 2790, 24024, 229080, 2399760, 27422640, 339696000, 4536362880, 64988179200, 994447238400, 16190733081600, 279499828608000
DIFFERENCES OF FACTORIAL NUMBERS. REF JRAM 198 61 57.

1981 1, 9, 54, 273, 1260, 5508, 23256, 95931, 389367, 1562275, 6216210, 24582285, 96768360, 379629720, 1485507600, 5801732460, 22626756594, 88152205554
LAPLACE TRANSFORM COEFFICIENTS. REF QAM 14 407 56.

1982 1, 9, 56, 300, 1485, 7007, 32032, 143208, 629850, 2735810, 11767536
PARTITIONS OF A POLYGON BY NUMBER OF PARTS. REF CAY 13 95.

1983 1, 9, 61, 381, 2332, 14337, 89497, 569794, 3704504, 24584693, 166335677, 1145533650, 8017098273, 56928364553, 409558170361, 2981386305018
PERMUTATIONS BY SUBSEQUENCES. REF MTAC 22 390 68.

1984 1, 9, 64, 326, 1433, 5799, 22224, 81987, 293987
PARTIALLY LABELED ROOTED TREES. REF R1 134.

1985 1, 9, 66, 450, 2955, 18963, 119812, 748548, 4637205, 28537245
SPHEROIDAL HARMONICS. REF MES 54 75 24.

1986 1, 9, 67, 525, 4651, 47229
EXPANSION OF AN INTEGRAL. REF SKA 11 95 28.

1987 1, 9, 70, 571, 4820, 44676, 450824, 4980274, 59834748
ASYMMETRIC PERMUTATIONS. REF LU1 1 222.

1988 1, 9, 71, 580, 5104, 48860, 509004, 5753736, 70290936, 924118272, 13020978816, 195869441664, 3134328981120, 53180752331520, 953884282141440
GENERALIZED STIRLING NUMBERS. REF PEF 77 7 62.

1989 1, 9, 72, 600, 5400, 52920, 564480, 6531840, 81648000, 1097712000, 15807052800
COEFFICIENTS OF LAGUERRE POLYNOMIALS. REF LA4 519. AS1 799.

1990 1, 9, 74, 638, 5944, 60216, 662640, 7893840, 101378880, 1397759040, 20606463360, 323626665600, 5395972377600, 95218662067200, 1773217155225600
GENERALIZED STIRLING NUMBERS. REF PEF 77 44 62.

1991 1, 9, 77, 1224, 7888, 202124, 1649375
CONSECUTIVE RESIDUES. REF MTAC 18 397 64.

1992 1, 9, 81, 729, 6561, 59049, 531441, 4782969, 43046721, 387420489, 3486784401, 31381059609, 282429536481, 2541865828329, 22876792454961
POWERS OF NINE. REF BA1.

1993 1, 9, 81, 835, 9990, 137466, 2148139, 37662381, 733015845, 15693217705, 366695853876, 9289111077324, 253623142901401, 7425873460633005
FROM BESSEL POLYNOMIALS. REF RCI 77. RI1.

1994 1, 9, 95, 420, 1225, 2834, 5652, 10165, 16940, 26625, 39949, 57722, 80835, 110260, 147050, 192339, 247342, 313355, 391755, 484000, 591629, 716262, 859600
DISCORDANT PERMUTATIONS. REF SMA 20 23 54.

1995 1, 9, 108, 3420, 114480, 7786800
GENERATORS FOR SYMMETRIC GROUP. JCT 9 111 70.

1996 1, 9, 120, 2100, 45360, 1164240, 34594560, 1167566400, 44108064000, 1843717075200, 84475764172800
COEFFICIENTS OF ORTHOGONAL POLYNOMIALS. REF MTAC 9 174 55.

1997 1, 9, 225, 11025, 893025, 108056025, 18261468225, 4108830350625, 1187451971330625, 428670161650355625, 189043541287806830625
CENTRAL FACTORIAL NUMBERS. REF RCI 217.

1998 1, 9, 259, 1974, 8778, 28743, 77077, 179452, 375972, 725781, 1312311, 2249170, 3686670, 5818995, 8892009, 13211704, 19153288, 27170913, 37808043
CENTRAL FACTORIAL NUMBERS. REF RCI 217.

1999 1, 9, 475, 36799, 2082753, 262747265
FROM HIGHER ORDER BERNOULLI NUMBERS. REF NO1 461.

2000 1, 9, 1375, 114562, 9458775
FROM HIGHER ORDER BERNOULLI NUMBERS. REF NO1 461.

SEQUENCES BEGINNING 1, 10 THROUGH 1, 19

2001 1, 10, 17, 106, 437, 20480, 44707, 1068404
SUMS OF LOGARITHMIC NUMBERS. REF MST 31 77 63.

2002 1, 10, 25, 37, 42, 48, 79, 145, 244, 415, 672, 1100, 1722, 2727, 4193, 6428, 9658, 14478, 21313, 31304, 45329, 65311, 93074, 132026, 185413, 259242, 359395, 495839
A GENERALIZED PARTITION FUNCTION. REF PNISI 17 236 51.

2003 1, 10, 30, 20, 10, 12, 20, 40, 90, 220, 572, 1560, 4920, 12920, 38760, 118864, 371450, 1179900
BINOMIAL COEFFICIENTS. REF TH1 164. FMR 1 55.

2004 1, 10, 34, 58, 73, 79, 86, 152, 265, 457, 763, 1268, 2058, 3308, 5236, 8220, 12731, 19546, 29685, 44702, 66714, 98806, 145154, 211756, 306667, 441249, 630771
A GENERALIZED PARTITION FUNCTION. REF PNISI 17 236 51.

2005 1, 10, 34, 206, 1351, 10543, 92708
HIT POLYNOMIALS. REF RI3.

2006 1, 10, 35, 84, 165, 286, 455, 680, 969, 1330, 1771, 2300, 2925, 3654, 4495, 5456, 6545, 7770, 9139, 10660, 12341, 14190, 16215, 18424, 20825, 23426, 26235, 29260
CENTRAL FACTORIAL NUMBERS. REF AMS 2 358 31 (DIVIDED BY 2). CC1 742. RCI 217.

2007 1, 10, 35, 271, 29821
SWITCHING NETWORKS. REF JFI 276 317 63.

2008 1, 10, 40, 110, 245, 476, 840, 1380, 2145, 3190, 4576, 6370, 8645, 11480, 14960, 19176, 24225, 30210, 37240, 45430, 54901, 65780, 78200, 92300, 108225, 126126
4-DIMENSIONAL FIGURATE NUMBERS. REF BE3 195.

2009 1, 10, 40, 315, 2464, 22260, 222480, 2447445, 29369120, 381798846, 5345183480, 80177752655, 1282844041920, 21808348713320, 392550276838944
RENCONTRES NUMBERS. REF R1 65.

2010 1, 10, 45, 120, 200, 162, 160, 810, 1530, 1730, 749, 1630, 4755, 7070, 6700, 2450, 5295, 14070, 20010, 19350, 10157, 6290, 25515, 40660, 44940, 34268, 9180, 24510
COEFFICIENTS OF A MODULAR FORM. REF JLMS 39 435 64.

2011 1, 10, 46, 556, 160948
SWITCHING NETWORKS. REF JFI 276 317 63.

2012 1, 10, 50, 385, 3130, 28764, 291900
FROM MENAGE POLYNOMIALS. REF R1 198.

2013 1, 10, 55, 220, 715, 2002, 5005, 11440, 24310, 48620, 92378, 167960, 293930, 497420, 817190, 1307504, 2042975, 3124550, 4686825, 6906900, 10015005, 14307150
FIGURATE NUMBERS OR BINOMIAL COEFFICIENTS C(N, 9). REF DI2 2 7. RS1. BE3 196. AS1 828.

2014 1, 10, 55, 385, 3025, 25333, 220825, 1978405, 18080425, 167731333, 1574304985, 14914341925, 142364319625, 1367428536133, 13202860761145
1N + 2**N + ··· + 10**N. REF AS1 813.**

2015 1, 10, 55, 1996, 11756666
SWITCHING NETWORKS. REF JFI 276 317 63.

2016 1, 10, 56, 234, 815, 2504, 7018, 18336
ARRAYS OF DUMBBELLS. REF JMP 11 3098 70.

2017 1, 10, 60, 462, 3920, 36954, 382740
KINGS PROBLEM. REF AMS 38 1253 67.

2018 1, 10, 65, 350, 1701, 7770, 34105, 145750, 611501, 2532530, 10391745, 42355950, 171798901, 694337290, 2798806985, 11259666950, 45232115901
STIRLING NUMBERS OF SECOND KIND. REF AS1 835. DKB 223.

2019 1, 10, 70, 420, 2310, 12012, 60060, 291720
PRODUCT OF BINOMIAL COEFFICIENTS. REF JO2 449.

2020 1, 10, 76, 8416, 268496896
SWITCHING NETWORKS. REF JFI 276 317 63.

2021 1, 10, 80, 365, 1246, 3535
SEQUENCES BY NUMBER OF INCREASES. REF JCT 1 372 66.

2022 1, 10, 85, 735, 6769, 67284, 723680, 8409500, 105258076, 1414014888, 20313753096, 310989260400, 5056995703824, 87077748875904, 1583313975727488
STIRLING NUMBERS OF FIRST KIND. REF AS1 833. DKB 226.

2023 1, 10, 88, 6616, 91666432
SWITCHING NETWORKS. REF JFI 276 317 63.

2024 1, 10, 91, 651, 4026, 22737
COEFFICIENTS FOR EXTRAPOLATION. REF SE2 93.

2025 1, 10, 91, 820, 7381, 66430, 597871, 5380840
CENTRAL FACTORIAL NUMBERS. REF TH1 36. FMR 1 112. RCI 217.

2026 1, 10, 99, 1024, 11234, 132269, 1670481, 22586759, 326098984, 5012274595
PERMUTATIONS BY LENGTH OF RUNS. REF DKB 262 (DIVIDED BY 2).

2027 1, 10, 99, 1024, 11304, 133669, 1695429, 23023811, 333840443, 5153118154, 84426592621, 1463941342191, 26793750988542, 516319125748337
PERMUTATIONS BY LENGTH OF RUNS. REF DKB 261.

2028 1, 10, 105, 1260, 17325, 270270, 4729725, 91891800, 1964187225, 45831035250, 1159525191825, 31623414322500, 924984868933125, 28887988983603750
ASSOCIATED STIRLING NUMBERS. REF TOH 37 259 33. JO2 152. DB1 296. CO1 2 98. (DIVIDED BY 2.)

2029 1, 10, 180, 4620, 152880, 6168960, 293025600
DISSECTIONS OF A BALL. REF CMA 2 25 70.

2030 1, 10, 199, 3970, 79201, 1580050, 31521799, 628855930, 12545596801, 250283080090, 4993116004999, 99612037019890, 1987247624392801
A(N) = 20A(N - 1) - A(N - 2). REF NCM 4 167 1378. MTS 65(4, SUPPLEMENT) 8 56.

2031 1, 10, 259, 12916, 1057221, 128816766, 21878089479, 4940831601000, 1432009163039625, 518142759828635250, 228929627246078500875
CENTRAL FACTORIAL NUMBERS. REF RCI 217.

2032 1, 10, 297, 13756, 925705
RESTRICTED PERMUTATIONS. REF R1 187.

2033 1, 10, 378, 16576, 819420
FINITE AUTOMATA. REF IC 10 507 67.

2034 1, 10, 438, 5893028544
POST FUNCTIONS. REF JCT 4 298 68.

2035 1, 10, 3330, 178981952
GROUPOIDS. REF PAMS 17 736 66. JL2 246.

2036 1, 11, 5, 137, 7, 363, 761, 7129, 671, 83711, 6617, 1145993, 1171733, 1195757, 143327, 42142223, 751279, 275295799
COEFFICIENTS FOR NUMERICAL DIFFERENTIATION. REF PHM 33 11 42. BAMS 48 922 42.

2037 1, 11, 7, 389, 1031, 19039, 24457, 1023497
SUMS OF LOGARITHMIC NUMBERS. REF MST 31 78 63.

2038 1, 11, 13, 19, 29, 37, 41, 53, 59, 61, 67, 71, 79, 83, 101, 107, 131, 139, 149, 163, 173, 179, 181, 191, 197, 199, 211, 227, 239, 251, 269, 271, 293, 311, 317, 347, 349
SOLUTION OF A CONGRUENCE. REF KR1 1 64.

2039 1, 11, 23, 83, 131, 179, 191, 239, 251, 359, 419, 431, 443, 491, 659, 683, 719, 743, 911, 1019, 1031, 1103, 1223, 1439, 1451, 1499, 1511, 1559, 1583, 1811, 1931, 2003
FROM LUCASIAN NUMBERS. REF BA3 564 1894. DI2 1 27.

2040 1, 11, 26, 39, 47, 53, 61, 67, 76, 83, 89, 104, 106, 109, 116, 118, 121, 139, 147, 152, 155, 170, 186, 191, 200, 207, 211, 212, 214, 219, 222, 233, 236, 244, 249, 262, 277
ELLIPTIC CURVES. REF JRAM 212 24 63.

2041 1, 11, 30, 77, 162, 323, 589, 1043, 1752, 2876, 4571, 7128, 10860, 16306, 24051, 35040, 50355, 71609, 100697, 140349, 193784, 265505, 360889, 487214, 653243
BIPARTITE PARTITIONS. REF NI1 1.

2042 1, 11, 31, 151, 911, 5951, 40051, 272611, 1863551, 12760031, 87424711, 599129311, 4106261531, 28144128251, 192901135711, 1322159893351
RELATED TO FACTORS OF FIBONACCI NUMBERS. REF JA2 20.

2043 1, 11, 31, 161, 601, 2651, 10711, 45281, 186961, 781451, 3245551, 13524161, 56258281, 234234011, 974792551, 4057691201, 16888515361, 70296251531
A(N) = 2A(N - 1) + 9A(N - 2). REF MQET 1 11 16.

2044 1, 11, 61, 181, 421, 461, 521, 991, 1621, 1871, 3001, 4441, 4621, 6871, 9091, 9931, 12391, 13421, 14821, 19141, 25951, 35281, 35401, 55201, 58321, 61681, 62071
QUINTAN PRIMES. REF CU1 2 201.

2045 1, 11, 61, 231, 681, 1683, 3653, 7183, 13073, 22363, 36365, 56695, 85305, 124515, 177045, 246047, 335137, 448427, 590557, 766727, 982729, 1244979, 1560549
A SQUARE RECURRENCE. REF SIAMR 12 277 70.

2046 1, 11, 66, 286, 1001, 3003, 8008, 19448, 43758, 92378, 184756, 352716, 646646, 1144066, 1961256, 3268760, 5311735, 8436285, 13123110, 20030010, 30045015
FIGURATE NUMBERS OR BINOMIAL COEFFICIENTS C(N, 10). REF DI2 2 7. RS1. BE3 196. AS1 828.

2047 1, 11, 66, 302, 1191, 4293, 14608, 47840, 152637, 478271, 1479726, 4537314, 13824739, 41932745
EULERIAN NUMBERS. REF R1 215. DB1 151. JCT 1 351 66. DKB 260. CO1 2 84.

2048 1, 11, 77, 440, 2244, 10659, 48279, 211508, 904475, 3798795, 15737865, 64512240, 262256280, 1059111900, 4254603804, 17018415216, 67837293986
LAPLACE TRANSFORM COEFFICIENTS. REF QAM 14 407 56.

2049 1, 11, 85, 575, 3661, 22631, 137845, 833375, 5019421, 30174551
DIFFERENCES OF RECIPROCALS OF UNITY. REF DKB 228.

2050 1, 11, 87, 693, 5934, 55674, 572650, 6429470, 78366855, 1031378445, 14583751161, 220562730171, 3553474061452
PERMUTATIONS BY NUMBER OF PAIRS. REF DKB 264.

2051 1, 11, 101, 781, 5611, 39161, 270281, 1857451, 12744061, 87382901, 599019851, 4105974961, 28143378001, 192899171531, 1322154751061, 9062194370461
RELATED TO FACTORS OF FIBONACCI NUMBERS. REF JA2 20.

2052 1, 11, 107, 1066, 11274, 127860, 1557660, 20355120, 284574960, 4243508640, 67285058400, 1131047366400, 20099588140800, 376612896038400
GENERALIZED STIRLING NUMBERS. REF PEF 77 61 62.

2053 1, 11, 113, 1099, 11060, 118484, 1366134, 16970322, 226574211, 3240161105, 49453685911, 802790789101
PERMUTATIONS BY NUMBER OF PAIRS. REF DKB 263.

2054 1, 11, 121, 1331, 14641, 161051, 1771561, 19487171, 214358881, 2357947691, 25937424601, 285311670611, 3138428376721, 34522712143931
POWERS OF ELEVEN. REF BA1.

2055 1, 11, 188, 2992, 51708, 930436, 17131724
SPECIFIC HEAT FOR CUBIC LATTICE. REF JMP 3 187 62.

2056 1, 11, 191, 2497, 14797, 92427157, 36740617, 61430943169, 23133945892303, 16399688681447
COEFFICIENTS FOR NUMERICAL INTEGRATION. REF OP1 545. PHM 35 263 44.

2057 1, 11, 301, 15371, 1261501, 151846331, 25201039501, 5515342166891, 1538993024478301, 533289474412481051, 224671379367784281901
MULTIPLES OF GLAISHERS I NUMBERS. REF PLMS 31 232 1899. FMR 1 76.

2058 1, 11, 309, 5805, 95575, 1516785, 24206055, 396475975, 6733084365, 119143997490
PERMUTATIONS BY NUMBER OF PAIRS. REF DKB 264.

2059 1, 11, 361, 24611, 2873041, 512343611, 129570724921, 44110959165011, 19450718635716001, 1078405256112570481 1, 7342627959965776406281
GENERALIZED TANGENT NUMBERS. REF QJM 45 202 14. MTAC 21 690 67.

2060 1, 11, 1230, 47093135946
POST FUNCTIONS. REF JCT 4 296 68.

2061 1, 12, 4, 129, 72, 1332, 960, 13419, 11372, 132900
SUSCEPTIBILITY FOR TRIANGULAR LATTICE. REF PHA 28 934 62.

2062 1, 12, 4, 360, 40, 20160, 12096, 259200, 604800, 239500800, 760320, 43589145600, 217945728000, 1494484992000, 697426329600, 3201186852864000
FROM GENERALIZED BERNOULLI NUMBERS. REF JM2 23 211 44.

2063 1, 12, 6, 180, 10, 560, 1260, 12600, 1260, 166320, 13860, 2522520, 2702700, 2882880, 360360, 110270160, 2042040, 775975200
COEFFICIENTS FOR NUMERICAL DIFFERENTIATION. REF PHM 33 11 42. BAMS 48 922 42.

2064 1, 12, 14, 135, 276, 1520, 4056, 17778, 54392, 213522, 700362, 2601674, 8836812, 31925046, 110323056
SUSCEPTIBILITY FOR CUBIC LATTICE. REF JCP 38 811 63.

2065 1, 12, 24, 60, 180, 588, 1968, 6840, 24240, 87252, 318360, 1173744, 4366740, 16370700, 61780320, 234505140
WALKS ON A TRIANGULAR LATTICE. REF JCP 46 3481 67.

2066 1, 12, 24, 168, 1440, 24480, 297024, 28017216, 533681664, 41156316672
SPECIFIC HEAT FOR CUBIC LATTICE. REF PRV 164 801 67.

2067 1, 12, 48, 16, 414, 960, 672, 4800, 2721, 9064, 8880, 6912, 2398, 13440, 29280, 30976, 10878, 57228, 9360, 252384, 53760, 177600, 113952, 107520, 436131, 16488
RELATED TO REPRESENTATION AS SUMS OF SQUARES. REF QJM 38 325 07.

2068 1, 12, 48, 540, 4320, 42240, 403200, 4038300, 40958400
WALKS ON A CUBIC LATTICE. REF AIP 9 345 60.

2069 1, 12, 54, 88, 99, 540, 418, 648, 594, 836, 1056, 4104, 209, 4104, 594, 4256, 6480, 4752, 298, 5016, 17226, 12100, 5346, 1296, 9063, 7128, 19494, 29160, 10032, 7668
COEFFICIENTS OF A MODULAR FORM. REF QJM 38 56 07. KNAW 59 207 56.

2070 1, 12, 60, 210, 630, 1736, 4536, 11430
LABELED TREES OF DIAMETER 3. REF IBMJ 4 478 60.

2071 1, 12, 66, 220, 483, 660, 252, 1320, 4059, 6644, 6336, 240, 12255, 27192, 35850, 27972, 2343, 50568, 99286, 122496, 96162, 11584, 115116, 242616, 315216, 283800
COEFFICIENTS OF A MODULAR FORM. REF JLMS 39 436 64.

2072 1, 12, 66, 245, 715, 1768, 3876, 7752, 14421, 25300, 42287, 67860, 105183, 158224, 231880, 332112, 466089, 642341, 870922, 1163580, 1533939, 1997688
FERMAT COEFFICIENTS. REF MMAG 27 141 54.

2073 1, 12, 78, 364, 1365, 4368, 12376, 31824, 75582, 167960, 352716, 705432, 1352078, 2496144, 4457400, 7726160, 13037895, 21474180, 34597290, 54627300
FIGURATE NUMBERS OR BINOMIAL COEFFICIENTS C(N, 11). REF DI2 2 7. RS1. BE3 196. AS1 828.

2074 1, 12, 84, 468, 2332, 11068, 51472, 237832, 1095384
WALKS ON A CUBIC LATTICE. REF PCPS 58 99 62.

2075 1, 12, 90, 560, 3150, 16632, 84084, 411840, 1969110, 9237800, 42678636, 194699232, 878850700, 3931426800, 17450721000
COEFFICIENTS FOR NUMERICAL DIFFERENTIATION. REF OP1 21. SE2 92. JM2 22 120 43. LA4 514.

2076 1, 12, 110, 945, 8092, 70756
GENERALIZED STIRLING NUMBERS OF SECOND KIND. REF FQ 5 366 67.

2077 1, 12, 119, 1175, 12154, 133938, 1580508, 19978308, 270074016, 3894932448, 59760168192, 972751628160, 16752851775360, 304473528961920
GENERALIZED STIRLING NUMBERS. REF PEF 77 26 62.

2078 1, 12, 120, 720, 3360, 13440, 48384, 161280, 506880, 1520640
COEFFICIENTS OF HERMITE POLYNOMIALS. REF AS1 801.

2079 1, 12, 120, 1200, 12600, 141120, 1693440, 21772800, 299376000, 4390848000, 68497228800, 1133317785600, 19833061248000, 366148823040000
LAH NUMBERS. REF R1 44. CO1 1 166.

2080 1, 12, 132, 847, 3921, 14506, 45402, 124707, 308407, 699766
STOCHASTIC CUBIC ARRAYS. REF CJ1 13 283 70.

2081 1, 12, 132, 1404, 14652, 151116, 1546332, 15734460, 159425580, 1609987708, 16215457188, 162961837500, 1634743178420
SUSCEPTIBILITY FOR CUBIC LATTICE. REF SSP 3 268 70.

2082 1, 12, 132, 1404, 14700, 152532, 1573716, 16172148, 165697044, 1693773924, 17281929564, 176064704412, 1791455071068
WALKS ON A CUBIC LATTICE. REF JCP 46 3481 67.

2083 1, 12, 137, 1602, 19710, 257400, 3574957, 52785901, 827242933, 13730434111, 240806565782, 4452251786946, 86585391630673
PERMUTATIONS BY LENGTH OF RUNS. REF DKB 261.

2084 1, 12, 144, 1728, 20736, 248832, 2985984, 35831808, 429981696, 5159780352, 61917364224, 743008370688, 8916100448256, 106993205379072
POWERS OF TWELVE. REF BA1.

2085 1, 12, 144, 1750, 23420, 303240, 3641100, 46113200, 575360400, 7346545000
POWERS OF TEN WRITTEN IN BASE 8. REF AS1 1017.

2086 1, 12, 148, 2568, 53944
PARTITION FUNCTION FOR CUBIC LATTICE. REF PHM 2 745 57.

2087 1, 12, 180, 2800, 44100, 698544, 11099088, 176679360, 2815827300, 44914183600, 716830370256, 11445589052352, 182811491808400, 2920656969720000
REMAINDER IN GAUSSIAN QUADRATURE. REF MTAC 1 53 43.

2088 1, 12, 180, 3360, 75600, 1995840, 60540480, 2075673600, 79394515200, 3352212864000, 154872234316800, 7771770303897600, 420970891461120000
COEFFICIENTS OF HERMITE POLYNOMIALS. REF MTAC 3 168 48.

2089 1, 12, 240, 4032, 34560, 101376, 50319360
FROM HIGHER ORDER BERNOULLI NUMBERS. REF NO1 459.

2090 1, 12, 240, 6624, 234720, 10208832, 526810176, 31434585600, 2127785025024, 161064469168128
SUSCEPTIBILITY FOR CUBIC LATTICE. REF PRV 164 801 67.

2091 1, 12, 288, 51840, 2488320, 209018880, 75246796800, 902961561600, 86684309913600, 514904800886784000, 86504006548979712000
EXPANSION OF GAMMA FUNCTION. REF MTAC 22 619 68.

2092 1, 12, 360, 20160, 1814400, 239500800, 43589145600, 10461394944000, 3201186852864000, 1216451004088320000, 562000363888803840000
COEFFICIENTS FOR CENTRAL DIFFERENCES. REF JM2 42 162 63.

2093 1, 12, 720, 60480, 3628800, 95800320, 2615348736000, 4483454976000, 32011868528640000, 51090942171709440000, 152579284313702400000
COEFFICIENTS FOR NUMERICAL INTEGRATION. REF OP1 545. PHM 35 263 44.

2094 1, 12, 10206, 2148007936
POST FUNCTIONS. REF JCT 4 295 68.

2095 1, 13, 11, 1093, 757, 3851, 797161, 4561, 34511, 363889, 368089, 1001523179, 391151, 8209, 20381027, 4404047, 2413941289, 2644097031, 17189128703, 7333
LARGEST FACTOR OF 3(2N + 1) − 1. REF KR1 2 28.**

2096 1, 13, 19, 37, 61, 109, 157, 193, 241, 283, 367, 373, 379, 397, 487
RELATED TO KUMMERS CONJECTURE. REF HA3 482.

2097 1, 13, 41, 671, 73, 597871, 7913, 28009, 792451, 170549237, 19397633, 317733228541, 9860686403
COEFFICIENTS FOR CENTRAL DIFFERENCES. REF JM2 42 162 63.

2098 1, 13, 47, 73, 2447, 16811
RELATED TO WEBER FUNCTIONS. REF KNAW 66 751 63.

2099 1, 13, 51, 601, 4806, 39173, 775351
LABELED SERIES-REDUCED TREES. REF RI1.

2100 1, 13, 61, 73, 193, 241, 541, 601, 1021, 1801, 1873, 1933, 2221, 3121, 3361, 4993, 5521, 6481, 8461, 9181, 9901, 10993, 11113, 12241, 12541, 13633, 14173, 17761, 20593
SEXTAN PRIMES. REF CU1 1 256.

2101 1, 13, 72, 595, 4096, 39078, 379760, 4181826, 49916448, 647070333, 9035216428, 135236990388, 2159812592384, 36658601139066, 658942295734944
DISCORDANT PERMUTATIONS. REF SMA 20 23 54.

2102 1, 13, 85, 377, 1289, 3653, 8989, 19825, 40081, 75517, 134245, 227305, 369305, 579125, 880685, 1303777, 1884961, 2668525, 3707509, 5064793, 6814249
A SQUARE RECURRENCE. REF SIAMR 12 277 70.

2103 1, 13, 87, 4148, 153668757
SWITCHING NETWORKS. REF JFI 276 317 63.

2104 1, 13, 104, 663, 3705, 19019, 92092, 427570, 1924065, 8454225, 36463440, 154969620, 650872404, 2707475148, 11173706960, 45812198536, 186803188858
LAPLACE TRANSFORM COEFFICIENTS. REF QAM 14 407 56.

2105 1, 13, 109, 193, 433, 769, 1201, 1453, 2029, 3469, 3889, 4801, 10093, 12289, 13873, 18253, 20173, 21169, 22189, 28813, 37633, 43201, 47629, 60493, 63949, 65713
CUBAN PRIMES. REF CU1 1 259.

2106 1, 13, 158, 66336, 122544034314
SWITCHING NETWORKS. REF JFI 276 317 63.

2107 1, 13, 169, 2197, 28561, 371293, 4826809, 62748517, 815730721, 10604499373, 137858491849, 1792160394037, 23298085122481, 302875106592253
POWERS OF THIRTEEN. REF BA1.

2108 1, 13, 181, 2521, 35113, 489061, 6811741, 94875313, 1321442641, 18405321661, 256353060613, 3570537526921, 49731172316281, 692665874901013
FROM THE SOLUTION TO A PELLIAN. REF AMM 56 174 49.

2109 1, 13, 192, 1085, 3880, 10656, 24626, 50380, 94128, 163943, 270004, 424839, 643568, 944146, 1347606, 1878302, 2564152, 3436881, 4532264, 5890369, 7555800
DISCORDANT PERMUTATIONS. REF SMA 20 23 54.

2110 1, 13, 237, 356026, 2932175712336
SWITCHING NETWORKS. REF JFI 276 317 63.

2111 1, 13, 252, 3740, 51300, 685419, 9095856, 120872850, 1614234960, 21697730849
C-NETS. REF JCT 4 275 68.

2112 1, 13, 273, 4641, 85085, 1514513, 27261234, 488605194, 8771626578
FROM FIBONACCI IDENTITIES. REF FQ 6 82 68.

2113 1, 14, 21, 25, 30, 33, 38, 41, 43, 48, 50, 53, 56, 59, 61, 65, 67, 70, 72, 76, 77, 79, 83, 85, 87, 89, 92, 95, 96, 99, 101, 104, 105, 107, 111, 112, 114, 116, 119, 121, 123, 124
ZEROS OF RIEMANN ZETA FUNCTION. REF RS7 58.

2114 1, 14, 70, 140, 70, 28, 28, 40, 70, 140, 308, 728, 1820, 4760, 12920, 36176, 104006, 305900
BINOMIAL COEFFICIENTS. REF TH1 164. FMR 1 55.

2115 1, 14, 120, 825, 5005, 28028, 148512, 755820, 3730650, 17978180
PARTITIONS OF A POLYGON BY NUMBER OF PARTS. REF CAY 13 95.

2116 1, 14, 126, 630, 2310, 6930, 18018, 42042
RELATED TO BINOMIAL MOMENTS. REF JO2 449.

2117 1, 14, 135, 5478, 165826, 13180268, 834687179
COEFFICIENTS OF ELLIPTIC FUNCTIONS. REF TM1 4 92.

2118 1, 14, 147, 1408, 13013, 118482
CENTRAL FACTORIAL NUMBERS. REF TH1 35. FMR 1 112. RCI 217.

2119 1, 14, 155, 1665, 18424, 214676, 2655764, 34967140, 489896616, 7292774280, 115119818736, 1922666722704, 33896996544384, 629429693586048
GENERALIZED STIRLING NUMBERS. REF PEF 77 7 62.

2120 1, 14, 196, 2744, 38416, 537824, 7529536, 105413504, 1475789056, 20661046784, 289254654976, 4049565169664, 56693912375296, 793714773254144
POWERS OF FOURTEEN. REF BA1.

2121 1, 14, 273, 7645, 296296, 15291640, 1017067024, 84865562640, 8689315795776, 1071814846360896, 156823829909121024, 26862299458337581056
CENTRAL FACTORIAL NUMBERS. REF RCI 217.

2122 1, 14, 386, 5868, 65954, 614404, 5030004, 37460376
ROOTED PLANAR MAPS. REF BAMS 74 74 68.

2123 1, 14, 560, 11200, 197568, 3378944, 57573888
ALMOST CUBIC MAPS. REF PL2 1 292 70.

2124 1, 14, 818, 141, 13063, 16774564, 1057052, 4651811, 778001383, 1947352646, 1073136102266, 72379420806883
DOUBLE SUMS OF RECIPROCALS. REF RO2 316. FMR 1 117.

2125 1, 15, 17, 24, 37, 43, 57, 63, 65, 73, 79, 89, 101, 106, 122, 129, 131, 142, 145, 148, 151, 161, 164, 168, 171, 186, 195, 197, 198, 204, 217, 222, 223, 225, 229, 232, 233, 248
ELLIPTIC CURVES. REF JRAM 212 24 63.

2126 1, 15, 21, 33, 35, 39, 51, 55, 57, 65, 69, 77, 85, 87, 91, 93, 95, 115, 119, 133, 143, 145, 155, 161, 187, 203, 209, 217, 221, 247, 253, 299, 319, 323, 341, 377, 391, 403, 437
RELATED TO LIOUVILLES FUNCTION. REF JIMS 7 71 43.

2127 1, 15, 29, 12, 26, 12, 26, 9, 23, 7, 21, 4, 18, 2, 16, 30, 13, 27, 10, 24, 8, 22, 5, 19, 3, 17, 31, 14, 28, 11, 25, 11, 25, 8, 22, 6, 20, 3, 17, 1, 15, 29, 12, 26, 9, 23, 7, 21, 4, 18, 2
DAYS AT FORTNIGHTLY INTERVALS FROM JANUARY 1. REF EUR 13 11 50.

2128 1, 15, 42, 90, 126, 165, 273, 612, 630, 855, 1020, 1404, 1512, 1518, 1950, 1980, 2457, 3045, 3720, 5040, 6327, 6480, 7280, 8184, 8610, 9933, 12972, 14700, 15600
NON-CYCLIC SIMPLE GROUPS. REF DI1 309. JL2 137. (DIVIDED BY 4.)

2129 1, 15, 45, 118, 257, 522, 975, 1752, 2998, 4987, 8043, 12693, 19584, 29719, 44324, 65210, 94642, 135805, 192699, 270822, 377048, 520624, 713123, 969784
BIPARTITE PARTITIONS. REF NI1 1.

2130 1, 15, 51, 97, 127, 145, 152, 160, 273, 481, 811, 1372, 2250, 3692, 5924, 9472, 14887, 23310, 36005, 55314, 84042, 126998, 190138, 283108, 418175, 614429, 896439
A GENERALIZED PARTITION FUNCTION. REF PNISI 17 236 51.

2131 1, 15, 60, 450, 4500, 55125, 793800, 13097700
EXPANSION OF AN INTEGRAL. REF CO1 1 176.

2132 1, 15, 70, 630, 5544, 55650, 611820, 7342335, 95449640, 1336295961, 20044438050, 320711010620, 5452087178160, 98137569209940, 1864613814984984
RENCONTRES NUMBERS. REF R1 65.

2133 1, 15, 72, 609, 4960, 46188, 471660, 5275941, 64146768, 842803767, 11902900380, 179857257960, 2895705788736, 49491631601635, 895010868095256
DISCORDANT PERMUTATIONS. REF SMA 20 23 54.

2134 1, 15, 73, 143, 208, 244, 265, 273, 282, 490, 838, 1426, 2367, 3908, 6356, 10246, 16327, 25812, 40379, 62748, 96660, 147833, 2, 24446, 338584, 507293, 755612
A GENERALIZED PARTITION FUNCTION. REF PNISI 17 235 51.

2135 1, 15, 76, 275, 720, 1666, 3440, 6129, 11250, 17545, 28896, 41405, 65072, 85950, 128960, 162996, 238545, 286995, 404600, 482160, 662112, 756470, 1042560
RELATED TO THE DIVISOR FUNCTION. REF SMA 19 39 53.

2136 1, 15, 90, 350, 1050, 2646, 5880, 11880, 22275, 39325, 66066, 106470, 165620, 249900, 367200, 527136, 741285, 1023435, 1389850, 1859550, 2454606, 3200450
STIRLING NUMBERS OF SECOND KIND. REF AS1 835. DKB 223.

2137 1, 15, 99, 429, 1430, 3978, 9690, 21318, 43263, 82225, 148005, 254475, 420732, 672452, 1043460, 1577532, 2330445, 3372291, 4790071, 6690585, 9203634
FERMAT COEFFICIENTS. REF MMAG 27 141 54.

2138 1, 15, 105, 490, 1918, 6825, 22935, 74316, 235092, 731731, 2252341, 6879678, 20900922, 63259533
ASSOCIATED STIRLING NUMBERS. REF R1 76. DB1 296. CO1 2 58.

2139 1, 15, 113, 575, 2241, 7183, 19825, 48639, 108545, 224143, 433905, 795455, 1392065, 2340495, 3800305, 5984767, 9173505, 13726991, 20103025, 28875327
A SQUARE RECURRENCE. REF SIAMR 12 277 70.

2140 1, 15, 140, 1050, 6930, 42042, 240240
PRODUCT OF BINOMIAL COEFFICIENTS. REF JO2 449.

2141 1, 15, 140, 1050, 6951, 42525, 246730, 1379400, 7508501, 40075035, 210766920, 1096190550, 5652751651, 28958095545, 147589284710, 749206090500
STIRLING NUMBERS OF SECOND KIND. REF AS1 835. DKB 223.

2142 1, 15, 175, 1960, 22449, 269325, 3416930, 45995730, 657206836, 9957703756, 159721605680, 2706813345600, 48366009233424, 909299905844112
STIRLING NUMBERS OF FIRST KIND. REF AS1 833. DKB 226.

2143 1, 15, 179, 2070, 24574, 305956, 4028156, 56231712, 832391136, 13051234944, 216374987520, 3785626465920, 69751622298240, 1350747863435520
GENERALIZED STIRLING NUMBERS. REF PEF 77 44 62.

2144 1, 15, 200, 2672, 37600, 554880, 8514560, 134864640
ALMOST CUBIC MAPS. REF PL2 1 292 70.

2145 1, 15, 210, 2380, 26432, 303660, 3678840, 47324376, 647536032, 9418945536, 145410580224, 2377609752960, 41082721413120, 748459539843840
ASSOCIATED STIRLING NUMBERS. REF R1 75. CO1 2 98.

2146 1, 15, 210, 3150, 51975, 945945, 18918900
VALUES OF BESSEL POLYNOMIALS. REF RCI 77. RI1.

2147 1, 15, 225, 3375, 50625, 759375, 11390625, 170859375, 2562890625, 38443359375, 576650390625, 8649755859375, 129746337890625, 1946195068359375
POWERS OF FIFTEEN. REF BA1.

2148 1, 15, 528, 3990, 232305, 4262895, 128928632, 1420184304, 186936865290
COEFFICIENTS FOR STEP-BY-STEP INTEGRATION. REF JACM 11 231 64.

2149 1, 15, 575, 46760, 6998824, 1744835904, 673781602752, 381495483224064, 303443622431870976
DIFFERENCES OF RECIPROCALS OF UNITY. REF DKB 228.

2150 1, 16, 0, 256, 1054, 0, 0, 4096, 6561, 16864, 0, 0, 478, 0, 0, 65536, 63358, 104976, 0, 269824, 0, 0, 0, 0, 720291, 7648, 0, 0, 1407838, 0, 0, 1048576, 0, 1013728, 0
RELATED TO REPRESENTATION AS SUMS OF SQUARES. REF QJM 38 304 07.

2151 1, 16, 18, 0, 252, 576, 519, 3264, 12468, 20568, 26662, 215568, 528576, 164616, 3014889, 10894920
SUSCEPTIBILITY FOR CUBIC LATTICE. REF JCP 38 811 63.

2152 1, 16, 25, 33, 49, 52, 64, 73, 100, 121, 148, 169, 177
SUMS OF SQUARES. REF MMAG 40 198 67.

2153 1, 16, 80, 1056, 320416
SWITCHING NETWORKS. REF JFI 276 317 63.

2154 1, 16, 81, 256, 625, 1296, 2401, 4096, 6561, 10000, 14641, 20736, 28561, 38416, 50625, 65536, 83521, 104976, 130321, 160000, 194481, 234256, 279841, 331776
FOURTH POWERS. REF BA1.

2155 1, 16, 104, 320, 260, 1248, 3712, 1664, 6890, 7280, 5568, 4160, 33176, 4640, 74240, 29824, 14035, 54288, 27040, 142720, 1508, 110240, 289536, 222720, 380770
COEFFICIENTS OF A MODULAR FORM. REF KNAW 59 207 56.

2156 1, 16, 125, 680, 3135, 13155, 51873, 195821
PARTIALLY LABELED TREES. REF R1 138.

2157 1, 16, 144, 984, 5756, 30760, 155912, 766424
WALKS ON A CUBIC LATTICE. REF PCPS 58 99 62.

2158 1, 16, 150, 926, 4788, 22548, 100530, 433162, 1825296, 7577120, 31130190, 126969558
PERMUTATIONS BY LENGTH OF RUNS. REF DKB 260.

2159 1, 16, 160, 13056, 183305216
SWITCHING NETWORKS. REF JFI 276 317 63.

2160 1, 16, 177, 5548, 39615, 2236440, 40325915, 1207505768, 13229393814, 1737076976040
COEFFICIENTS FOR STEP-BY-STEP INTEGRATION. REF JACM 11 231 64.

2161 1, 16, 181, 1821, 17557, 167449, 1604098, 15555398, 153315999, 1538907304, 15743413076, 164161815768, 1744049683213, 18865209953045
PERMUTATIONS BY SUBSEQUENCES. REF MTAC 22 390 68.

2162 1, 16, 192, 2016, 20160, 197940, 1930944
WALKS ON A CUBIC LATTICE. REF PCPS 58 100 62.

2163 1, 16, 200, 2400, 29400, 376320, 5080320, 72576000, 1097712000, 17563392000
COEFFICIENTS OF LAGUERRE POLYNOMIALS. REF LA4 519. AS1 799.

2164 1, 16, 256, 4096, 65536, 1048576, 16777216, 268435456, 4294967296, 68719476736, 1099511627776, 17592186044416, 281474976710656
POWERS OF SIXTEEN. REF BA1.

2165 1, 16, 272, 2880, 24576, 185856, 1304832, 8728576, 56520704, 357888000, 2230947840, 13754155008
PERMUTATIONS BY LENGTH OF RUNS. REF DKB 261.

2166 1, 16, 272, 3968, 56320, 814080, 12207360
GENERALIZED TANGENT NUMBERS. REF TOH 42 152 36.

2167 1, 16, 361, 3362, 16384, 55744, 152166, 355688, 739328, 1415232, 2529614
GENERALIZED TANGENT NUMBERS. REF MTAC 21 690 67.

2168 1, 16, 435, 7136, 99350
CARD MATCHING. REF R1 193.

2169 1, 16, 1280, 249856
GENERALIZED EULER NUMBERS. REF MTAC 21 689 67.

2170 1, 16, 19683, 4294967296, 298023223876953125, 10314424798490535546171949056, 2569235775210588780886114772242356213216O7
N(N**2). REF ELM 3 20 48.**

2171 1, 16, 7625597484987 (THE NEXT TERM HAS 155 DIGITS)
N(N**N). REF ELM 3 20 48.**

2172 1, 17, 27, 33, 52, 73, 82, 83, 103, 107, 137, 153, 162, 217, 219, 227, 237, 247, 258, 268, 271, 282, 283, 302, 303, 313, 358, 383, 432, 437, 443, 447, 502, 548, 557, 558, 647
NOT THE SUM OF 4 TETRAHEDRALS. REF MTAC 12 142 58.

2173 1, 17, 31, 1, 5461, 257, 73, 1271, 60787, 241, 22369621, 617093
COEFFICIENTS FOR CENTRAL DIFFERENCES. REF JM2 42 162 63.

2174 1, 17, 31, 691, 5461, 929569, 3202291, 221930581, 4722116521, 968383680827, 14717667114151, 2093660879252671, 86125672563301143
RELATED TO GENOCCHI NUMBERS. REF AMP 26 5 1856. QJM 46 38 14. FMR 1 73.

2175 1, 17, 73, 241, 1009, 2641, 8089, 18001, 53881, 87481, 117049, 515761, 1083289, 3206641, 3818929, 9257329, 22000801, 48473881, 175244281, 427733329
PSEUDO-SQUARES. REF MTAC 8 241 54, 24 434 70.

2176 1, 17, 73, 241, 1009, 2689, 8089, 33049, 53881, 87481, 483289, 515761, 1083289, 3818929, 9257329, 22000801, 48473881, 175244281, 427733329, 898716289
PRIMES WITH LARGE LEAST NONRESIDUES. REF RS5 XV.

2177 1, 17, 82, 273, 626, 1394, 2402, 4369, 6643, 10642, 14642, 22386, 28562, 40834, 51332, 69905, 83522, 112931, 130322, 170898, 196964, 248914, 279842, 358258
SUM OF 4TH POWERS OF DIVISORS OF N. REF AS1 827.

2178 1, 17, 97, 257, 337, 641, 881, 1297, 2417, 2657, 3697, 4177, 4721, 6577, 10657, 12401, 14657, 14897, 15937, 16561, 28817, 38561, 39041, 49297, 54721, 65537, 65617
QUARTAN PRIMES. REF CU1 1 253.

2179 1, 17, 98, 354, 979, 2275, 4676, 8772, 15333, 25333, 39974, 60710, 89271, 127687, 178312, 243848, 327369, 432345, 562666, 722666, 917147, 1151403, 1431244
SUMS OF FOURTH POWERS. REF AS1 813.

2180 1, 17, 257, 241, 65537, 61681, 673, 15790321, 6700417, 38737, 4278255361, 2931542417, 22253377, 308761441, 54410972897, 4562284561, 67280421310721
LARGEST FACTOR OF 16N + 1. REF KR1 2 88.**

2181 1, 17, 259, 2770, 27978, 294602, 3331790, 40682144, 535206440, 7557750635, 114101726625, 1834757172082
PERMUTATIONS BY NUMBER OF PAIRS. REF DKB 264.

2182 1, 17, 289, 4913, 83521, 1419857, 24137569, 410338673, 6975757441, 118587876497, 2015993900449, 34271896307633, 582622237229761
POWERS OF SEVENTEEN. REF BA1.

2183 1, 17, 367, 27859, 1295803, 5329242827, 25198857127, 11959712166949, 11153239773419941, 31326450596954510807
COEFFICIENTS FOR NUMERICAL INTEGRATION. REF OP1 545. PHM 36 217 45.

2184 1, 17, 1835, 195013, 3887409, 58621671097
FROM HIGHER ORDER BERNOULLI NUMBERS. REF NO1 463.

2185 1, 18, 23, 28, 32, 35, 39, 42, 46, 49, 52, 55, 58, 60, 63, 66, 68, 71, 74, 76, 79, 81, 84, 86, 88, 91, 93, 95, 98, 100, 102, 104, 107, 109, 111, 113, 115, 118, 120, 122, 124, 126
THE GRAM POINTS. REF RS7 58.

2186 1, 18, 45, 69, 96, 120, 147, 171
RATIONAL POINTS IN A QUADRILATERAL. REF JRAM 227 49 67.

2187 1, 18, 108, 180, 5040, 162000, 14565600, 563253408, 17544639744, 750651187968
SPECIFIC HEAT FOR CUBIC LATTICE. REF PRV 164 801 67.

2188 1, 18, 126, 420, 630, 252, 84, 72, 90, 140, 252, 504, 1092, 2520, 6120, 15504, 40698, 110124
BINOMIAL COEFFICIENTS. REF TH1 164. FMR 1 55.

2189 1, 18, 160, 1120, 6912, 39424, 212992, 1105920, 5570560, 27394048, 132120576
COEFFICIENTS OF CHEBYSHEV POLYNOMIALS. REF LA4 516.

2190 1, 18, 245, 3135, 40369, 537628, 7494416, 109911300, 1698920916, 27679825272, 474957547272, 8572072384512, 162478082312064, 3229079010579072
GENERALIZED STIRLING NUMBERS. REF PEF 77 26 62.

2191 1, 18, 251, 3325, 44524, 617624, 8969148, 136954044, 2201931576, 37272482280, 663644774880, 12413008539360, 243533741849280, 5003753991174720
GENERALIZED STIRLING NUMBERS. REF PEF 77 61 62.

2192 1, 18, 324, 5832, 104976, 1889568, 34012224, 612220032, 11019960576, 198359290368, 3570467226624, 64268410079232, 1156831381426176
POWERS OF EIGHTEEN. REF BA1.

2193 1, 18, 2862, 158942078604
POST FUNCTIONS. REF JCT 4 297 68.

2194 1, 19, 3, 863, 275, 33953, 8183, 3250433, 4671, 13695779093, 2224234463, 132282840127, 2639651053, 111956703448001, 50188465, 2334028946344463
NUMERATORS OF LOGARITHMIC NUMBERS. REF JM2 22 49 43. PHM 38 336 47. MTAC 20 465 66.

2195 1, 19, 43, 67, 163, 77683, 1333963, 2404147, 20950603, 36254563, 51599563, 96295483, 114148483, 269497867, 585811843, 52947440683, 71837718283
SEQUENCE OF PRESCRIBED QUADRATIC CHARACTER. REF MTAC 24 440 70.

2196 1, 19, 145, 100, 2191, 8592, 14516, 29080, 114575, 320417, 615566, 1125492, 2139700, 3664750, 5997448, 10103304, 15992719, 23857290, 36059435, 53341900
RELATED TO REPRESENTATION AS SUMS OF SQUARES. REF QJM 38 349 07.

2197 1, 19, 205, 1795, 14221, 106819
CONNECTED RELATIONS. REF CR 268 579 69.

2198 1, 19, 361, 6859, 130321, 2476099, 47045881, 893871739, 16983563041, 322687697779, 6131066257801, 116490258898219, 2213314919066161
POWERS OF NINETEEN. REF BA1.

2199 1, 19, 1513, 315523, 136085041, 105261234643, 132705221399353, 254604707462013571, 705927677520644167681, 2716778010767155313771539
GENERALIZED EULER NUMBERS. REF AMM 36 649 35.

2200 1, 19, 4315, 237671, 9751299, 150653570023
FROM HIGHER ORDER BERNOULLI NUMBERS. REF NO1 461.

SEQUENCES BEGINNING 1, 20 THROUGH 1, 49

2201 1, 20, 74, 24, 157, 124, 478, 1480, 1198, 3044, 480, 184, 2351, 1720, 3282, 5728, 2480, 1776, 10326, 9560, 8886, 9188, 11618, 23664, 16231, 23960
RELATED TO REPRESENTATION AS SUMS OF SQUARES. REF QJM 38 56 07.

2202 1, 20, 74, 186, 388, 721, 1236, 1995, 3072, 4554, 6542, 9152, 12516, 16783, 22120, 28713, 36768, 46512, 58194, 72086, 88484, 107709, 130108, 156055, 185952
POWERS OF ROOTED TREE ENUMERATOR. REF R1 150.

2203 1, 20, 80, 144, 610, 448, 1120, 2240, 3423, 12200, 14800, 29440, 5470, 6272, 48800, 81664, 73090, 68460, 15600, 87840, 139776, 82880, 189920, 474112, 18525
RELATED TO REPRESENTATION AS SUMS OF SQUARES. REF QJM 38 311 07.

2204 1, 20, 154, 1676, 14292, 155690, 1731708, 21264624, 280260864, 3970116255, 60113625680, 969368687752, 16588175089420, 300272980075896
DISCORDANT PERMUTATIONS. REF SMA 20 23 54.

2205 1, 20, 220, 23932, 2390065448
SWITCHING NETWORKS. REF JFI 276 317 63.

2206 1, 20, 295, 4025, 54649, 761166, 11028590, 167310220, 2664929476, 44601786944, 784146622896, 14469012689040, 279870212258064, 5667093514231200
GENERALIZED STIRLING NUMBERS. REF PEF 77 7 62.

2207 1, 20, 300, 4200, 58800, 846720, 12700800, 199584000, 3293136000, 57081024000, 1038874636800, 19833061248000, 396661224960000
LAH NUMBERS. REF R1 44. CO1 1 166.

2208 1, 20, 371, 2588, 11097, 35645, 94457, 218124, 454220, 872648, 1571715, 2684936, 4388567, 6909867, 10536089, 15624200, 22611330, 32025950, 44499779
DISCORDANT PERMUTATIONS. REF SMA 20 23 54.

2209 1, 20, 484, 497760, 1957701217328
SWITCHING NETWORKS. REF JFI 276 317 63.

2210 1, 20, 784, 52480, 5395456, 791691264, 157294854144, 40683662475264, 13288048674471936
CENTRAL FACTORIAL NUMBERS. REF OP1 7. FMR 1 110. RCI 217.

2211 1, 20, 1120, 3200, 3942400, 66560000, 10035200000
ASYMPTOTIC EXPANSION OF AN INTEGRAL. REF MTAC 19 114 65.

2212 1, 21, 31, 6257, 10293, 279025, 483127, 435506703, 776957575, 22417045555, 40784671953
COEFFICIENTS OF JACOBI NOME. REF HE1 477. MTAC 3 234 48.

2213 1, 21, 42, 65, 86, 109, 130, 151, 174, 195, 218, 239, 262, 283, 304, 327, 348, 371, 392, 415, 436, 457, 480, 501, 524, 545, 568, 589, 610, 633, 654, 677, 698, 721, 742, 763
RELATED TO POWERS OF 3. REF AMM 64 367 57.

2214 1, 21, 60, 90, 182, 378, 861, 1737, 3458, 6717, 13377, 25877, 49949, 95085, 180254, 338003, 631124, 1168226, 2151409, 3934674, 7159108, 12948649, 23307439
SOLID PARTITIONS. REF PNISI 26 135 60.

2215 1, 21, 266, 2646, 22827, 179487, 1323652, 9321312, 63436373, 420693273, 2734926558, 17505749898, 110687251039, 693081601779, 4306078895384
STIRLING NUMBERS OF SECOND KIND. REF AS1 835. DKB 223.

2216 1, 21, 322, 4536, 63273, 902055, 13339535, 206070150, 3336118786, 56663366760, 1009672107080, 18861567058880, 369012649234384
STIRLING NUMBERS OF FIRST KIND. REF AS1 833. DKB 226.

2217 1, 21, 378, 6930, 135135, 2837835
VALUES OF BESSEL POLYNOMIALS. REF RCI 77. RI1.

2218 1, 21, 21000, 101, 121, 1101, 1121, 21121, 101101, 101121, 121121, 1101121, 1121121, 21121121, 101101121, 101121121, 121121121, 1101121121, 1121121121
SMALLEST NUMBER REQUIRING N WORDS IN ENGLISH.

2219 1, 22, 67, 181, 401, 831, 1576, 2876, 4987, 8406, 13715, 21893, 34134, 52327, 78785, 116982, 171259, 247826, 354482, 502090, 704265, 979528, 1351109, 1849932
BIPARTITE PARTITIONS. REF NI1 1.

2220 1, 22, 79, 107, 311, 432, 487, 665, 692, 1044, 1536, 13909, 15204, 29351, 44332, 66287, 70877
CLASS NUMBERS OF QUADRATIC FIELDS. REF MTAC 24 441 70.

2221 1, 22, 328, 4400, 58140, 785256
FROM SUMS OF PRODUCTS OF POWERS. REF NMT 7 16 59.

2222 1, 22, 355, 5265, 77224, 1155420, 17893196, 288843260, 4876196776, 86194186584, 1595481972864, 30908820004608, 626110382381184
GENERALIZED STIRLING NUMBERS. REF PEF 77 44 62.

2223 1, 23, 11, 563, 1627, 88069, 1423, 1593269, 7759469, 31730711, 46522243
COEFFICIENTS FOR NUMERICAL DIFFERENTIATION. REF PHM 33 13 42.

2224 1, 23, 47, 73, 97, 103, 157, 167, 193, 263, 277, 307, 383, 397, 433, 503, 577, 647, 673, 683, 727, 743, 863, 887, 937, 967, 983, 1033, 1093, 1103, 1153, 1163, 1223, 1367
PRIMES WITH 5 AS SMALLEST PRIMITIVE ROOT. REF KR1 1 57. AS1 864.

2225 1, 23, 65, 261, 1370, 8742, 65304, 557400, 5343120
BINOMIAL COEFFICIENT SUMS. REF CJM 22 26 70.

2226 1, 23, 71, 311, 479, 1559, 5711, 10559, 18191, 31391, 307271, 366791, 2155919, 6077111, 98538359, 120293879, 131486759, 508095719, 2570169839
NEGATIVE PSEUDO-SQUARES. REF MTAC 24 436 70.

2227 1, 23, 263, 133787, 157009, 16215071, 2689453969, 26893118531, 5600751928169
COEFFICIENTS FOR REPEATED INTEGRATION. REF JM2 28 56 49.

2228 1, 23, 1681, 257543, 67637281
GLAISHERS T NUMBERS. REF QJM 29 76 1897. FMR 1 76.

2229 1, 24, 12, 640, 1920, 107520, 1792, 2064384, 10321920, 43253760, 64880640
COEFFICIENTS FOR NUMERICAL DIFFERENTIATION. REF PHM 33 13 42.

2230 1, 24, 26, 0, 0, 72, 378, 1080, 665, 384, 1968, 2016, 25698, 39552, 3872, 20880, 65727, 379856, 1277646
SUSCEPTIBILITY FOR CUBIC LATTICE. REF JCP 38 811 63.

2231 1, 24, 44, 80, 144, 260, 476, 872, 1600, 2940, 5404, 9936, 18272, 33604, 61804, 113672, 209072, 384540, 707276, 1300880, 2392688, 4400836, 8094396, 14887912
RESTRICTED PERMUTATIONS. REF CMB 4 32 61.

2232 1, 24, 44, 80, 144, 264, 484, 888, 1632, 3000, 5516, 10144, 18656, 34312, 63108, 116072, 213488, 392664, 722220, 1328368, 2443248, 4493832, 8265444
PERMANENTS OF CYCLIC (0, 1) MATRICES. REF CMB 7 262 64. JCT 7 315 69.

2233 1, 24, 144, 984, 7584, 65304, 622704, 6523224
BINOMIAL COEFFICIENT SUMS. REF CJM 22 26 70.

2234 1, 24, 154, 580, 1665, 4025, 8624, 16884, 30810, 53130, 87450, 138424, 211939, 315315, 457520, 649400, 903924, 1236444, 1664970, 2210460, 2897125, 3752749
GENERALIZED STIRLING NUMBERS. REF PEF 77 7 62.

2235 1, 24, 240, 1560, 8400, 40824, 186480, 818520, 3498000, 14676024, 60780720, 249401880, 1016542800, 4123173624, 16664094960, 67171367640
DIFFERENCES OF ZERO. REF VO1 31. DA2 2 212. R1 33.

2236 1, 24, 240, 2520, 26880, 304080, 3671136
3-LINE LATIN RECTANGLES. REF R1 210.

2237 1, 24, 252, 1472, 4830, 6048, 16744, 84480, 113643, 115920, 534612, 370944, 577738, 401856, 1217160, 987136, 6905934, 2727432, 10661420, 7109760, 4219488
RAMANUJAN TAU FUNCTION. REF PLMS 51 4 50. MTAC 24 495 70.

2238 1, 24, 264, 3312, 48240, 762096, 12673920
WALKS ON A CUBIC LATTICE. REF JCP 34 1537 61.

2239 1, 24, 274, 1624, 6769, 22449, 63273, 157773, 357423, 749463, 1474473, 2749747, 4899622, 8394022, 13896582, 22323822, 34916946, 53327946, 79721796
STIRLING NUMBERS OF FIRST KIND. REF AS1 833. DKB 226.

2240 1, 24, 282, 2008, 10147, 40176, 132724, 381424, 981541, 2309384, 5045326, 10356424
STOCHASTIC MATRICES OF INTEGERS. REF ST2. CJ1 13 283 70.

2241 1, 24, 300, 3360, 38850, 475776, 6231960, 87530400
LABELED TREES OF HEIGHT 3. REF IBMJ 4 478 60.

2242 1, 24, 600, 10800, 176400, 2822400, 45722880, 762048000, 13172544000, 237105792000
COEFFICIENTS OF LAGUERRE POLYNOMIALS. REF AS1 799.

2243 1, 24, 640, 7168, 294912, 2883584, 54525952, 167772160, 36507222016, 326417514496
COEFFICIENTS FOR NUMERICAL DIFFERENTIATION. REF OP1 23. PHM 33 14 42.

2244 1, 24, 924, 26432, 705320, 18858840, 520059540, 14980405440, 453247114320, 14433720701400, 483908513388300, 17068210823664000, 632607429473019000
ASSOCIATED STIRLING NUMBERS. REF TOH 37 259 33. JO2 152. CO1 2 98.

2245 1, 24, 936, 56640, 4968000, 598328640, 94916183040, 19200422062080, 4826695329792000, 1476585999504000000, 540272647694971699200
CONNECTED GRAPHS BY POINTS AND LINES. REF AMS 30 748 59.

2246 1, 24, 1920, 322560, 92897280, 40874803200, 25505877196800, 21424936845312000, 23310331287699456000, 31888533201572855808000
COEFFICIENTS FOR CENTRAL DIFFERENCES. REF JM2 42 162 63.

2247 1, 24, 2040, 297200, 68938800, 24046189440, 12025780892160, 8302816499443200, 7673688777463632000, 9254768770160124288000
STOCHASTIC MATRICES OF INTEGERS. REF ST2.

2248 1, 24, 3852, 18534400
PERMANENTS OF PROJECTIVE PLANES. REF RYS 124.

2249 1, 24, 5760, 322560, 51609600, 13624934400, 19837904486400, 2116043145216, 20720294477955072, 15747423803245854720
COEFFICIENTS FOR NUMERICAL DIFFERENTIATION. REF OP1 23. PHM 33 14 42.

2250 1, 24, 5760, 967680, 464486400, 122624409600, 2678117105664000, 64274810535936000, 14985212970663936 0000, 669659197233029971968000
COEFFICIENTS FOR NUMERICAL INTEGRATION. REF PHM 36 217 45.

2251 1, 25, 421, 6105, 83029, 1100902, 14516426, 192422979, 2579725656, 35098717902, 485534447114, 6835409506841, 97966603326993, 1429401763567226
PERMUTATIONS BY SUBSEQUENCES. REF MTAC 22 390 68.

2252 1, 25, 445, 7140, 111769, 1767087, 28699460, 483004280, 8460980836, 154594537812, 2948470152264, 58696064973000, 1219007251826064
GENERALIZED STIRLING NUMBERS. REF PEF 77 26 62.

2253 1, 25, 450, 7350, 117600, 1905120, 31752000, 548856000, 9879408000
COEFFICIENTS OF LAGUERRE POLYNOMIALS. REF LA4 519. AS1 799.

2254 1, 25, 490, 9450, 190575, 4099095, 94594500, 2343240900
ASSOCIATED STIRLING NUMBERS. REF AJM 56 92 34. DB1 296.

2255 1, 26, 302, 2416, 15619, 88234, 455192, 2203488, 10187685, 45533450, 198410786, 848090912, 3572085255
EULERIAN NUMBERS. REF R1 215. DB1 151. JCT 1 351 66. DKB 260. CO1 2 84.

2256 1, 26, 485, 8175, 134449, 2231012, 37972304, 668566300, 12230426076, 232959299496, 4623952866312, 95644160132976, 2060772784375824
GENERALIZED STIRLING NUMBERS. REF PEF 77 61 62.

2257 1, 27, 184, 875, 2700, 7546, 17600, 35721, 72750, 126445, 223776, 353717, 595448, 843750, 1349120, 1827636, 2808837, 3600975, 5306000, 6667920, 9599172
RELATED TO THE DIVISOR FUNCTION. REF SMA 19 39 53.

2258 1, 27, 378, 4536, 48600
CARD MATCHING. REF R1 193.

2259 1, 27, 511, 8624, 140889, 2310945, 38759930, 671189310, 12061579816, 225525484184, 4392554369840, 89142436976320, 1884434077831824
GENERALIZED STIRLING NUMBERS. REF PEF 77 7 62.

2260 1, 27, 10206, 1271126683458
POST FUNCTIONS. REF JCT 4 295 68.

2261 1, 28, 2, 8, 6, 992, 1, 3, 2, 16256, 2, 16, 16
DIFFERENTIAL STRUCTURES (ASSUMING POINCARE CONJECTURE). REF ICM 50 62. ANN 77 504 63.

2262 1, 28, 153, 496, 1225, 2556, 4753, 8128, 13041, 19900, 29161, 41328, 56953, 76636, 101025, 130816, 166753, 209628, 260281, 319600, 388521, 468028, 559153
SUMS OF CUBES OF ODD NUMBERS. REF CC1 742.

2263 1, 28, 462, 5880, 63987, 627396, 5715424, 49329280, 408741333, 3281882604, 25708104786, 197462483400, 1492924634839, 11143554045652
STIRLING NUMBERS OF SECOND KIND. REF AS1 835. DKB 223.

2264 1, 28, 546, 9450, 157773, 2637558, 44990231, 790943153, 14409322928, 272803210680, 5374523477960, 110228466184200, 2353125040549984
STIRLING NUMBERS OF FIRST KIND. REF AS1 834. DKB 226.

2265 1, 30, 97, 267, 608, 1279, 2472, 4571, 8043, 13715, 22652, 36535, 57568, 89079, 135384, 202747, 299344, 436597, 629364, 897970, 1268634, 1776562, 2466961
BIPARTITE PARTITIONS. REF NI1 2.

2266 1, 30, 625, 11515, 203889, 3602088, 64720340, 1194928020, 22800117076, 450996059800, 9262414989464, 197632289814960, 4381123888865424
GENERALIZED STIRLING NUMBERS. REF PEF 77 44 62.

2267 1, 30, 630, 11760, 211680, 3810240, 69854400, 1317254400, 25686460800, 519437318400, 10908183686400, 237996734976000, 5394592659456000
LAH NUMBERS. REF R1 44. CO1 1 166.

2268 1, 30, 840, 1197504000, 60281712691200
STEINER TRIPLE SYSTEMS. REF CO1 2 153.

2269 1, 30, 1023, 44473, 2475473, 173721912, 15088541896, 1593719752240, 201529405816816, 30092049283982400, 5242380158902146624
CENTRAL FACTORIAL NUMBERS. REF RCI 217.

2270 1, 30, 1260, 75600, 6237000, 681080400, 95351256000, 16672848192000, 3563821301040000, 914714133933600000, 277707211062240960000
RELATED TO EULER NUMBERS. REF QJM 47 110 16. FMR 1 112. DA2 2 283.

2271 1, 31, 223, 433, 439, 457, 727, 919, 1327, 1399, 1423, 1471, 1831, 1999, 2017, 2287, 2383, 2671, 2767, 2791, 2953, 3271, 3343, 3457, 3463, 3607, 3631, 3823, 3889
2 IS A 6-TH POWER RESIDUE MODULO P. REF KR1 1 59.

2272 1, 31, 301, 1701, 6951, 22827, 63987, 159027, 359502, 752752, 1479478, 2757118, 4910178, 8408778, 13916778, 22350954, 34952799, 53374629, 79781779
STIRLING NUMBERS OF SECOND KIND. REF AS1 835. DKB 223.

2273 1, 31, 304, 4230, 43880, 547338, 6924960, 94714620, 1375878816, 21273204330, 348919244768, 6056244249682, 110955673493568, 2140465858763844
DISCORDANT PERMUTATIONS. REF SMA 20 23 54.

2274 1, 31, 307, 643, 5113, 21787, 39199, 360007, 4775569, 10318249, 65139031
PRIMES WITH LARGE LEAST NONRESIDUES. REF RS5 XVI.

2275 1, 31, 696, 5823, 29380, 108933, 327840, 848380, 1958004, 4130895, 8107024, 14990889, 26372124, 44470165, 72305160, 113897310, 174496828, 260846703
DISCORDANT PERMUTATIONS. REF SMA 20 23 54.

2276 1, 31, 3661, 1217776, 929081776, 1413470290176, 3878864920694016, 17810567950611972096
DIFFERENCES OF RECIPROCALS OF UNITY. REF DKB 228.

2277 1, 32, 243, 1024, 3125, 7776, 16807, 32768, 59049, 100000, 161051, 248832, 371293, 537824, 759375, 1048576, 1419857, 1889568, 2476099, 3200000, 4084101
FIFTH POWERS. REF BA1.

2278 1, 33, 79, 107, 311, 487, 665, 857, 2293, 3523, 13909, 26713, 29351, 59801, 66287, 70877
CLASS NUMBERS OF QUADRATIC FIELDS. REF MTAC 24 441 70.

2279 1, 33, 244, 1057, 3126, 8052, 16808, 33825, 59293, 103158, 161052, 257908, 371294, 554664, 762744, 1082401, 1419858, 1956669, 2476100, 3304182, 4101152
SUM OF 5TH POWERS OF DIVISORS OF N. REF AS1 827.

2280 1, 33, 276, 1300, 4425, 12201, 29008, 61776, 120825, 220825, 381876, 630708, 1002001, 1539825, 2299200, 3347776, 4767633, 6657201, 9133300, 12333300
SUMS OF FIFTH POWERS. REF AS1 813.

2281 1, 33, 524, 2322, 81912, 214181, 1182276, 3736614, 9972264, 24622002, 51265020, 106396576, 202547304, 357914103
RELATED TO REPRESENTATION AS SUMS OF SQUARES. REF QJM 38 305 07.

2282 1, 35, 835, 17360, 342769, 6687009, 131590430, 2642422750, 54509190076, 1159615530788, 25497032420496, 580087776122400, 13662528306823824
GENERALIZED STIRLING NUMBERS. REF PEF 77 61 62.

2283 1, 35, 966, 24970, 631631, 15857205, 397027996
CENTRAL FACTORIAL NUMBERS. REF TH1 36. FMR 1 112. RCI 217.

2284 1, 35, 1974, 172810, 21967231, 3841278805, 886165820604, 261042753755556, 95668443268795341, 42707926241367380631, 22821422608929422854674
CENTRAL FACTORIAL NUMBERS. REF RCI 217.

2285 1, 36, 330, 22060, 920737780
SWITCHING NETWORKS. REF JFI 276 317 63.

2286 1, 36, 666, 384112, 735192450952
SWITCHING NETWORKS. REF JFI 276 317 63.

2287 1, 36, 820, 7645, 44473, 191620, 669188, 1999370, 5293970, 12728936, 28285400, 58856655, 115842675, 217378200, 391367064, 679524340, 1142659012
CENTRAL FACTORIAL NUMBERS. REF RCI 217.

2288 1, 36, 841, 16465, 296326, 5122877, 87116283, 1477363967, 25191909848, 434119587475, 7583461369373, 134533482045389, 2426299018270338
PERMUTATIONS BY SUBSEQUENCES. REF MTAC 22 390 68.

2289 1, 36, 882, 18816, 381024, 7620480, 153679680, 3161410560
COEFFICIENTS OF LAGUERRE POLYNOMIALS. REF LA4 519. AS1 799.

2290 1, 36, 1072, 2100736, 17592201773056
SWITCHING NETWORKS. REF JFI 276 317 63.

2291 1, 36, 1225, 41616, 1413721, 48024900, 1631432881, 55420693056, 1882672131025, 63955431761796, 2172602007770041, 73804512832419600
BOTH TRIANGULAR AND SQUARE. REF DI2 2 10. MAG 47 237 63. BE3 193. FQ 9 95 71.

2292 1, 37, 59, 67, 101, 103, 131, 149, 157, 233, 257, 263, 271, 283, 293, 307, 311, 347, 353, 379, 389, 401, 409, 421, 433, 461, 463, 467, 491, 523, 541, 547, 557, 577, 587, 593
IRREGULAR PRIMES. REF PNAS 40 31 54. BO1 430.

2293 1, 41, 109, 151, 229, 251, 271, 367, 733, 761, 971, 991, 1069, 1289, 1303, 1429, 1471, 1759, 1789, 1811, 1879, 2411, 2441, 2551, 2749, 2791, 3061, 3079, 3109, 3229
PRIMES WITH 6 AS SMALLEST PRIMITIVE ROOT. REF KR1 1 57. AS1 864.

2294 1, 41, 313, 353, 1201, 3593, 4481, 7321, 8521, 10601, 14281, 14321, 14593, 21601, 26513, 32633, 41761, 41801, 42073, 42961, 49081, 56041, 66361, 67073, 72481
HALF-QUARTAN PRIMES. REF CU1 1 254.

2295 1, 42, 139, 392, 907, 1941, 3804, 7128, 12693, 21893, 36535, 59521, 94664, 147794, 226524, 342006, 508866, 747753, 1085635, 1559725, 2218272, 3126541
BIPARTITE PARTITIONS. REF NI1 2.

2296 1, 42, 462, 2772, 12012, 42042, 126126
RELATED TO BINOMIAL MOMENTS. REF JO2 449.

2297 1, 42, 1176, 28224, 635040, 13970880, 307359360, 6849722880, 155831195520, 3636061228800, 87265469491200, 2157837063782400, 55024845126451200
LAH NUMBERS. REF R1 44. CO1 1 166.

2298 1, 42, 1586, 31388, 442610, 5030004, 49145460
ROOTED PLANAR MAPS. REF BAMS 74 74 68.

2299 1, 43, 109, 157, 229, 277, 283, 307, 499, 643, 691, 733, 739, 811, 997, 1021, 1051, 1069, 1093, 1459, 1579, 1597, 1627, 1699, 1723, 1789, 1933, 2179, 2203, 2251, 2341
2 IS A CUBIC RESIDUE MODULO P. REF KR1 1 59.

2300 1, 44, 432, 1136, 610, 5568, 6048, 11456, 3423, 26840, 79920, 768, 5470, 77952, 263520, 61696, 73090, 150612, 84240, 692960, 139776, 1030080, 1025568
RELATED TO REPRESENTATION AS SUMS OF SQUARES. REF QJM 38 329 07.

2301 1, 44, 4940800, 564083990621761115783168
DISCRIMINANTS OF SHAPIRO POLYNOMIALS. REF PAMS 25 115 70.

2302 1, 48, 264, 1680, 11640, 86352, 673104, 5424768, 44828400, 377810928, 3235366752, 28074857616, 246353214240
WALKS ON A CUBIC LATTICE. REF JCP 46 3481 67.

2303 1, 48, 1152, 30720, 1152000, 65630208
COLORED GRAPHS. REF CJM 12 412 60.

2304 1, 49, 1513, 38281, 874886, 18943343, 399080475, 8312317976, 172912977525, 3615907795025, 76340522760097, 1631788075873114, 35378058306185002
PERMUTATIONS BY SUBSEQUENCES. REF MTAC 22 390 68.

SEQUENCES BEGINNING 1, 50, 1, 51, ...

2305 1, 50, 1660, 46760, 1217776, 30480800, 747497920, 18139003520, 437786795776
DIFFERENCES OF RECIPROCALS OF UNITY. REF DKB 228.

2306 1, 52, 472, 3224, 18888, 101340, 511120, 2465904
SERIES-PARALLEL NUMBERS. REF R1 142.

2307 1, 55, 150, 210, 280, 580, 1275, 2905, 5350, 9985, 17965, 33665, 62895, 117287, 214610, 389805, 700720, 1259890, 2250405, 4008717, 7092366, 12497237
SOLID PARTITIONS. REF PNISI 26 135 60.

2308 1, 56, 1120, 18592, 300288, 4877824, 80349696, 1344154112
ALMOST CUBIC MAPS. REF PL2 1 292 70.

2309 1, 56, 1918, 56980, 1636635, 47507460, 1422280860
ASSOCIATED STIRLING NUMBERS. REF AJM 56 92 34. DB1 296.

2310 1, 57, 1191, 15619, 156190, 1310354, 9738114, 66318474, 423281535, 2571742175, 15041229521, 85383238549
EULERIAN NUMBERS. REF R1 215. DB1 151. JCT 1 351 66. DKB 260. CO1 2 84.

2311 1, 60, 168, 360, 504, 660, 1092, 2448, 2520, 3420, 4080, 5616, 6048, 6072, 7800, 7920, 9828, 12180, 14880, 20160, 25308, 25920, 29120, 32736, 34440, 39732, 51888
NON-CYCLIC SIMPLE GROUPS. REF DI1 309. JL2 137.

2312 1, 60, 720, 6090, 47040, 363384, 2913120
LABELED TREES OF DIAMETER 4. REF IBMJ 4 478 60.

2313 1, 61, 841, 7311, 51663, 325446, 1910706, 10715506, 58258210, 309958755, 1623847695
PERMUTATIONS BY LENGTH OF RUNS. REF DKB 260.

2314 1, 61, 1385, 19028, 206276, 1949762, 16889786, 137963364, 1081702420, 8236142455, 61386982075
PERMUTATIONS BY LENGTH OF RUNS. REF DKB 260.

2315 1, 61, 2763, 38528, 249856, 1066590, 3487246, 9493504, 22634496, 48649086, 96448478
GENERALIZED CLASS NUMBERS. REF MTAC 21 689 67, 22 698 68.

2316 1, 63, 22631, 30480800, 117550462624, 1083688832185344, 21006340945438768128
DIFFERENCES OF RECIPROCALS OF UNITY. REF DKB 228.

2317 1, 64, 625, 4016, 21256, 100407, 439646, 1823298
PARTIALLY LABELED ROOTED TREES. REF R1 134.

2318 1, 64, 729, 4096, 15625, 46656, 117649, 262144, 531441, 1000000, 1771561, 2985984, 4826809, 7529536, 11390625, 16777216, 24137569, 34012224, 47045881
SIXTH POWERS. REF BA1.

2319 1, 64, 960, 135040, 14333211520
SWITCHING NETWORKS. REF JFI 276 317 63.

2320 1, 64, 1024, 12480, 137472, 1443616
WALKS ON A CUBIC LATTICE. REF PCPS 58 100 62.

2321 1, 64, 2304, 2928640, 11745443774464
SWITCHING NETWORKS. REF JFI 276 317 63.

2322 1, 65, 794, 4890, 20515, 67171, 184820, 446964, 978405, 1978405, 3749966, 6735950, 11562759, 19092295, 30482920, 47260136, 71397705, 105409929, 152455810
SUMS OF SIXTH POWERS. REF AS1 813.

2323 1, 65, 1795, 36317, 636331
CONNECTED RELATIONS. REF CR 268 579 69.

2324 1, 70, 19355, 11180820, 11555272575
CUBIC GRAPHS. REF RE4.

2325 1, 71, 239, 241, 359, 431, 499, 599, 601, 919, 997, 1051, 1181, 1249, 1439, 1609, 1753, 2039, 2089, 2111, 2179, 2251, 2281, 2341, 2591, 2593, 2671, 2711, 2879, 3119
PRIMES WITH 7 AS SMALLEST PRIMITIVE ROOT. REF KR1 1 58. AS1 864.

2326 1, 73, 241, 1009, 2641, 8089, 18001, 53881, 87481, 117049, 515761, 1083289, 3206641, 3818929, 9257329, 22000801, 48473881, 175244281, 427733329, 898716289
PSEUDO-SQUARES. REF MTAC 24 434 70.

2327 1, 82, 707, 3108, 9669, 24310, 52871, 103496, 187017, 317338, 511819, 791660, 1182285, 1713726, 2421007, 3344528, 4530449, 6031074, 7905235, 10218676
SUMS OF FOURTH POWERS OF ODD NUMBERS. REF AMS 2 358 31 (DIVIDED BY 2). CC1 742.

2328 1, 88, 326, 1631, 10112, 74046, 622704, 5900520
BINOMIAL COEFFICIENT SUMS. REF CJM 22 26 70.

2329 1, 90, 1260, 13230, 126720, 1171170, 10663380, 96461910, 870123240, 7838973450, 70582218300, 635365793790, 5718795460560, 51471172410930
RELATED TO EULER NUMBERS. REF QJM 47 110 16. FMR 1 112. DA2 2 283.

2330 1, 96, 1776, 43776, 1237920
WALKS ON A CUBIC LATTICE. REF PRV 114 53 59.

2331 1, 97, 139, 151, 199, 211, 331, 433
RELATED TO KUMMERS CONJECTURE. REF HA3 482.

2332 1, 113, 281, 353, 577, 593, 617, 1033, 1049, 1097, 1153, 1193, 1201, 1481, 1601, 1889, 2129, 2273, 2393, 2473, 3049, 3089, 3137, 3217, 3313, 3529, 3673, 3833, 4001
2 IS A QUARTIC RESIDUE MODULO P. REF KR1 1 59.

2333 1, 120, 265, 579, 1265, 2783, 6208, 13909, 31337, 70985, 161545, 369024, 845825, 1944295, 4480285, 10345391, 23930320, 55435605, 128577253, 298529333
PERMANENTS OF CYCLIC (0, 1) MATRICES. REF CMB 7 262 64. JCT 7 315 69.

2334 1, 120, 1800, 16800, 126000, 834120, 5103000, 29607600, 165528000, 901020120, 4809004200, 25292030400, 131542866000, 678330198120, 3474971465400
DIFFERENCES OF ZERO. REF VO1 31. DA2 2 212. R1 33.

2335 1, 120, 2520, 43680, 757680, 13747104, 264181680
LABELED TREES OF HEIGHT 4. REF IBMJ 4 478 60.

2336 1, 120, 4293, 88234, 1310354, 15724248, 162512286, 1505621508, 12843262863, 102776998928, 782115518299
EULERIAN NUMBERS. REF R1 215. DB1 151. JCT 1 351 66. DKB 260. CO1 2 84.

2337 1, 120, 4320, 105840, 2257920, 45722880, 914457600, 18441561600, 379369267200
COEFFICIENTS OF LAGUERRE POLYNOMIALS. REF AS1 799.

2338 1, 120, 7308, 303660, 11098780, 389449060, 13642629000, 486591585480, 17856935296200, 678103775949600, 26726282654771700
ASSOCIATED STIRLING NUMBERS. REF TOH 37 259 33. JO2 152. CO1 2 98.

2339 1, 121, 12321, 1234321, 123454321, 12345654321, 1234567654321, 123456787654321, 12345678987654321, 1234567900987654321
WONDERFUL DEMLO NUMBERS. REF MST 6 68 38.

2340 1, 125, 1296, 8716, 47787, 232154, 1040014
PARTIALLY LABELED TREES. REF R1 138.

2341 1, 128, 2187, 16384, 78125, 279936, 823543, 2097152, 4782969, 10000000, 19487171, 35831808, 62748517, 105413504, 170859375, 268435456, 410338673
SEVENTH POWERS. REF BA1.

2342 1, 128, 12758, 5134240
THRESHOLDS OF COMPLETENESS. REF JL2 367.

2343 1, 129, 2316, 18700, 96825, 376761, 1200304, 3297456, 8080425, 18080425, 37567596, 73399404, 136147921, 241561425, 412420800, 680856256, 1091194929
SUMS OF SEVENTH POWERS. REF AS1 815.

2344 1, 130, 1270932917454
ALGEBRAS. REF PAMS 17 737 66.

2345 1, 132, 1716, 12012, 60060, 240240
RELATED TO BINOMIAL MOMENTS. REF JO2 450.

2346 1, 136, 64573605
ALGEBRAS. REF PAMS 17 736 66.

2347 1, 139, 571, 163879, 5246819, 534703531, 4483131259, 432261921612371, 6232523202521089, 25834629665134204969, 1579029138854919086429
EXPANSION OF GAMMA FUNCTION. REF MTAC 22 619 68.

2348 1, 151, 431, 6581, 67651, 241981, 2081921, 3395921
PRIMES WITH LARGE LEAST NONRESIDUES. REF RS5 XXIII.

2349 1, 157, 262, 367, 412, 577, 682, 787, 877, 892, 907, 997, 1072, 1207, 1237, 1312, 1402
RELATED TO EULERS TOTIENT FUNCTION. REF AMM 54 332 47.

2350 1, 163, 907, 2683, 5923, 10627, 15667, 20563, 34483, 37123, 38707, 61483, 90787, 93307, 103387, 166147, 133387, 222643, 210907, 158923, 253507, 296587
PRIMES BY CLASS NUMBER. REF MTAC 24 492 70.

2351 1, 211, 281, 421, 461, 521, 691, 881, 991, 1031, 1151, 1511, 1601, 1871, 1951, 2221, 2591, 3001, 3251, 3571, 3851, 4021, 4391, 4441, 4481, 4621, 4651, 4691, 4751
ARTIADS. REF PLMS 24 256 1893. JM1 15 118 66.

2352 1, 220, 1184, 2620, 5020, 6232, 10744, 12285, 17296, 63020, 66928, 67095, 69615, 79750, 100485, 122265, 122368, 141664, 142310, 171856, 176272, 185368
AMICABLE NUMBERS. REF MTAC 21 242 67.

2353 1, 244, 3126, 16808, 59293, 161052, 371294, 762744, 1419858, 2476100, 4101152, 6436344, 9768751, 14408200, 20511150, 28629152, 39296688, 52541808
A DIVISOR FUNCTION. REF QJM 38 56 07.

2354 1, 244, 3369, 20176, 79225, 240276, 611569, 1370944, 2790801, 5266900, 9351001, 15787344, 25552969, 39901876, 60413025, 89042176, 128177569, 180699444
SUMS OF FIFTH POWERS OF ODD NUMBERS. REF CC1 742.

2355 1, 247, 14608, 455192, 9738114, 162512286, 2275172004, 27971176092, 311387598411, 3207483178157
EULERIAN NUMBERS. REF R1 215. DB1 151. JCT 1 351 66. DKB 260. CO1 2 84.

2356 1, 251, 571, 971, 1181, 1811, 2011, 2381, 2411, 3221, 3251, 3301, 3821, 4211, 4861, 4931, 5021, 5381, 5861, 6221, 6571, 6581, 8461, 8501, 9091, 9461
2 IS A QUINTIC RESIDUE MODULO P. REF KR1 1 59.

2357 1, 256, 6561, 65536, 390625, 1679616, 5764801, 16777216, 43046721, 100000000, 214358881, 429981696, 815730721, 1475789056, 2562890625, 4294967296
EIGHTH POWERS. REF BA1.

2358 1, 257, 6818, 72354, 462979, 2142595, 7907396, 24684612, 67731333, 167731333, 382090214, 812071910, 1627802631, 3103591687, 5666482312
SUMS OF EIGHTH POWERS. REF AS1 815.

2359 1, 263, 293, 368, 578, 683, 743, 788, 878, 893, 908, 998, 1073, 1103, 1208, 1238, 1268, 1403, 1418
RELATED TO EULERS TOTIENT FUNCTION. REF AMM 54 332 47.

2360 1, 272, 7936, 137216, 1841152, 21253376, 222398464, 2174832640, 20261765120, 182172651520
PERMUTATIONS BY LENGTH OF RUNS. REF DKB 261.

2361 1, 272, 24611, 515086, 4456448, 23750912, 93241002, 296327464, 806453248, 1951153920, 4300685074
GENERALIZED TANGENT NUMBERS. REF MTAC 21 690 67.

2362 1, 274, 48076, 6998824, 929081776, 117550462624, 14500866102976, 1765130436471424
DIFFERENCES OF RECIPROCALS OF UNITY. REF DKB 228.

2363 1, 284, 1210, 2924, 5564, 6368, 10856, 14595, 18416, 76084, 66992, 71145, 87633, 88730, 124155, 139815, 123152, 153176, 168730, 176336, 180848, 203432
AMICABLE NUMBERS. REF MTAC 21 242 67.

2364 1, 331, 39139, 253243, 4397207, 21587171
PRIMES WITH LARGE LEAST NONRESIDUES. REF RS5 XXIV.

2365 1, 341, 561, 645, 1105, 1387, 1729, 1905, 2047, 2465, 2701, 2821, 3277, 4033, 4369, 4371, 4681, 5461, 6601, 7957, 8321, 8481, 8911, 10261, 10585, 11305, 12801
PSEUDO-PRIMES. REF SPH 8 45 38. LE1 48. SI1 215.

2366 1, 502, 47840, 2203488, 66318474, 1505621508, 27971176092, 447538817472, 6382798925475
EULERIAN NUMBERS. REF R1 215. DB1 151. JCT 1 351 66. DKB 260. CO1 2 84.

2367 1, 504, 270648, 144912096, 77599626552, 41553943041744, 22251789971649504, 11915647845248387520, 6380729991419236488504
EXPANSION OF A MODULAR FUNCTION. REF RAM 317.

2368 1, 512, 19683, 262144, 1953125, 10077696, 40353607, 134217728, 387420489, 1000000000, 2357947691, 5159780352, 10604499373, 20661046784
NINTH POWERS. REF BA1.

2369 1, 720, 1854, 4738, 12072, 30818, 79118, 204448, 528950, 1370674, 3557408, 9244418, 24043990, 62573616, 162925614, 424377730, 1105703640, 2881483458
PERMANENTS OF CYCLIC (0, 1) MATRICES. REF CMB 7 262 64. JCT 7 317 69.

2370 1, 720, 15120, 191520, 1905120, 16435440, 129230640, 953029440, 6711344640, 45674188560, 302899156560, 1969147121760, 12604139926560
DIFFERENCES OF ZERO. REF VO1 31. DA2 2 212. R1 33.

2371 1, 725, 1125, 1600, 2000, 2048, 2225, 2304, 2525, 2624, 3600, 4205, 4225, 4352, 4400, 4525, 4752, 4913, 5125, 5225, 5725, 6125, 7056, 7168, 7225, 7232, 7488, 7600
DISCRIMINANTS OF REAL QUARTIC FIELDS. REF JLMS 31 484 56.

2372 1, 744, 196884, 21493760, 864299970, 20245856256, 333202640600, 4252023300096, 44656994071935, 401490886656000, 3176440229784420
COEFFICIENTS OF A MODULAR INVARIANT. REF KNAW 56 398 53. MTAC 8 77 54.
TA CARD

BIBLIOGRAPHY

AB1 *Annales de la Société Scientifique de Bruxelles.*

ACA *Acta Arithmetica.*

ACP *Advances in Chemical Physics.*

AENS *Annales Scientifiques de l'Ecole Normale Supérieure, Paris.*

AFAS *Association Française pour l'Advancement des Sciences, Comptes Rendus.*

AFI *American Federation of Information Processing Societies, Conference Proceedings.*

AFM *Arkiv för Matematik.*

AH1 W. Ahrens, "Mathematische Unterhaltungen und Spiele," Vol. 1, 3rd ed. Teubner, Leipzig, 1921.

AH2 W. Ahrens, "Mathematische Unterhaltungen und Spiele," Vol. 2, 2nd ed. Teubner, Leipzig, 1918.

AIEE *Transactions of the American Institute of Electrical Engineers.*

AIP *Advances in Physics.*

AJM *American Journal of Mathematics.*

AMM *The American Mathematical Monthly.*

AMP *Archiv der Mathematik und Physik.*

AMS *Annals of Mathematical Statistics.*

AM1 *Acta Mathematica (Uppsala).*

AM2 *Acta Mathematica (Budapest).*

AM3 *Archiv for Mathematik og Naturvidenskab.*

ANN *Annals of Mathematics.*

ANY *Annals of the New York Academy of Sciences.*

AS1 M. Abramowitz and I. A. Stegun, "Handbook of Mathematical Functions." National Bureau of Standards, Washington, D. C., 1964; republished by Dover, New York, 1965.

AT1 A. O. L. Atkin and B. J. Birch, editors, "Computers in Number Theory." Academic Press, New York, 1971.

BAMS *Bulletin of the American Mathematical Society.*

BAT P. M. Batchelder, "An Introduction to Finite Difference Equations." Harvard Univ. Press, Cambridge, Massachusetts, 1927.

BA1 British Association Mathematical Tables, Vol. 9, "Tables of Powers," by J. W. L. Glaisher and others. Cambridge Univ. Press, London and New York, 1940.

BA2 British Association Mathematical Tables, Vol. 8, "Number-Divisor Tables," by J. W. L. Glaisher. Cambridge Univ. Press, London and New York, 1940.
BA3 *Reports of the British Association for the Advancement of Science.*
BA4 *Reports of the British Association for the Advancement of Science, Transactions Section.*
BA5 British Association Mathematical Tables, Vol. 6, "Bessel Functions, Part 1, Functions of Orders Zero and Unity." Cambridge Univ. Press, London and New York, 1937.
BER *Chemische Berichte* (Formerly *Berichte der Deutschen Chemischen Gesellschaft).*
BE1 R. S. Berkowitz, editor, "Modern Radar." Wiley, New York, 1965.
BE2 E. R. Berlekamp, "Algebraic Coding Theory." McGraw-Hill, New York, 1968.
BE3 A. H. Beiler, "Recreations in the Theory of Numbers." Dover, New York, 1964.
BE4 *Sitzungsberichte der Königlich Preussischen Akadamie der Wissenschaften, Berlin.*
BE5 C. Berge, "Principles of Combinatorics." Academic Press, New York, 1971.
BE6 E. F. Beckenbach, editor, "Applied Combinatorial Mathematics." Wiley, New York, 1971.
BIT *Nordisk Tidskrift for Informationsbehandling.*
BI1 G. Birkhoff, "Lattice Theory." American Mathematical Society, Colloquium Publications, Vol. 25, 3rd ed., Providence, Rhode Island, 1967.
BMB *The Bulletin of Mathematical Biophysics.*
BMG *Bulletin de la Société Mathématique de Grèce.*
BO1 Z. I. Borevich and I. R. Shafarevich, "Number Theory." Academic Press, New York, 1966.
BO3 M. Boll, "Tables Numériques Universelles." Dunod, Paris, 1947.
BO4 C. J. Bouwkamp, personal communication.
BSM *Bulletin de la Société Mathématique de Belgique*
BSMF *Bulletin de la Société Mathématique de France.*
BS1 R. G. Busacker and T. L. Saaty, "Finite Graphs and Networks." McGraw-Hill, New York, 1965.
BU2 *Bulletin of the Calcutta Mathematical Society.*
CACM *Communications of the Association for Computing Machinery.*
CAU A. Cauchy, "Oeuvres Complètes." Gauthier-Villars, Paris, 1882–1938.
CAY A. Cayley, "Collected Mathematical Papers," Vols. 1–13. Cambridge Univ. Press, London and New York, 1889–1897.
CA1 R. D. Carmichael, "Introduction to the Theory of Groups of Finite Order." Republished by Dover, New York, 1956.
CA3 D. G. Cantor, personal communication.
CC1 F. E. Croxton and D. J. Cowden, "Applied General Statistics," 2nd ed. Prentice- Hall, Englewood Cliffs, New Jersey, 1955.
CH1 M. S. Cheema, personal communication.
CJM *Canadian Journal of Mathematics.*
CJ1 *The Computer Journal.*
CL1 "Tables of the Modified Hankel Functions of Order One-Third and of Their Derivatives." Annals of the Computation Laboratory of Harvard University, Vol. 2. Harvard Univ. Press, Cambridge, Massachusetts, 1945.
CMA "Combinatorial Mathematics and Its Applications," Vol. 1. *Proceedings of a Conference held at University of North Carolina, Chapel Hill, April 1967*

(R. C. Bose and T. A. Dowling, eds.). University of North Carolina Press, Chapel Hill, 1969. "Combinatorial Mathematics and Its Applications," Vol. 2. *Proceedings of 2nd Conference held at University of North Carolina, Chapel Hill, May 1970.* University of North Carolina, Chapel Hill, 1970.

CMB *Canadian Mathematical Bulletin.*

CM1 H. S. M. Coxeter and W. O. J. Moser, "Generators and Relations for Discrete Groups," 2nd ed. Springer-Verlag, Berlin and New York, 1965.

CO1 L. Comtet, "Analyse Combinatoire," Vols. 1 and 2. Presses Universitaires de France, Paris, 1970.

CPM *Časopis pro Pestování Matematiky.*

CR *Comptes Rendus Hebdomadaires des Séances de l'Académie des Sciences, Paris, Série A.*

CSA "Combinatorial Structures and Their Applications" *(Proceedings Calgary Conference).* Gordon & Breach, New York, 1970.

CU1 A. J. C. Cunningham, "Binomial Factorisations," Vols. 1–9. Hodgson, London, 1923–1929.

CU2 A. J. C. Cunningham, "Quadratic Partitions." Hodgson, London, 1904.

CU3 A. J. C. Cunningham, "Quadratic and Linear Tables." Hodgson, London, 1927.

DA1 H. T. Davis, "The Summation of Series." Principia Press of Trinity Univ., San Antonio, Texas, 1962.

DA2 H. T. Davis, "Tables of the Mathematical Functions," Vols. 1 and 2, 2nd ed., 1963, Vol. 3 (with V. J. Fisher), 1962. Principia Press of Trinity Univ., San Antonio, Texas.

DB1 F. N. David and D. E. Barton, "Combinatorial Chance." Hafner, New York, 1962.

DE1 C. F. Degen, "Canon Pellianus." Hafniae, Copenhagen, 1817.

DI1 L. E. Dickson, "Linear Groups with an Exposition of the Galois Field Theory." Republished by Dover, New York, 1958.

DI2 L. E. Dickson, "History of the Theory of Numbers." Carnegie Institute Publications 256, Washington, D.C., Vol 1, 1919; Vol. 2, 1920; Vol. 3, 1923.

DKB F. N. David, M. G. Kendall and D. E. Barton, "Symmetric Function and Allied Tables." Cambridge Univ. Press, London and New York, 1966.

DMJ *Duke Mathematical Journal.*

DO1 A. P. Domoryad, "Mathematical Games and Pastimes." Macmillan, New York, 1964.

DVSS *Det Kongelige Danske Videnskabernes Selskabs Skrifter.*

ELM *Elemente der Mathematik.*

EMN *The Edinburgh Mathematical Notes.*

EMS *Proceedings of the Edinburgh Mathematical Society.*

ESB *Éditions Spéciales de l'Institut Mathématique de Belgrade.*

EUL L. Euler, "Opera Omnia." Teubner, Leipzig, 1911.

EUR *Eureka, the Journal of the Archimedeans (Cambridge University Mathematical Society).*

FE1 W. Feller, "An Introduction to Probability Theory and Its Applications," Vol. 1, 2nd ed., 1960; Vol. 2, 1966. Wiley, New York.

FI1 T. Fine, "An Introduction to Theories of Probability." Academic Press, New York, 1973.

FI2 R. A. Fisher, "Contributions to Mathematical Statistics." Wiley, New York, 1950.

FMR A. Fletcher, J. C. P. Miller, L. Rosenhead, and L. J. Comrie, "An Index of Mathematical Tables," Vols. 1 and 2, 2nd ed. Blackwell, Oxford and Addison-Wesley, Reading, Massachusetts, 1962.

FQ *The Fibonacci Quarterly.*

FVS *Finska Vetenskaps-Societeten, Commentationes Physico-Mathematicae.*

FW1 Federal Works Agency, Work Projects Administration for the City of New York, "Tables of the Exponential Function." National Bureau of Standards, Washington, D.C., 1939.

FY1 R. A. Fisher and F. Yates, "Statistical Tables for Biological, Agricultural and Medical Research," 6th ed. Hafner, New York, 1963.

GA1 M. Gardner, "The 2nd Scientific American Book of Mathematical Puzzles and Diversions." Simon and Schuster, New York, 1961.

GO1 S. W. Golomb, "Shift Register Sequences." Holden-Day, San Francisco, California, 1967.

GO2 S. W. Golomb, "Polyominoes." Scribners, New York, 1965.

GO3 H. W. Gould, Exponential Binomial Coefficient Series. Technical Report 4. Department of Mathematics, West Virginia Univ., Morgantown, West Virginia, September 1961.

GO4 H. W. Gould, "Research Bibliography of Two Special Number Sequences." *Mathematica Monongaliae* 12, 1971.

GR1 D. A. Grave, "Traktat z Algebrichnogo Analizu (Monograph on Algebraic Analysis)," Vol. 2. Vidavnitstvo Akademiia Nauk, Kiev, 1938.

GR2 B. Grünbaum, "Convex Polytopes." Wiley, New York, 1967.

GU1 R. K. Guy, Dissecting a polygon into triangles. Research Paper 9. Department of Mathematics, Univ. of Calgary, Calgary, Alberta, January 1967.

GU2 R. K. Guy, The crossing number of the complete graph. Research Paper 8. Department of Mathematics, Univ. of Calgary, Calgary, Alberta, January 1967.

GU3 R. K. Guy and P. A. Kelly, The no-three-in-line problem. Research Paper 33. Department of Mathematics, Univ. of Calgary, Calgary, Alberta, January 1968.

GU4 R. K. Guy, A theorem in partitions. Research Paper 11. Department of Mathematics, Univ. of Calgary, Calgary, Alberta, January 1967.

GU5 R. K. Guy, personal communication.

GU6 R. C. Gunning, "Lectures on Modular Forms." Princeton Univ. Press, Princeton, New Jersey, 1962.

GU7 R. K. Guy and J. L. Selfridge, The nesting and roosting habits of the laddered parenthesis. Research Paper 127. Department of Mathematics, Univ. of Calgary, Calgary, Alberta, June 1971.

GU8 R. K. Guy, Sedláček's conjecture of disjoint solutions of $x + y = z$. Research Paper 129. Department of Mathematics, Univ. of Calgary, Calgary, Alberta, June 1971.

HAR G. H. Hardy, "Collected Papers," Vols. 1–. Oxford Univ. Press, London and New York, 1966–.

HA1 F. Harary, editor, "Graph Theory and Theoretical Physics." Academic Press, New York, 1967.

HA2 M. A. Harrison, "Introduction to Switching and Automata Theory." McGraw Hill, New York, 1965.

HA3 H. Hasse, "Vorlesungen über Zahlentheorie." Springer-Verlag, Berlin and New York, 1964.

HA4 K. Hayashi, "Sieben- und Mehrstellige Tafeln der Kreis- und Hyperbelfunktionen und deren Produkte Sowie der Gammafunktion." Springer-Verlag, Berlin and New York, 1926.

HA5 F. Harary, "Graph Theory." Addison-Wesley, Reading, Massachusetts, 1969.

HE1 C. Hermite, "Oeuvres," Vol. 4. Gauthier-Villars, Paris, 1917.
HM1 B. H. Hannon and W. L. Morris, Tables of arithmetical functions related to the Fibonacci numbers. Report ORNL-4261. Oak Ridge National Laboratory, Oak Ridge, Tennessee, June 1968.
HO1 V. E. Hoggatt, Jr., "Fibonacci and Lucas Numbers." Houghton, Boston, Massachusetts, 1969.
HO2 R. Honsberger, "Ingenuity in Mathematics." Random House, New York, 1970.
HO3 L. Hogben, "Choice and Chance by Cardpack and Chessboard," Vol. 1, Chanticleer Press, New York, 1950.
HPR H. P. Robinson, personal communication.
HSG F. Harary and L. Beineke, editors, "A Seminar on Graph Theory." Holt, New York, 1967.
HS1 M. Hall, Jr. and J. K. Senior, "The Groups of Order 2^n ($n \leq 6$)." Macmillan, New York, 1964.
HUR A. Hurwitz, "Mathematische Werke," Vols. 1 and 2. Birkhaeuser, Basel, 1962–1963.
HW1 G. H. Hardy and E. M. Wright, "An Introduction to the Theory of Numbers," 3rd ed. Oxford Univ. Press, London and New York, 1954.
IAS *Proceedings of the Indian Academy of Sciences, Section A.*
IBMJ *IBM Journal of Research and Development.*
IC *Information and Control.*
ICM *Proceedings of the International Congress of Mathematicians.*
IDM *L'Intermédiaire des Mathématiciens.*
IJM *Illinois Journal of Mathematics.*
IJ1 *Indian Journal of Mathematics.*
JACM *Journal of the Association for Computing Machinery.*
JACS *Journal of the American Chemical Society.*
JA1 *The Journal of the Australian Mathematical Society.*
JA2 D. Jarden, "Recurring Sequences." Riveon Lematematika, Jerusalem, 1966.
JCP *The Journal of Chemical Physics.*
JCPM J. C. P. Miller, personal communication.
JCT *Journal of Combinatorial Theory.*
JDM *Journal de Mathématiques Pures et Appliquées.*
JEP *Journal de l'École Polytechnique, Paris.*
JFI *Journal of the Franklin Institute.*
JIA *Journal of the Institute of Actuaries.*
JIMS *The Journal of the Indian Mathematical Society.*
JLMS *Journal of the London Mathematical Society.*
JL2 J. Leech, editor, "Computational Problems in Abstract Algebra." Pergamon, Oxford, 1970.
JMP *Journal of Mathematical Physics.*
JMSJ *Journal of the Mathematical Society of Japan.*
JM1 *Journal of Mathematical Analysis and Applications.*
JM2 *Studies in Applied Mathematics* (Formerly *The Journal of Mathematics and Physics).*
JNSM *The Journal of Natural Sciences and Mathematics.*
JO1 L. B. W. Jolley, "Summation of Series," 2nd ed. Republished by Dover, New York, 1961.
JO2 C. Jordan, "Calculus of Finite Differences." Budapest, 1939.
JPC *The Journal of Physical Chemistry.*
JRAM *Journal für die Reine und Angewandte Mathematik.*

JRM *Journal of Recreational Mathematics.*

JSIAM *SIAM Journal on Applied Mathematics* (formerly *Journal of the Society for Industrial and Applied Mathematics).*

KNAW *Proceedings of the Koninklijke Nederlandse Akademie van Wetenschappen, Series A.*

KN1 D. E. Knuth, "The Art of Computer Programming," Vols. 1– . Addison-Wesley, Reading, Massachusetts, 1968–.

KN2 D. E. Knuth, personal communication.

KO1 R. R. Korfhage, personal communication.

KR1 M. B. Kraitchik, "Recherches sur la Théorie des Nombres." Gauthiers-Villars, Paris, Vol. 1, 1924, Vol. 2, 1929.

KU1 H. Kumin, The enumeration of trees by height and diameter. M. A. Thesis, Department of Mathematics, Univ. of Texas, Austin, Texas, January 1964 [See *Mathematics of Computation*, Vol. 25, p. 632 (1971).]

KYU *Memoirs of the Faculty of Science, Kyusyu University, Series A.*

LA2 C. A. Laisant, "Recueil de Problèmes de Mathématiques," Vols. 1–3. Gauthier-Villars, Paris, 1893.

LA3 H. Langman, "Play Mathematics." Hafner, New York, 1962.

LA4 C. Lanczos, "Applied Analysis." Prentice-Hall, Englewood Cliffs, New Jersey, 1956.

LEM *L'Enseignement Mathématique.*

LE1 D. H. Lehmer, Guide to tables in the theory of numbers. Bulletin No. 105. National Research Council, Washington, D.C., 1941.

LE2 H. Levy and F. Lessman, "Finite Difference Equations." Pitman, London, 1959.

LE3 W. J. Leveque, "Topics in Number Theory," Vol. 1. Addison-Wesley, Reading, Massachusetts, 1956.

LE4 R. S. Lehman, A study of regular continued fractions. Report 1066. Ballistic Research Laboratories, Aberdeen Proving Ground, Maryland, February 1959.

LF1 A. V. Lebedev and R. M. Fedorova, "A Guide to Mathematical Tables." Pergamon, Oxford, 1960.

LF2 N. M. Burunova, "A Guide to Mathematical Tables," Supplement No. 1. Pergamon, Oxford, 1960.

LIN *Atti della Reale Accademia Nazionale dei Lincei, Memorie della Classe di Scienze Fisiche, Matematiche e Naturali.*

LINR *Atti della Reale Accademia Nazionale dei Lincei, Rendiconti della Classe di Scienze Fisiche, Matematiche e Naturali.*

LI2 Shen Lin, personal communication

LO1 J. S. Lomont, "Applications of Finite Groups." Academic Press, New York, 1961.

LU1 E. Lucas, "Théorie des Nombres," Vol. 1. Gauthier-Villars, Paris, 1891.

LU2 W. F. Lunnon, personal communication.

LU5 W. F. Lunnon, Counting hexagonal and triangular polyominoes, *in* "Graph Theory and Computing" (R. C. Read, ed.). Academic Press, New York, 1972.

LU6 W. F. Lunnon, Symmetry of cubical and general polyominoes, *in* "Graph Theory and Computing" (R. C. Read, ed.). Academic Press, New York, 1972.

MAG *The Mathematical Gazette.*

MAN *Mathematische Annalen.*

MAT *Mathematika.*

MA1 V. Mangulis, "Handbook of Series for Scientists and Engineers." Academic Press, New York, 1965.

MA2 P. A. MacMahon, "Combinatory Analysis." Cambridge Univ. Press, London and New York, Vol. 1, 1915 and Vol. 2, 1916.

MA3 C. L. Mallows, personal communication.

MA4 *Mathematical Algorithms.*

MES *The Messenger of Mathematics.*

ME1 N. Metropolis, M. L. Stein, and P. R. Stein, On finite limit sets for transformations on the unit interval. *Journal of Combinatorial Theory*, **A15**, 25-44 (1973).

MFC *Matematický Časopis* (formerly *Matematicko-Fyzikálny Časopis).*

MFM *Monatshefte für Mathematik.*

MFS M. F. Sykes, personal communication.

MI1 *Massachusetts Institute of Technology, Research Laboratory of Electronics, Quarterly Progress Reports.*

MLG *The Mathematical Magazine.*

MMAG *Mathematics Magazine.*

MNAS *Memoirs of the National Academy of Sciences.*

MNR *Mathematische Nachrichten.*

MOD *Atti del Seminario Matematico e Fisico dell' Università di Modena.*

MO1 J. W. Moon, "Topics on Tournaments." Holt, New York, 1968.

MO2 P. Montel, "Leçons sur les Récurrences et leurs Applications." Gauthier-Villars, Paris, 1957.

MQET *Mathematical Questions and Solutions from the Educational Times.*

MST *The Mathematics Student.*

MTAC *Mathematics of Computation* (formerly *Mathematical Tables and Other Aids to Computation*).

MTS *Mathesis.*

MT1 L. M. Milne-Thomson, "The Calculus of Finite Differences." Macmillan, New York, 1933.

MU1 T. Muir, "The Theory of Determinants in the Historical Order of Development," Vols. 1–4. Macmillan, New York, 1906–1923.

MU2 T. Muir, "A Treatise on the Theory of Determinants." Republished by Dover, New York, 1960.

MU3 A. Mukhopadhyay, editor, "Recent Developments in Switching Theory." Academic Press, New York, 1971.

MZT *Mathematische Zeitschrift.*

NAM *Nouvelles Annales de Mathématiques.*

NAMS *Notices of the American Mathematical Society.*

NBS *Journal of Research of the National Bureau of Standards.*

NCM *Nouvelles Correspondance Mathématique.*

NET E. Netto, "Lehrbuch der Combinatorik," 2nd ed. Teubner, Leipzig, 1927.

NI1 National Institute of Sciences of India, Mathematical Tables, Vol. 1, "Tables of Partitions of Gaussian Integers," by M. S. Cheema and H. Gupta, New Delhi, 1956.

NMT *Nordisk Matematisk Tidskrift.*

NO1 N. E. Nörlund, "Vorlesungen über Differenzenrechnung." Springer-Verlag, Berlin and New York, 1924.

NVSF *Det Kongelige Norske Videnskabers Selskabs, Forhandlinger.*

NZ1 I. Niven and H. S. Zuckerman, "An Introduction to the Theory of Numbers," 2nd ed. Wiley, New York, 1966.

OB1 W. Oberschelp, Strukturzahlen in Endlichen Relationssystemen, "Contributions to Mathematical Logic" (*Proceedings 1966 Hanover Colloquium*), pp. 199–213. North-Holland Publ., Amsterdam, 1968.

OP1 T. R. Van Oppolzer, "Lehrbuch zur Bahnbestimmung der Kometen und Planeten," Vol. 2. Engelmann, Leipzig, 1880.

PAMS *Proceedings of the American Mathematical Society.*

PA1 T. R. Parkin, L. J. Lander, and D. R. Parkin, Polyomino enumeration results, *SIAM Fall Meeting, Santa Barbara, California, 1967.*

PCPS *Proceedings of the Cambridge Philosophical Society.*

PEF *Publications de la Faculté d'Électrotechnique de l'Université à Belgrade.*

PE1 O. Perron, "Die Lehre von den Kettenbrüchen," 2nd ed. Teubner, Leipzig, 1929.

PE2 J. T. Peters, "Ten-Place Logarithm Table," Vols. 1 and 2, rev. ed. Ungar, New York, 1957.

PE3 J. K. Percus, "Combinatorial Methods." Courant Institute of Mathematical Sciences, New York, 1969.

PGEC *IEEE Transactions on Computers* (formerly *IEEE Transactions on Electronic Computers*).

PGIT *IEEE Transactions on Information Theory.*

PHA *Physica.*

PHM *The Philosophical Magazine.*

PIEE *Proceedings of the Institution of Electrical Engineers.*

PJM *Pacific Journal of Mathematics.*

PLMS *Proceedings of the London Mathematical Society.*

PL1 R. J. Plemmons, Cayley tables for all semigroups of order less than 7. Department of Mathematics, Auburn Univ., Auburn, Alabama, 1965.

PL2 *Proceedings Louisiana Conferences on Combinatorics, Graph Theory, and Computing.*

PNAS *Proceedings of the National Academy of Sciences of the United States of America.*

PNISI *Proceedings of the Indian National Science Academy* (formerly *Proceedings of the National Institute of Sciences of India*), *Part A.*

PO1 L. Poletti, "Tavole di Numeri Primi Entro Limiti Diversi e Tavole Affini." Heopli, Milan, 1920.

PO2 P. Poulet, "La Chasse aux Nombres," Vol. 2. Librairie du Sphinx, Brussels, 1934.

PPS *Proceedings of the Physical Society.*

PRSE *Proceedings of the Royal Society of Edinburgh, Section A.*

PRV *Physical Review.*

PR1 G. Prévost, "Tables de Fonctions Sphériques." Gauthier-Villars, Paris, 1933.

PSAM *Proceedings of Symposia in Applied Mathematics,* American Mathematical Society, Providence, Rhode Island.

PSPM *Proceedings of Symposia in Pure Mathematics,* American Mathematical Society, Providence, Rhode Island.

PURB *Research Bulletin of the Panjab University* (*Science Section*).

QAM *Quarterly of Applied Mathematics.*

QJM *The Quarterly Journal of Pure and Applied Mathematics.*

RAM Srinivasa Ramanujan, "Collected Papers." Cambridge Univ. Press, London and New York, 1927.

RA1 The Rand Corporation, "A Million Random Digits with 100,000 Normal Deviates." The Free Press, New York, 1955.

RCI J. Riordan, "Combinatorial Identities." Wiley, New York, 1968.

REC *Recreational Mathematics Magazine.*

RE1 K. B. Reid, personal communication.

RE2 K. G. Reuschle, "Tafeln Complexer Primzahlen." Königl. Akademie der Wissenschaften, Berlin, 1875.

RE3 R. C. Read, personal communication.

RE4 R. C. Read, Some enumeration problems in graph theory. Ph.D. Thesis, Department of Mathematics, London Univ., London, 1958.

RG1 I. M. Ryshik and I. S. Gradstein, "Tables of Series, Products, and Integrals." Deutscher Verlag der Wissenschaften, Berlin, 1957.

RI1 J. Riordan, personal communication.

RI3 J. Riordan, The enumeration of permutations with three-ply staircase restrictions. Unpublished memorandum. Bell Telephone Laboratories, Murray Hill, New Jersey, October 1963.

RLG R. L. Graham, personal communication.

RLM *Riveon Lematematika.*

RMM C. R. Rao, S. K. Mitra, and A. Matthai, editors, "Formulae and Tables for Statistical Work." Statistical Publishing Society, Calcutta, India, 1966.

RO1 P. Rosenthiehl, editor, "Theory of Graphs" (*International Symposium, Rome, 1966*). Gordon & Breach, New York, 1967.

RO2 H. A. Rothe, *in* "Sammlung Combinatorisch-Analytischer Abhandlungen" (C. F. Hindenburg, ed.), Vol. 2, Chapter XI. Fleischer, Leipzig, 1800.

RO3 I. Rosenberg, The number of maximal closed classes in the set of functions over a finite domain. *Journal of Combinatorial Theory* (*A*) **14**, 1–7 (1973).

RS1 Royal Society Mathematical Tables, Vol. 3, "Table of Binomial Coefficients" (J. C. P. Miller, ed.). Cambridge Univ. Press, London and New York, 1954.

RS2 Royal Society Mathematical Tables, Vol. 4, "Tables of Partitions," by H. Gupta and others. Cambridge Univ. Press, London and New York, 1958.

RS3 *Philosophical Transactions of the Royal Society of London, Series A.*

RS4 Royal Society Mathematical Tables, Vol. 8, "Tables of Natural and Common Logarithms," by W. E. Mansell. Cambridge Univ. Press, London and New York, 1964.

RS5 Royal Society Mathematical Tables, Vol. 9, "Tables of Indices and Primitive Roots," by A. E. Western and J. C. P. Miller. Cambridge Univ. Press, London and New York, 1968.

RS6 *Proceedings of the Royal Society of London, Series A.*

RS7 Royal Society Mathematical Tables, Vol. 6, "Tables of the Riemann Zeta Function," by C. B. Haselgrove and J. C. P. Miller. Cambridge Univ. Press, London and New York, 1960.

RYS H. J. Ryser, "Combinatorial Mathematics," Mathematical Association of America, Carus Mathematical Monograph Number 14, 1963.

R1 J. Riordan, "An Introduction to Combinatorial Analysis." Wiley, New York, 1958.

SA1 M. Fiedler, editor, "Theory of Graphs and Its Applications" (*Proceedings of the Symposium, Smolenice, Czechoslovakia, 1963*). Academic Press, New York, 1964.

SA2 A. V. Aho and N. J. A. Sloane, Some doubly exponential sequences. *Fibonacci Quarterly,* to be published.

SC1 I. J. Schwatt, "An Introduction to the Operations with Series," 2nd ed. Chelsea, Bronx, New York, 1961.

SE1 J. L. Selfridge, personal communication.

SE2 J. Ser, "Les Calculs Formels des Séries de Factorielles." Gauthier-Villars, Paris, 1933.

SE3 J. J. Seidel, personal communication.

SIAMR *SIAM Review.*

SI1 W. Sierpiński, "Elementary Theory of Numbers." Państwowe Wydawnictwo Naukowe, Warsaw, 1964.

SI2 W. Sierpiński, "A Selection of Problems in the Theory of Numbers." Macmillan, New York, 1964.

SKA *Skandinavisk Aktuarietidskrift.*

SL1 M. A. Sainte-Laguë, "Les Réseaux (ou Graphes)," Mémorial des Sciences Mathématiques, Fasc. 18. Gauthier-Villars, Paris, 1926.

SMA *Scripta Mathematica.*

SMD *Soviet Mathematics-Doklady.*

SMH *Studia Scientiarum Mathematicarum Hungarica.*

SPH *Sphinx.*

SPS Space Programs Summary. Jet Propulsion Laboratory, California Institute of Technology, Pasadena, California.

SSP *Solid State Physics* (*Journal of Physics C*).

ST1 M. L. Stein and P. R. Stein, Enumeration of linear graphs and connected linear graphs up to $p = 18$ points. Report LA-3775. Los Alamos Scientific Laboratory of the University of California, Los Alamos, New Mexico, October 1967.

ST2 M. L. Stein and P. R. Stein, Enumeration of stochastic matrices with integer elements. Report LA-4434. Los Alamos Scientific Laboratory of the University of California, Los Alamos, New Mexico, June 1970.

ST3 P. R. Stein, personal communication.

ST4 M. L. Stein and P. R. Stein, Tables of the number of binary decompositions of all even numbers less than 200,000 into prime numbers and lucky numbers. Report LA-3106. Los Alamos Scientific Laboratory of the University of California, Los Alamos, New Mexico, September 1964.

SY1 J. J. Sylvester, "Collected Mathematical Papers," Vols. 1–4. Cambridge Univ. Press, London and New York, 1904–1912.

TAMS *Transactions of the American Mathematical Society.*

TA1 P. G. Tait, "Scientific Papers," Cambridge Univ. Press, London and New York, Vol. 1, 1898, and Vol. 2, 1900.

TCPS *Transactions of the Cambridge Philosophical Society.*

TH1 T. N. Thiele, "Interpolationsrechnung." Teubner, Leipzig, 1909.

TH2 V. Thébault, "Les Récréations Mathématiques." Gauthier-Villars, Paris, 1952.

TI1 P. Erdös and G. Katona, editors, "Theory of Graphs" (*Proceedings of the Colloquium, Tihany, Hungary.*) Academic Press, New York, 1968.

TM1 J. Tannery and J. Molk, "Éléments de la Théorie des Fonctions Elliptiques," Vols. 1–4. Gauthier-Villars, Paris, 1893–1902.

TOH *Tôhoku Mathematical Journal.*

UL1 S. M. Ulam, "Problems in Modern Mathematics." Wiley, New York, 1960.

UWP *University of Washington Publications in Mathematics.*

VO1 A. H. Voigt, "Theorie der Zahlenreihen und der Reihengleichungen." Goschen, Leipzig, 1911.

WA1 G. L. Watson, personal communication.

WCC W. A. Whitworth, "Choice and Chance." Hafner, New York, 1959.

WE1 D. J. A. Welsh, editor, "Combinatorial Mathematics and Its Applications." Academic Press, New York, 1971.

WE2 M. B. Wells, "Elements of Combinatorial Computing." Pergamon, Oxford, 1971.
WH1 N. L. White, personal communication.
WIEN *Sitzungsberichte der Kaiserlichen Akademie der Wissenschaften in Wien, Mathematisch-Naturwissenschaftlichen Klasse.*
WO1 E. Wolman, personal communication.
WR1 J. A. Wright, personal communication.
ZFK *Zeitschrift für Kristallographie.*
ZML *Zeitschrift für Mathematische Logik und Grundlagen der Mathematik.*
ZMP *Zeitschrift für Mathematik und Physik.*

INDEX

An asterisk * denotes the principal sequence of its type

Q

R